BEACHES

FORM & PROCESS

TITLES OF RELATED INTEREST

Deep marine environments
K. Pickering et al.

Experiments in physical sedimentology
J. R. L. Allen

Geomorphological techniques
A. S. Goudie (ed.)

Free-surface hydraulics
J. M. Townson

The history of geomorphology
K. J. Tinkler (ed.)

Hydraulic structures
P. Novak et al.

Hydraulics in civil engineering
A. Chadwick & J. Morfett

Introduction to theoretical geomorphology
C. Thorn

Marine geochemistry
R. Chester

Marine geotechnics
H. Poulos

Mathematics in geology
J. Ferguson

Models in geomorphology
M. J. Woldenberg (ed.)

Principles of physical sedimentology
J. R. L. Allen

BEACHES

FORM & PROCESS

numerical experiments with monochromatic waves on the orthogonal profile

by

J. Hardisty
University of Hull

London
UNWIN HYMAN
Boston   Sydney   Wellington

This volume has been typeset, proofed and passed for press by the author.

Published by the Academic Division of
Unwin Hyman Ltd
15/17 Broadwick Street, London W1V 1FP, UK

Unwin Hyman Inc.
955 Massachusetts Avenue, Cambridge, MA 02139, USA

Allen & Unwin (Australia) Ltd
8 Napier Street, North Sydney, NSW 2060, Australia

Allen & Unwin (New Zealand) Ltd
in association with the Port Nicholson Press Ltd
Compusales Building, 75 Ghuznee Street, Wellington 1, New Zealand

First published in 1990

British Library Cataloguing in Publication Data

Applied for

ISBN 0-04-445219-5

Library of Congress Cataloging in Publication Data

Applied for

ISBN 0-04-445219-5

Typeset by the author using an Apple Macintosh SE system running Microsoft Word in 10 on
12 pt Times Roman with 12 pt section headers and 14 pt chapter headers and utilising Microsoft
Excel and Cricket Graph. A range of micro-computer packages is available containing programs
which simulate the processes described in this book. The packages operate on either MS DOS or
Macintosh machines and form part of the *GeoSystems Library* which is referenced in the text.
Full details may be obtained from Unico Geosystems Ltd, School of Earth Resources,
University of Hull, HU6 7RX, U.K.
Printed in Great Britain by Cambridge University Press

CONTENTS

SECTION A. INTRODUCTION

SECTION B. HYDRODYNAMICS

SECTION C. SEDIMENT DYNAMICS

SECTION D. MORPHODYNAMICS

SECTION E. NUMERICAL MODELLING

Abbreviated Figure Captions

PLATE CAPTIONS

Plate I   Preparing to position an ultrasonic current meter beneath surging breakers on a gravel beach in south west Wales. (page 1).

Plate II   The complexities of the shoaling transformations are apparent in this view on a stormy winters day at Spurn Head on the North Sea coast of Northern England. (page 45).

Plate III   The plunging breaker. (page 111).

Plate IV   This steep, coarse sand beach at Argeles on the micro-tidal Mediterranean coast of southern France responds constantly to changes in wave conditions. (page 187).

Plate V   The SLOPES numerical model is being tested in the British Beach and Nearshore Dynamics Project using, in this photograph, electromagnetic current meters at Llangenith on the Bristol Channel. (page 239).

This book is dedicated to the memory of W.E.H.Culling. Listening to astro-physicists arguing about the second decimal place in the velocity of a nebula which they could not see, he wondered why geomorphologists were not predicting the landforms over which they could walk. Herein will be found some predictions of surface elevation in microns.

PREFACE

Isaac Newton (c1674)

> *I do not know what I may seem to the world but to myself I am as a small boy playing on the shore whose attention is now and then diverted by a rounder pebble or shinier stone while the whole ocean of truth lies undiscovered before me.*

W.F.Tanner (1974)

> *The equilibrium idea is that an energetic wave will establish in due time and barring too many complications, a delicately adjusted balance among activity, three-dimensional geometry and sediment transport such that the system will tend to correct short or minor interference.*

This book has two simple aims. Firstly it organises, presents and tests existing theories for the processes which operate on a beach. Secondly it integrates the best fit theories with some morphological ideas in order to erect and to examine a model for the two-dimensional form of the orthogonal profile.

If these objectives have been achieved it is largely due to the guidance which I have received from a number of colleagues and students. The colleagues include Doug Hamilton, Colin Jago, Keith Dyer, Tony Heathershaw, Nick Langhorne, John Pethick, Duncan McGregor, Chris Green, David Huntley, Rob Ferguson, Bill Carter, Anne Hinton and Sarah Metcalfe. The post-graduate students include R.J.S.Whitehouse, A.W.Evans, D.P.Horn, J.M.Woodruff, J.P.Lowe, K.A.Jagger and J.P.Hoad who helped to clarify many aspects of the problem; and Mark Davidson and Paul Russell who led me into random waves. That the book was written at all is due to the encouragement of J.B.Thornes who seemed to expect Section D some years before it was formulated. That there is any data in here at all is due to Mike Overs, Bill Miller and Robin Powell for they were not only good company on many research cruises in the North Atlantic but they also convinced me that, despite the immiscible nature of seawater and electricity, meaningful measurements can be made in the marine environment.

J.Hardisty
North Ferriby

SYMBOLS

A	coefficient in the Pierson-Moskowitz spectrum	
A_a	attenuation coefficient for seabed currents	
A_b	slope coefficient for bedload rate parameter	
b	packing coefficient in the linear concentration	
b_b	depth of bedload layer	m
B	coefficient in the Pierson-Moskowitch spectrum	
B_b	slope coefficient for threshold velocity	
B_{hy}	hydraulic bore parameter	
B_t	turbulence factor in the longshore currents	
C	sediment concentration on the bed	$cm^3 \ cm^{-2}$
C_a	sediment concentration at height a above bed	$cm^3 \ cm^{-2}$
C_{Dz}	drag coefficient at a height z above bed	
C_o	condenser capacitance *in vacuuo*	
C_m	condenser capacitance in material	
C_z	sediment concentration at height z above bed	$g \ cm^{-3}$
C_∞	deep water wave celerity	$m \ s^{-1}$
D	sediment grain diameter	m
D_{50} etc	Grain size percentile	ϕ
$E(\omega)$	spectral density function	
f_w	wave friction factor	
F_D	drag force on a particle	N
F_G	gravitational force on a particle	N
F_L	lift force on a particle	N
G	coefficient in grain pemeability formula	
	or the universal gravitational constant	
h	water depth	m
ΔH_f	change in wave height due to friction	m
ΔH_r	change in wave height due to refraction	m
H	local wave height	m
H_b	breaking wave height	m
H_i	incident wave height	m
H_r	reflected wave height	m
H_{rms}	root mean squared wave height	m
H_s	shoaled wave height	m
H_∞	deep water wave height	m
$i_b(t)$	instantaneous bedload rate	$kg \ m^{-1} \ s^{-1}$
i_s	suspended load transport rate	$kg \ m^{-1} \ s^{-1}$
I_s	net suspended load transport per wave	$kg \ m^{-1}$
k	wave number $= 2\pi/L$	
k_b	bedload rate parameter	$kg \ m^{-4} \ s^{-2}$
k_e	edge wave number	
K	permeability *or* elliptic function in cnoidal waves	

K_r	refraction coefficient	
K_{rf}	reflection coefficient	
K_s	shoaling coefficient	
L	local wave length	m
L_∞	deep water wave length	m
m	modal number in edge wave equations	
n	shoaling parameter and	
	number of sinusoids in a random wave distribution	
n_e	edge wave mode number	
M_ϕ	mean grain diameter	ϕ
$M_{d\phi}$	modal grain diameter	ϕ
M_2	amplitude of lunar semi-diurnal tide	m
p_1	coefficient in the longshore current equations	
p_2	coefficient in the longshore current equations	
P	wave power	$J\ m^{-1}$
P_c	horizontal eddy coefficient for longshore currents	
P_H	hydrostatic pressure at depth h	$N\ m^2$
P_t	breaker phase difference parameter	
Re_w	wave Reynolds Number	
R_h	ratio in shallow water limit of wave equations	
R_y	frictional drag in longshore current equations	
s	packing coefficient in the linear grain concentration	
s_h	local inter-orthogonal spacing	
sk_ϕ	sediment skewness	ϕ
s_L	coefficient in longshore current equations	
S_∞	deep water inter-orthogonal spacing	
$S^o/_{oo}$	salinity of seawater	$gm\ kg^{-1}$
S_2	amplitude of solar semi-diurnal tide	m
S_{xx}	radiation stress across x=constant and along x	
S_{yy}	radiation stress across y=constant and along y	
t°	tidal phase	$^\circ$
T	wave period	s
T_f	turbulence factor in the threshold equations	
T_{fp}	freezing point of water	$^\circ C$
u_{cr}	threshold velocity	$m\ s^{-1}$
u_{zcr}	threshold velocity at a height z above the bed	$m\ s^{-1}$
u(t)	seabed current	$m\ s^{-1}$
u_*	shear velocity	$m\ s^{-1}$
U(h)	drift current velocity profile	ms^{-1}
U_r	Ursell number	
U_s	drift current velocity	$m\ s{-1}$
U_{so}	drift current at the bed	
U_o and U_1	Flow velocities in Bowen's model (15.2)	ms^{-1}
U_z	Velocity of second order drift current	$m\ s^{-1}$

x	distance offshore	m
z	depth below sea surface (positive upwards)	m
z_t	sea surface elevation	m
Z_o	datum height	m
α	(alpha) local wave orthogonal direction	°
α_∞	deep water wave orthogonal direction	°
β	(beta) seabed gradient	
δ	(delta) thickness of wave induced boundary layer	m
ΔH_f	wave height loss due to seabed friction	m
ΔH_r	wave height loss (or gain) due to refraction	m
ΔH_b	wave height loss due to breaking	m
ε	(epsilon) surf scaling parameter	
ε_e	dielectric constant	
ε_L	coefficient in a longshore current equations	
ε_R	beach reflectivity parameter	
γ	(gamma) surface tension and	dyn cm^{-1}
	concentration parameter in the Rouse equation	
γ_b	breaker coefficient	
η	(eta) surface elevation or set up	m
λ	(lambda) linear grain concentration	
λ_e	edge wave length	m
κ	(kappa) von Karman's constant	
K	(kappa) elliptic function in cnoidal wave theory	
μ	(mu) dynamic viscosity of water	N s m^{-2}
μ_e	eddy coefficient for surf zone mixing	
ν	(nu) kinematic viscosity of water	m^2 s^{-1}
ϕ	(phi) angle of internal friction of sediment and units of grain diameter	
θ	(theta) Shields parameter	
θ_{cr}	threshold Shields parameter	
ρ	(rho) fluid density	kg m^{-3}
ρ_b	sediment bulk density	kg m^{-3}
ρ_s	sediment density	kg m^{-3}
σ_ϕ	standard deviation of sediment grain size	ϕ
τ	(tau) bed shear stress	N m^{-2}
ω	(omega) wave radian frequency ($=2\pi/T$)	
ω_e	edge wave radian frequency	
ψ_w	surface stream function in the surf zone	
ξ	(xi) surf similarity parameter	
ζ	(zeta) wave set up coefficient	
max	local, maximum values of periodic parameters	
∞	deep water value of various parameters	

SECTION A

INTRODUCTION

chapter one

BEACHES : AN INTRODUCTION

1.1 Geomorphology

Geomorphology is one of the earth sciences along with geology, sedimentology, hydrology, oceanography, meteorology and so forth. In common with the other sciences, geomorphology attempts firstly to explain and secondly to predict particular aspects of the physical and chemical world. In common with the other earth sciences, geomorphology is broadly concerned with planet earth; with the solid rock making up the lithosphere, with the river and ocean waters which constitute the hydrosphere and with the gases of the atmosphere. In this sense each of the earth sciences has identified and researched its own particular aspect of the natural world, but each draws upon and contributes to the other disciplines. Geomorphology is no exception, for although this science seeks to explain and to predict the surface form of the solid earth, it achieves that aim by drawing upon the other disciplines, and in turn offering its own successes to their work. This book makes much use of the abstract shorthand methods of mathematics. It is equally dependent upon the basic principles of physics for it examines the forces which move individual grains of sand across the beach surface and must account for the work which is done and the energy which is consumed during these movements. Again, chemistry is required to understand the cohesive nature of the finer sedimentary particles, astronomy for the tides, meteoreology and oceanography for the waves, geology for the rocks, and not a little electronics and computer science for the instrumentation and simulations.

Geomorphology combines essential aspects of all of these various fields in order to explain and to predict the evolution and preservation of the surface form of the solid earth. Scheidegger (1970 pv) separates the subject into two categories, the first of which comprises those forms that are due to processes occuring inside the solid earth (endogenetic features from the Greek *endon* within), and the second, those that are due to processes occuring outside the solid earth (exogenetic features from the Greek *exo* without). We are here concerned with features and forms due to exogenetic, earth surface processes.

A casual perusal of any of the general geomorphological textbooks shows that the subject also separates easily into a number of environmental subdivisions. Typically Embleton and Thornes (1979) suggest geomorphologies dealing with hillslope, fluvial, glacial, nival (snow covered), aeolian and marine landforms. Although we are clearly concerned here with marine, and specifically with shallow marine landforms, such a subdivision has little significance beyond a convenient classification of the geography of the earth's surface features. This is because the processes by which form is evolved and maintained in a state of static or more usually dynamic equilibrium with its environment are those of sediment mass transport. The basic physics of the sediment mass transport processes is common to most environments so that the laws governing observed features, once deduced, are applicable to very many environments.

Recent developments have moved away from the traditional schools of geomorphological research which separated process from form and we now acknowledge that a complex structure of feedbacks is in operation whereby processes transport sediment and sculpt the landform, but the landform severely influences the operation of the processes. Such is the approach taken in this book and detailed in the following chapters.

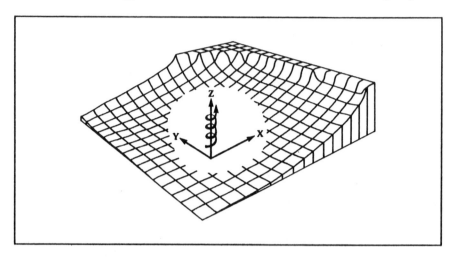

Figure 1.1 Definition diagram of the co-ordinate axes using the right hand screw rule, along with an example of the geomorphological form function.

1.2 The Form Problem

The following is a three dimensional treatment of the corresponding section in Scheidegger (1970, paragraph 1.20). It is necessary to begin by precisely defining the problem which we seek to address. We have seen that geomorphology is concerned with the surface form of the earth and using conventional co-ordinate axes this can be written with reference to the example shown in Figure 1.1 as:

$$z = f_1(x,y)$$
Eq.1.1

where z denotes a height above some datum which is a function (f_1) of the two horizontal coordinates. Here, and throughout this book, x,y and z are three mutually perpendicular axes and we shall use the *right-handed co-ordinate system* to define the positive directions of the three axes. Such a system takes its name from the fact that a right threaded screw (the normal type which is moved in by rotating the driver clockwise) rotated through $90°$ from the positive x direction to the positive y direction will advance along the positive z direction as shown in the lower part of Figure 1.1. This means that if x is to the North then y lies to the West and z is vertically upwards. We shall refer to the onshore direction as being the x axis so that the y axis extends along the coast to the left of an observer positioned offshore. The velocity of water currents also follows the same convention with the positive directions of the three flow components u, v and w being along x, y and z respectively. The same convention will also be applied to the sediment mass transport rate with the positive direction of the three transport components, i, j and k being along x, y and z respectively.

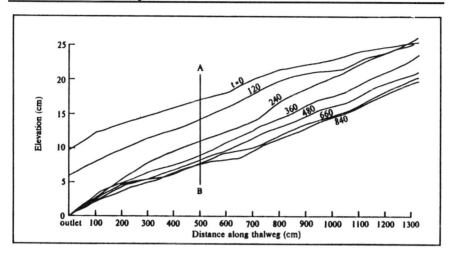

Figure 1.2 The evolution of a small scale stream bed through time in a laboratory experiment (after Schum et al., 1987).

The fourth dimension, time, is also of great importance in geomorphological work and must be included in any specification of the form. It is a fairly easy task to imagine the evolution of a landform, but it is unfortunately more difficult to represent this evolution on a flat printed page. An example is the type of step diagram shown in Figure 1.2, or by reducing the information contained within the diagram, the evolution of the elevation of the point A-B may be depicted as in Figure 1.3. In general terms the evolution of the form may be written by the function:

$$z = f_2(x,y,t,z_m)$$
Eq.1.2

where f_2 is a second function of x and y as before, but now also depends upon time, t, and upon some previous form z_m which is called the *geomorphic memory*. The complete definition of such as f_2 is the geomorphological problem, and we shall refer to it as the *form function*. The complete definition is difficult to obtain but, at least empirically, it is sometimes achieved through repeated ground surveys of an area, the result being a series of contour maps. To a certain extent this has been the object of much geomorphological work in efforts to determine the evolution of aspects of landscapes. This branch of the subject is called denudation geochronology. The result of such mapping does achieve the first of the two scientific objectives which were identified earlier, in that a certain predictive capability is obtained and suitable extrapolations can suggest the likely elevation of forms at some future or earlier date. It does not however offer any explanation and for this attention must be directed to the operative processes.

1.3 Parameters and Processes

Modern geomorphological research utilises the so called *scientific method* which involves proposing a hypothesis which formalises the relationship between variables and then executing experiments which are designed to disprove the hypothesis. If the experiments do not disprove or refute the hypothesis then the relationship is accepted until a better hypothesis or better experimental technique is developed. We shall investigate the types of hypotheses which have been erected for beaches in Section 1.5 and Chapter Two, but we

initially examine geomorphological science in general terms to clarify some of the ideas
and expressions with which the subject explains itself.

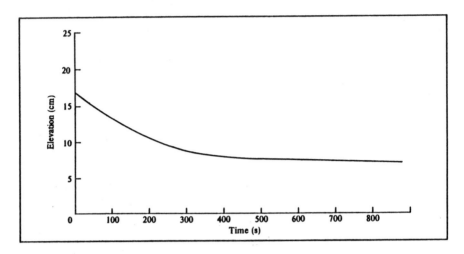

Figure 1.3 Evolution of the elevation of a point along A-B in Figure 1.2 .

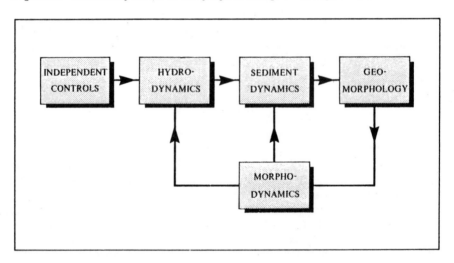

Figure 1.4 General overview of subaqueous geomorphological systems.

In the geomorphological parlance which is adopted here the variables in the scientific
method are called parameters, the relationships between them are called processes and the
whole structure is referred to as a model. The term model thus has a critical significance
for the conduct of the science, and yet definitions of the term are confusing and sometimes
confused. Haines-Young and Petch (1986) define a model as "a device used to generate
predictions" which is accurate but which obscures the differences between models and
simulations. It is important that the former includes the notion of refutability, whereas
this is not necessarily the case for the latter. Instead we shall use the working definition
that:

6

A model is a device used to generate predictions which can be tested by experimental methods.

The variables can usually be divided into dependent and independent parameters where the former controls the latter but the latter does not influence the former. Geomorphologically the ultimate dependent parameter is the landscape itself and we wish to explain and to predict the evolution of the three dimensional form in terms of the independent parameters. The ultimate independent parameters are the inputs to the system under consideration and these will usually include a manifestation of the energy input, and perhaps acknowledge the addition of sediment and water.

Consider the four upper. boxes in Figure 1.4. They represent a simple model of subaqueous geomorphological systems. It identifies four groups of parameters which are linked by three groups of processes. The input to the system is characterised by the independent parameters. The hydrodynamic processes transform these into the flow parameters and the sediment dynamic processes transform these into the transport parameters which characterise the movement of sediment within the system. Finally, the morphodynamic processes which usually involve little more than mass continuity transform the transport parameters into the form of the landscape which was characterised by the form function, Eq.1.2. This is called a process-response system and it is easy to envisage the application of this model to, for example, a river channel. Increased rainfall (the independent parameter) leads to a mean discharge (the flow parameter) which is competent to remove a certain load (the transport parameter) which results in erosion and a deeper channel (the morphological parameter).

Before developing this model for beaches, it is useful to define the geomorphological features which constitute the beach system.

1.4 Beach Geomorphology

Although the term *beach* is traditionally applied to a *'shore with a covering of sand or gravel'* (Shepard, 1973) we shall be dealing here with all of the associated problems of wave controlled coastal geomorphology and will therefore use the definition:

The orthogonal system is a coastal accumulation of non-cohesive sediment along the orthogonal path, the form and texture of which is controlled by wave dominated processes.

The orthogonal system (Figure 1.5) thus extends from the landward limit of the swash to the depth at which wave action ceases to be competent to transport non-cohesive seabed sediment. It should be noted that this usage of the orthogonal system encompasses the term shoreface which is used by e.g. Swift *et al.* (1985). Although no permanent lines can be drawn for the boundaries of the orthogonal system because both the landwards and the seawards limits will move continually under changing wave and tidal conditions, this definition does properly exclude rocky or muddy shorelines whilst leaving a geomorphology which is amenable to interesting, theoretical analysis.

The definition also represents a deliberate limitation of the scope of this book to a consideration of lines along the three dimensional beach and nearshore surface. This limitation has been imposed for the practical reason that a more comprehensive treatment of beach geomorphology would, at the present time, have necessitated a less rigorous methodology. The limitations of the orthogonal view are, nevertheless, discussed in more detail in Chapter Twenty.

The measured length of a coastline depends upon the interval over which measurements are made, and is usually an underestimate because straight line sections tend to cut across

Table 1.1 Glossary of Second Order Beach Features

Backshore. *The zone of the beach lying between the berm and the coastline. Shorenormal processes.*

Bar. *[Longshore bar, ball or ridge]. An elongate, slighly submerged sand ridge or ridges which may be exposed at low water. Shorenormal processes.*

Barrier. *A sand beach (barrier beach), island (barrier island) or spit (barrier spit) that extends roughly parallel with the general coastal trend but is separated from the mainland by a relatively narrow body of water or marsh. Shoreparallel processes.*

Beach face. *The sloping section of the beach below the berm. Shorenormal processes.*

Beach ridge. *[Storm beach, chenier] A low lengthy ridge of beach material piled up by storm waves landward of the berm. Shorenormal processes.*

Berm. *The nearly horizontal part of the beach landwards of the sloping foreshore. Shorenormal processes.*

Berm Crest. *The seaward limit of the berm. Shorenormal processes.*

Cusps. *One of a series of short ridges on the foreshore extending normal to the shoreline and recurring at more or less regular intervals. Shoreparallel processes.*

Foreshore. *The sloping part of the beach between the berm and the low tide level. Shorenormal processes.*

Low Tide Terrace. *The flat portion of the beach seawards of the beach face exposed at low water. Shorenormal processes.*

Offshore. *The zone seawards of the low tide mark. Shorenormal processes.*

Rip Channel. *Channel cut by seawards flow of rip currents which usually crosses the longshore bar. Shoreparallel and shorenormal processes.*

Trough. *[Longshore trough, runnel or low] Elongate depression or series of depressions along the lower beach or in the offshore zone which may be exposed at low water. Shorenormal processes.*

bays and headlands. Nevertheless Bird (1984) estimates that there are more than 500,000 km of coastline around the world ocean. Detailed studies by the U.S. Army Corps of Engineers (Shore Protection Manual, 1984) show that, of the 76,100 kms of U.S. shoreline exclusive of Alaska, about 19,550 kms are beaches (33%) and the remainder are rocky or urbanised. If one assumes that the same percentage applies to the rest of the world then there are more than 170,000 kms of beaches around the world ocean. All of these beaches exhibit a more or less smooth profile which is concave upwards and has a more or less straight or gently curved plan shape. This worldwide uniformity of form must suggest the action of common processes and it is the identification and analyses of these which leads to an understanding of the general features of beaches, and which is the subject of most of the succeeding chapters. We shall refer to these general features as *first order* forms, and find that the shape of the beach profile is generally due to *orthogonal processes*, that is those which operate along a vertical plane in the direction of wave advance which is roughly shorenormal, whilst the plan shape of the beach is generally due to *longshore processes* which operate in a shore-parallel or coastwise direction.

First order forms often have superimposed upon them smaller scale *second order* features and Table 1.1 details these with a note which suggests whether the feature is controlled by orthogonal or by longshore processes. The table is assembled from Wiegel (1964), Shepard (1973) and the Glossary of Geology (1972).

If the beach is examined in still finer detail then *third order* forms are revealed which range from wave generated ripples, swash and backwash marks and drainage channels to the individual grains and internal bedding structures which constitute the fabric of the beach deposit. We shall see that second and third order forms are very much a product of local beach conditions and are therefore not usually amenable to general analysis.

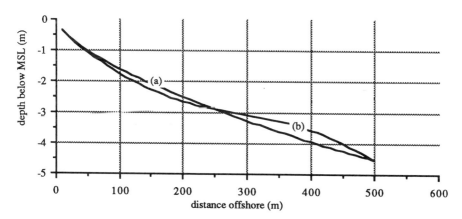

Figure 1.5 First and second order orthogonal features showing (a) the summer or swell wave, monotonic profile and (b) the winter or storm wave profile with sediment moved offshore to form a ridge and trough system. The profiles were simply generated using $h=ax^b$ for the monotonic profile and $h=ax^b + cx^b sin(2\pi x/500)$ for the barred profile, where a, b and c are appropriate constants. The form problem is therefore to explain these functions in terms of the processes which operate within the orthogonal system.

1.5 A Hierarchy of Models

The definition of a model in Section 1.3 represents both the overall framework within which geomorphologists seek to comprehend landform systems and also, at a different scale, the basic comprehension of each process or sub-process within that framework.

9

Thus, although Figure 1.4 is a model for the whole system, the relationship between molecular attractions and turbulence in the boundary layer is itself a model which forms a part of the hydrodynamic processes and thus of the whole. In later chapters we shall review some of the more important work which has taken place in beach geomorphology during the last fifty years, and rather than simply provide a chronological list of papers, the contributions have been ordered within a hierarchy of model types.

The hierarchy is based upon the way in which research is conducted and also upon the two, parallel aims of science, which are to provide on the one hand a predictive capability and on the other hand an explanation of natural phenomena. The hierarchy has three levels. The simplest approach is to formulate a *descriptive model* which is generally a map or a diagram or perhaps a few sentences linking the parameters of interest. Thus observation has shown that steep beaches tend to be associated with long waves. This is a descriptive model relating the two parameters, wave length and gradient. It is clear that descriptive models offer no explanation and very little predictive capability.

Table 1.2 A Hierarchy of Models		
Type of model	Explanation	Prediction
3. Theoretical	Good	Good
2. Empirical	Poor	Good
1. Descriptive	Poor	Poor

At the second level (Table 1.2) we erect *empirical models* by making measurements of the parameters and by then deducing a best fit or statistical relationship between them. Thus we shall see in Chapter Fourteen that King (1972) measured the wave length and gradient of a number of beaches and produced a graph of the results. The regression equation through the data is the empirical model and again it is clear that empirical models offer a reasonable predictive capability, but still very little explanation. The differences between descriptive and empirical models are often subtle (cf Section 2.1) and, therefore, we shall review both descriptive and empirical orthogonal system models in Chapter Fourteen.

The third type of model is theoretical and theoretical models are distinguished from empirical models because the model is erected *before* the data is collected. Thus a relationship is proposed between the parameters of interest and only then are field or laboratory data collected to test the model. Theoretical models offer both a predictive capability and, if the tests support the model (or rather do not refute its predictions), then it is assumed that the processes which are proposed in the theory provide an explanation of the phenomena. Thus we shall see in Chapter Fifteen that Inman and Bagnold (1963) proposed a theoretical relationship between the rate of percolation of water into the beach and the gradient of the beach surface. This model was later tested and whilst rather limited it does offer a reasonable explanation and a certain predictive capability.

Theoretical models are divided into process-response and systems models. Process-response models are those in which a given input generates a particular output through a specified process or transfer function. Alternatively systems models include a feedback process by which the parameters at any one part of the system can influence the processes at an earlier stage. The lower box in Figure 1.4 illustrates the principle by considering that the morphology generated by the various processes can affect both the hydrodynamics and sediment dynamics of the system. This type of approach is also known as non-linear modelling to distinguish it from linear, process-response models. Tanner (1974) recognised the application of a systems model to beaches when he wrote:

The equilibrium idea is that an energetic wave system will establish in due time and barring too many complications, a delicately adjusted balance among activity, three-dimensional geometry and sediment transport such that the system will tend to correct short or minor interference.

In this book we shall attempt to utilise the systems approach to develop a model for the form function in terms of the independent parameters in order to explain and to predict the geomorphological features of natural beaches. Unfortunately, it is often difficult to separate the theoretical process-response and systems models. However it is possible to classify both types of model on the mathematical techniques which are employed to solve the process equations. This classification is used later in the book and we shall review the orthogonal systems models which employ analytical techniques in Chapter Fifteen and those which employ numerical techniques in Chapter Sixteen. The following section introduces the computer based numerical techniques which are to be employed in later chapters.

1.6 Computer Modelling

The bulk of this book is concerned with the erection and testing of a systems model of the type which was introduced in the previous section. The model will contain one or more formulae for each of the processes, and the enumeration of more than thirty processes which are considered here is detailed in the following chapter. Although the construction of the model is only theoretically possible due to the research of the very many scientists listed in the bibliography, it is practically possible only due to the use of modern computers. The results have been compiled into a micro-computer model named SLOPES which is a geomorphological acronym for the Shoreline and Orthogonal Process Emulation System. SLOPES is written within the Microsoft Excel modelling environment and for the Apple Macintosh computers. This deliberate choice of modelling environment significantly affects the format and utility of this book, and the three main reasons for the rationale should be stated clearly.

Firstly, the beach system is complex and although some of the simple models which will be described in later Chapters have permitted the production of analytical solutions for certain aspects of the problem and although these do appear to offer a clear explanation, it has become apparent that the potential dangers of model development from such a narrow and strict deductive concept are particularly real (Anderson and Sambles, 1988). As Rescher (1962) argues, such a view 'may be buttressed by fond memories of what explanation used to be like in nineteenth-century physics'. A realistic model of the beach system cannot now be amenable to analytical solution, and the alternative is the use of numerical techniques and iterative solutions on the computer. However, once completed, the model itself will be examined (Chapters 16 to 19) to determine its sensitivity to both thge input parameters and the operative processes because herein lies modern geomorphological explanation. It is necessary then to computerise the modelling methodology in order to obtain numerical solutions to the process equations.

Secondly, computer models are not in themselves a new departure in geomorphological research (Chapter Sixteen), but have generally involved the construction of a long and complicated program running in one of the higher level languages such as Fortran or Pascal on a mainframe installation. This has inevitably resulted in a model which is designed to serve a very specific research project, and which is then difficult to adapt to different process functions as new results become available. The model is written, written up and then usually abandoned as the author moves into a different problem. Anderson and Sambles (1988) also argue, with good reason, against the continuation of this approach, and advocate instead an open programming architecture in which process modules are fully

11

visible and documented and from which they can easily be extracted and replaced as the model develops. The novel use of the Excel spreadsheet environment in SLOPES is well suited to this requirement as changes can be incorporated with great facility and with little programming experience. It is necessary then to present the computer model in such an open architecture programming environment.

Thirdly, the model running in Excel on a modern, high powered but reasonably priced micro-computer is extremely portable and can be used immediately by the reader with a Macintosh system running Excel or with a PC system running Lotus or a compatible program. The results of the modelling are reported here, but extension of those results or the investigation of other scenarios can easily be undertaken by the reader. The model is available from the address listed on page ii of the preface to this book. It is necessary to present the computer model in a common format so that the results reported here can be subjected to proper scrutiny.

1.7 The Beach Literature

Although the diversity of sciences which are utilised in beach research means that the student can usefully read textbooks from many disciplines, the subject has been specifically served for almost two decades by C.A.M.King's *Beaches and Coasts* and by Komar's *Beach Processes and Sedimentation*. In addition a number of other books include sections on beaches and beach processes. These include Pethick's *An Introduction to Coastal Geomorphology*, which is a general undergraduate text and Zenkovitch's *Processes of Coastal Development* which reviews the Russian ideas and literature. From an engineering standpoint there is Wiegel's *Oceanographical Engineering* which includes tables and graphs for the wave equations, the two volumes of the Coastal Engineering Research Centre's *Shore Protection Manual*, which is similar to, but more comprehensive than Wiegel and the two volumes of Silvester's *Coastal Engineering*. Each of these will justify a thorough reading, and recently five good texts have emerged which have detailed sections on beach science. These are Dyer's *Coastal and Estuarine Sediment Dynamics*, Sleath's *Sea Bed Mechanics*, Carter's *Coastal Environments*, Horikawa's *Coastal Enginnering* and, more recently, his *Nearshore Dynamics and Coastal Processes.* which describes the Japanese Nearshore Environment Research Center (NERC) Program. Finally there is the volume of papers entitled *Nearshore Sediment Transport* which is edited by R.J.Seymour and describes the results of the American Nearshore Sediment Transport Study (NSTS), and the volume edited by Lakhan and Trenhaile which describes a number of differing approaches to coastal modelling.

The number of journal papers on beach research is multiplying at a seemingly exponential rate each year, but the important references which are used in this book appear to be concentrated in selected journals. The geomorphologists and sedimentologists have a preference for *Earth Surface Processes and Landforms, Sedimentology* and *Marine Geology*, whilst the *Journal of Coastal Research* contains a broad spectrum of papers. Each edition of the *Journal of Geophysical Research* usually contains a paper on either shallow water waves or beach sediment transport, and again the engineers are served by *Coastal Engineering* or the *Journal of the Hydraulics Division of the American Society of Civil Engineers*.

chapter two

BEACH MODELLING AND MONITORING

2.1 Introduction

Figure 2.1 A physical model of waves in a laboratory flume

There are, in the literature, a number of classifications of models (e.g. Rivett, 1972, Highland, 1973; Gordon, 1978). A comprehensive account of the various types of models used in the earth sciences is given by Krumbein, (1968), Huggett (1980) and Woldenberg (1985) and in geomorphology by Chorley (1967) and by some of the papers in Anderson (1988). There are also two, readily available, classifications of coastal models. In the first (Fox, 1985), physical, statistical, probabilistic and deterministic coastal models are described. A second, more comprehensive, classification is given by Lakhan and Trenhaile (1989a) which divides coastal models into physical models and mathematical models. The latter group is detailed below, but some mention should be made of physical models for they often provide the experimental data on which more sophisticated analyses are based. A model is said to be physical when the representation is "physical and tangible with elements made of materials and hardware" (Lakhan and Trenhaile, 1989a). The most common example in coastal work is the scaled flume or tank model (Figures 2.1 and 2.2)

13

in which the operation of the prototype is simulated. Physical models have proven central to the design and construction of coastal engineering works in which the response of the prototype to new construction works is assessed in order to optimise design considerations, and numerous examples are given by papers in Dalrymple (1985) and Lakhan and Trenhaile (1989b). Svendsen (1985), working on physical model of coastal waves, suggested that there are three complimentary goals that can be pursued with a physical model. These are to seek insight into a new phenomena, to obtain measurements to verify (or disprove) a theoretical result and to obtain measurements for phenomena which are so complicated that they have not been theoretically analysed. We shall return to the use of physical models as data sources in later sections, but shall not otherwise consider this important field in more detail.

Figure 2.2 Large scale physical model of the Humber Estuary used to investigate the consequences of land reclamation. (Courtesy of Associated British Ports)

This book is, instead, concerned with the erection and testing of theoretical mathematical models for the orthogonal profile, and such models are classified by Lakhan and Trenhaile (1989a) according to different criteria, depending upon the types of model data, parameters, mathematical relationships, solution techniques, time related behaviour and structure. Although such a classification is useful, for it provides a hierarchical structure within which to set orthogonal systems models, it is somewhat complicated and has therefore been simplified and its scope has been extended for the present purposes. The resulting classification is shown in Table 2.1 and, with the exclusion of the physical models, divides orthogonal systems models into the two main groups which were introduced in Chapter One. These two groups represent the empirical and the theoretical approach and can be separated in so far as the former seeks to identify patterns from experimental data, whereas the latter seeks to employ experimental data to test a formal representation of the system. The empirical models are divided into descriptive and correlation models wherein the former include sentences or graphs whereas the latter involves the erection of (usually) statistical relationships between the independent parameters and the orthogonal profile form. The theoretical models are divided according to the mathematical procedures which are used to solve the equations and consist of analytical models (in which the equations are solved in a more or less simultaneous fashion to achieve formulae for the orthogonal profile) and numerical models (in which the equations are solved by numerical techniques, usually on a computer). The descriptive and correlation models are detailed in Chapter Fourteen, the analytical models are detailed in Chapter Fifteen and the numerical models are detailed in Chapter Sixteen.

Table 2.1
A Classification of Beach Models

A PHYSICAL

B EMPIRICAL

Word Model
Diagrams
Static form functions (h=f(x))
Static form functions (h=f(x, inputs))
Dynamic form functions (h=f(x, inputs, t))

C THEORETICAL	Analytical	Static	Stochastic
			Deterministic
	Analytical	Dynamic	Stochastic
			Deterministic
	Numerical	Static	Stochastic
			Deterministic
	Numerical	Dynamic	Stochastic
			Deterministic

The relationships between the data which is used to construct empirical models or to test the theoretical models and the generation of the models themselves is central to the ethos of coastal modelling and it is clear that the successful explanation or prediction developed by any model depends upon the quality and quantity of available data. Measurements of the processes which operate in the orthogonal system certainly date back to the nineteenth century, and it is likely that mariners were concerned with the form of the seabed close to the shore in far earlier times. It is not possible within this book, however, to cover all of the various techniques which are used to make field or laboratory measurements, and these details are in any case readily available elsewhere. Instead the present chapter will conclude with sections on some of those techniques so that the reader will be aware of the logistical and technical problems posed by the environment.

Much of the work in this book deals with the identification and presentation of algorithms for the individual hydrodynamic, sediment dynamic and morphodynamic processes which are to be included in that model. However, to begin immediately with a concentrated presentation of the processes would be rather like asking a patient to undertake a long and painful operation without telling him what it was for. In order to give some direction to more than two hundred pages of process functions, let us begin by considering the geomorphological results. There are already a number of models in existence which purport to explain and to predict the orthogonal system, and some of these have now been in the public domain for more than thirty five years. Firstly, however, this chapter presents a brief and selective review of the literature which deals with geomorphological models for the orthogonal system. The chapter is designed to provide an overview of the orthogonal system, and a more detailed treatment of the models is presented after the process sections in Chapters Fourteen to Sixteen. It is designed also to provide a framework for that overview by describing the important developments in the same hierarchical structure as was employed in Chapter One. That is descriptive models are followed by empirical models, and then by analytical and numerical process-response models. We begin in the seventeenth century, although Komar (1976) notes that Leonardo da Vinci understood many of the features of beach behaviour, as evidenced by his designs for reclamation of the Pontin Marshes in northern Italy.

2.2 The Orthogonal Form

Following the great steps forward in world science which resulted from the development of Newtonian physics in the seventeenth and eighteenth centuries and the calculus at about the same time, the nineteenth century was a time of consolidation as these powerful tools were applied to all manner of natural and man-made phenomena. Chemists were arguing about the inner structure of atoms, and the Challenger expedition began the science of oceanography (Linklater, 1972).

Beaches were of concern from a practical point of view to coastal engineers, charged in Europe with developing Victorian promenades, and throughout the world with improving harbour facilities. However the answers which they devised were generally along the lines of "make it massive and if it falls down make it more massive still". Gentlemen geologists were finding marine strata, in the Alps and elsewhere, at great heights above sealevel, and were asking geomorphologists for a better understanding of present day environments in order to comprehend these fossil examples. Gilbert's treatise *The Topographic Features of Lake Shores* and Geikie's *Outlines of Geology* published in 1896 began a series of essentially descriptive books which has already continued for almost a century and will no doubt survive longer. Though often beautifully illustrated there is little mention in such texts of the processes responsible for the observed landforms, and in the early days a considerable debate seems to have raged about the respective roles of waves and tidal currents.

The theoretical investigation of water waves was, however, proceeding apace in the nineteenth century, as mathematicians developed most of the wave theories with which we are familiar today. Thus the work of Gerstner (1802, referenced in Sleath, 1984), Russell (1844), Airy (1845), Stokes (1847, 1880), Boussinesq (1872, referenced in Sleath, 1984), Rayleigh (1876, 1877) and Reynolds (1877) provided the theoretical basis for much of the wave work which is here presented in Chapters Five and Six. It should however be remembered that these were, until recently, merely mathematical abstractions on the properties of ideal fluids, and it was not until almost a century later that any comparison with the real world was made and, only in the last few years, have we seen serious testing of the theories with the development of high speed current meters (Section 2.9). The achievement of any kind of confidence in the predictive capability of these wave equations is therefore only a relatively recent development.

The period between the turn of the century and the second world war saw the beginnings of the process approach to all geomorphological environments, not least to the beach system. The now classic work by D.W.Johnson, *Shore Processes and Shoreline Development* was published in 1919 and was perhaps the most important contribution during this period. It is a largely descriptive book, but certainly recognises that both waves and tidal currents are responsible for the movement of sediment in the nearshore region. It also offers the first of many classifications of shorelines, and of beaches in particular, dividing the forms into shorelines of emergence and of submergence. Johnson identifies a gradation from a young, cliffed coast, to an old peneplaned continental shelf upon which loose sediment forms a thin veneer. Interestingly, the bulk of Johnson's book considers the large scale shoreline features, in much the same way as we here concentrate on first order beach forms. The last two chapters of Johnson's book deal with bars and ridges and small scale features and include some of the earliest photographs of plaster casts of ripples and other bedforms.

The Second World War found the British and Americans with insufficient knowledge to conduct the amphibious landing operations required in the Pacific and in Europe and highlighted the need for a better understanding of the generation and propogation of ocean waves and their breaking characteristics in shallow water, of the character of water movements in the nearshore region and of the resulting sediment movements and the factors affecting the slope of the beach face (Komar, 1976). The result was a rapid expansion in all fields of beach research which has continued to the present day. This research has concerned many aspects of the coastal environment from meteorology, through waves, tides and sediment transport to the geomorphological and stratigraphic results of changing conditions over the longer time scales which include the effects of sealevel rise. Here, however,we are primarily interested in predicting and explaining the form of the orthogonal profile and will therefore concentrate on geomorphological models.

The first geomorphological result of these efforts was the formal confirmation of the cyclic nature of the beach profile. Shepard (1950) was working at the Scripps Institute on the Californian coast and surveyed local beach profiles over a period of years in the late nineteen forties and early fifties. He found that an offshore shift of sand occurred during storm conditions leading to the development of longshore bars at the expense of the beach face and the berm structure, whilst onshore sand transport and berm development at the expense of the bars followed upon the return to gentler wave conditions. This seasonal beach response has been observed on numerous occasions and typical results are shown in Figure 2.3.

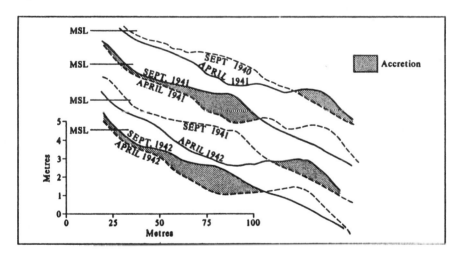

Figure 2.3 Profile changes at Scripps Pier, La Jolla, California, showing the tendency to shift from a more evenly sloping but steeper swell profile to a flatter storm profile (after Shepard, 1950).

These changes have become known as the *beach cycles* and the terms *winter, storm or destructive* were applied to conditions which lead to offshore sand transport and bar development whilst the terms *summer, swell or constructive* were applied to conditions which lead to onshore sand transport and berm development. Such cycles, either spatially or seasonally were often described from field measurements in the nineteen fifties by, for example, Shepard and LaFond (1940), Shepard and Inman (1950, 1951a) Inman (1953), Bascom (1954), Bruun (1954), Inman and Rusnak (1956), Ziegler *et al.* (1959), and more recently as a direct function of changing wave conditions by, for example, Rohrbough *et al.* (1964), Fox and Davis (1978) and Carr *et al.* (1982).

By the beginning of the second half of the twentieth century, mathematicians and fluid dynamicists had proposed various formulations for the wave equations, engineers were beginning to apply these equations to marine sediment transport and geomorphologists had obtained sufficient, long term records of form evolution to identify tentative process relationships.

2.3 The Orthogonal Processes

Wiegel (1964) was perhaps the first to offer a detailed list of the beach processes. He follows earlier authors by noting that it is the interaction of the waves and the beach which is responsible for the detailed morphology and then describes the first group of processes which begin to change the incident waves:

> *When waves enter shoaling waters along the coast and reach a depth of about one-half the wave length, the wave begins to "feel bottom"; they start to slow down, decrease in length, and after a bit, increase in height. When they reach a depth of approximately one to one and a half times their height, they break. On a steep beach, they break at the beach face; on a flat beach, they often break on an outer bar, then re-form as much lower waves and continue to shore, where they break again. In deep water, the water particles move in nearly closed circular orbits, with a small mass transport in the direction of wave propogation. The orbit's diameters decrease exponentially with distance below the surface until, at a depth equal to about one-half the wave length, there is effectively no motion. When the total depth of water is less than about one-half the wave length the waves begin to "feel bottom", i.e., there is motion at the bottom. In addition the particle motion becomes elliptical rather than circular. The more shallow the water, the flatter the ellipse; and the greater, the bottom velocity.*

This group are now referred to as the *hydrodynamic processes* (Figure 2.4). Wiegel continues:

> *A mechanism is thus available for moving sediments, the distance from the shore at which sediments can be put in motion depending upon the wave dimensions and sediment size; the motion, however, is relatively small. But the zone in which the waves break is one of great turbulence. The smaller materials are thrown into suspension, and once in suspension even currents of very low velocity are capable of transporting material in considerable amount. Sand moves along beaches as well as on and off them.*

This group is now referred to as the *sediment dynamic processes* (Figure 2.4) and it is the net transport of sediment in both the on/offshore and longshore directions which leads to the beach form. The application of continuity and mass balance principles to the transport of sediment in this way constitutes the *morphodynamic processes* (Figure 2.4). Wiegel's description of beach processes appears now, some quarter of a century later, to be

Table 2.2
BEACH PROCESSES

HYDRODYNAMIC PROCESSES

Deep Water Wave Parameters H_∞, T and α_∞	
Shoaling Transformations	
Shoaling Length	#5.4
Shoaling Height	#5.5
Frictional Height Reduction	#5.6(a)
Percolation Height Reduction	#5.6(b)
Refraction	#5.6(c)
Reflection	#5.6(d)
Breaking	#5.6(e)
Tidal Height	#8.4
Local Wave Parameters H,L,T and α	
Orthogonal Currents	
First order currents	#6.2(a)
Second order currents	#6.2(b)
Drift current	#6.4
Longshore Currents	#7.4
Local Flow Parameters $u(x,y,t)$, $v(x,y,t)$	

SEDIMENT DYNAMIC PROCESSES

Orthogonal Sediment Transport	
Bedload	#10.2
Suspended Load	#10.4
Bedslope Effects	#10.6
Grain size effects	#10.2-5
Longshore Sediment Transport	
Bedload	#11.3
Suspended Load	#11.3
Local Sediment Transport Parameters $i(x,y,t)$, $j(x,y,t)$	

MORPHODYNAMIC PROCESSES

Orthogonal Continuity	Ch.13
Longshore Continuity	Ch.13
Feedback Processes	
Hydrodynamic Feedback	Ch.17
Sediment Dynamic Feedback	Ch.17
Three Dimensional Beach Form Parameters $z(x,y,t)$	

A summary of the orthogonal system showing the input and output parameters for each group of processes. The numbers in the right hand column indicate the sections and chapters of the book which detail the corresponding functions.

simplistic but it is essentially correct and it was to take the next twenty five years to build and test the theories which would allow us to predict and quantitatively to evaluate their results.

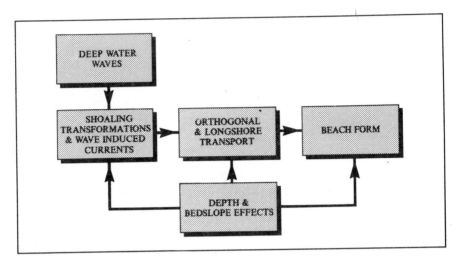

Figure 2.4 The orthogonal system.

The processes are summarised by the upper boxes in Figure 2.4 and by Table 2.2. We shall see (Chapter 5) that the wave form in deep water some distance from the shoreline is the energy source which drives the system, and is characterised by the wave height, H, the wave period T and by the direction of propogation of the waveform, α. The values of these independent variables may change as the wave progresses shorewards and the deep water values are therefore symbolised by the subscript ∞, thus H_∞ and α_∞.

The first group of three processes operate to bring the deep water wave into a local water depth, h, and to generate nearbed currents along both the wave orthogonal path and, shorewards of the breaker zone, in the longshore sense. This group is known as the hydrodynamic processes and is divided into the shoaling transformations which account for changes in the wave profile and the orthogonal and longshore current processes which generate nearbed flow fields from particular surface wave profiles. Table 2.2 indicates the sections in this book which deal with the equations which govern each process. The shoaling transformations commence with a deep water wave which has a wavelength which is related to its period and the wavelength reduces as the depth decreases (*shoaling length*). The wave height increases as the length decreases in order to maintain the potential energy of the wave (*shoaling height*). During shoaling wave energy is dissipated, and therefore wave height reduces due to *seabed friction*, the generation of percolation currents within the seabed (*percolation*), wave reflection from the seabed (*reflection*), alteration in wave direction (*refraction*) and wave *breaking* . It should be noted that wave refraction can cause either a divergence or a convergence of the wave paths and can therefore result in either an increase or a decreases in the wave height. The shoaling wave transformations depend upon the local seabed characteristics and upon the local water depth, and the latter may be constantly changing due to the *tidal amplitude* .

The action of the shoaling transformation processes on the deep water wave parameters leads to the second definition of variable values known as the *local wave parameters* which are T, H, α, L, and the water depth h. The remaining hydrodynamic processes operate to generate nearbed currents from the local wave parameters. The wave generates nearbed oscillatory currents with an onshore flow beneath the crest followed by an offshore flow beneath the trough. These are known as the *orthogonalal currents* u(x,t) and operate in the two dimensional vertical plane of the beach profile. Orthogonal currents are distinguished from *longshore currents* which are generated within the surf zone and operate in the two dimensional horizontal plane of the beach plan and which arc symbolised by v(x,t).

The effect of the hydrodynamic processes on the local wave parameters leads to the third definition of variable values known as the *local flow parameters* which are u(x,t) and v(x,t).

The third group of processes generates onshore, offshore and longshore movements of sediment as a result of the flows. These are called the *sediment dynamic processes*. Once the oscillatory currents become competent, then sediment is moved onshore and offshore by rolling and sliding along the seabed (*orthogonal bedload transport*) and in the body of the water (*orthogonal suspended load transport*) and both of these loads are severely affected by the influence of bedslope (*slope effects*). Conversely the longshore current may be competent to transport *bedload* and *suspended load* . The effect of the sediment dynamic processes on the flow parameters leads to the fourth definition of variable values known as the local transport parameters i(x,t) and j(x,t).

Finally it is the application of continuity and mass balance principles that define the relationships between the net sediment transport and the evolution of the morphological form. These are known as the morphodynamic processes and consist of *orthogonal continuity* and *longshore continuity* which together lead to local erosion and accretion and hence to the definition of the fifth and geomorphological set of parameters, the *three dimensional beach form.*

It becomes apparent that these processes can be linked together in an ordered fashion and at a fairly basic level this order must be: the offshore waves travel into the nearshore region, generate currents in both the shorenormal and longshore sense and these currents transport sediment. Nett sediment transport results in accretion or erosion and this leads to the three dimensional form of the beach. It will become apparent, however, that such an order, though initially perfectly adequate, is in fact insufficient because the beach form does itself affect the operation of the processes. The bathymetry controls the manner in which the deep water waves change and generate currents as they shoal: nearshore deep water brings the breakers shorewards, whereas an infilling moves them offshore. Again the slope and orientation of the seabed with respect to the shoaling waves affects the movement of seabed sediments: a steep seabed slope encourages downslope sediment movements whilst waves which break parallel to the shoreline cannot generate any overall coastwise movement of sediment.

It is necessary, therefore, to offer back the three dimensional beach form produced by the beach processes to the processes themselves, so that the process path is no longer a straight line. Instead it is a series of at least two and perhaps many more feedback loops of the type which were discussed in the preceeding chapter. The *feedback processes* are represented by the lower box in Figure 2.4 and operate to produce the three dimensional beach form as a result of the transport parameters. The effect of the morphodynamic processes leads to the fifth and final definition of variable values known as the *Form Function* (Chapter One).

Various attempts have been made at analysing the morphodynamics of the orthogonal system and these are dealt with in Section D of this book. There are, in principle, only two approaches to the problem. Firstly, no presumptions are made about the pre-existing forms and the process equations are solved directly through the specification of an equilibrium condition. The most common specification is that there is zero net sediment transport throughout the system. The result is then expressed in a manner which represents the stability of the system. This type of solution is known as analytical system modelling and is discussed further in Chapter Fifteen. Alternatively a model is constructed which simulates some or all of the processes across a given initial seabed bathymetry. The resulting sediment transport is then used to change the baythymetry and the model is re-run in a series of iterations. The iterations show that either a stable form evolves, or else the seabed continues to change with each new iteration. This leads to considerations of the state of equilibrium or otherwise of the form. This type of solution, which is known as numerical systems modelling, is discussed further in Chapter Sixteen.

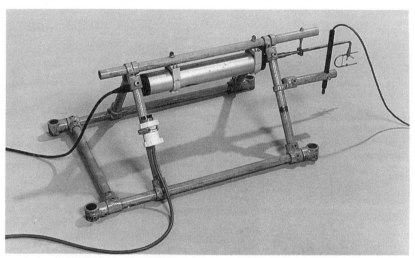

Figure 2.5 Photograph of a pressure transducer mounted on the galavanised frame of a beach rig. The transducer is powered along a single cable from a shorebased supply and converts varying pressure levels on the seabed into voltages. These signals are returned along the cable and recorded on chart recorders, paper or magnetic tape or, most recently directly through analogue to digital converter systems onto microcomputers for subsequent analysis.

2.4 The Measurement of Wave Profiles

Water depth and surface wave profiles have been measured with a variety of techniques which range from simple visual observations of floats (Paris, 1867 referenced in Defant, 1961) and staffs (Ingle, 1966) to the use of still cameras (Cote, 1960) and cine cameras (Suhayda and Pettigrew, 1977). However interest in wave research during the second World War, saw a development of the automated procedures which are described by Klebba (1945) for the Woods Hole Oceanographic Institution and Deacon (1946) for the British Admiralty Research Laboratories. Modern devices range from accelerometer buoys in deep water (Ewing, 1986) to upwards looking echo sounders and downwards looking lasers (Draper, 1970; Driver, 1980), resistance staffs (Koonitz and Inman, 1967) and capacitative or resistance gauges (Guza and Thornton, 1989). However the most rugged and practical device for shallow water work under wind waves, or general depth and tidal recording is the pressure transducer. Modern transducers include a piezo-resistive element which provides an electrical response proportional to the applied pressure and can therefore be calibrated to provide water depth and wave profile data. The construction and calibration of a shallow water pressure transducer system for wave recording on beaches is described in Hardisty (1990ad). A photograph of such a device being deployed as part of the British Beach and

Nearshore Dynamics Project is shown in Figure 2.5. The use of pressure transducers is discussed by Bishop and Donelan (1987) with special reference to pressure attenuation with depth.

2.5 The Measurement of Orthogonal Currents

There is now a large range of current meters, from traditional impellor or rotating cup devices, through acoustic and laser dopplers (Becker *et al.*, 1967), hot wire (Bourke et al., 1967) and hot bead (Bradbury and Castro, 1971) devices to electromagnetic (Hotta, 1981; Guza and Thornton, 1980, 1989) and acoustic (Tsuchiya *et al.*, 1983) instruments. However only the last two types are presently available in the form of a sufficiently robust and high sample rate instrument for the work described here. The electromagnetic (EM) devices are based upon the measurement of the electrical disturbance caused by water flow in a magnetic field generated by a small coil. The ultrasonic current meters (UCM) are based either on the Doppler shift or path time principle and is shown in Figure 2.6. An electromagnetic current meter array is shown deployed during the British Beach and Nearshore Dynamics Project in Plate V. A sled system used for towing an electromagnetic current meter rig across the surf zone is described in Sallenger *et al.* (1983).

Figure 2.6 Gravel movement under waves in a laboratory flume showing the ultrasonic current meter and small impellor used to monitor orthogonal currents and a twin hydrophone acoustic system listening to the gravel movements.

2.6 The Measurement of Orthogonal Transport

The measurement of the sediment transport rate, which is defined as the dry weight of solid material moving across unit width of the bed in unit time (Chapter Nine) has, until recently, lagged behind the measurement of the hydrodynamic parameters which were described in the preceeding sections. This lack of technical sophistication has been evident in both the accuracy of the transport rate measurements and the temporal resolution or frequency at which measurements are made. Nevertheless the sediment transport rate has been measured by techniques which range from morphological suveying (Langhorne, 1982), through traps (Hardisty and Hamilton, 1984), water samplers (Kana, 1976) and electromagnetic sensors to optical (Hanes and Huntley, 1986), acoustic devices (Thorne, 1986) and ultrasonic flux meters (Jensen and Sorenson, 1972; Vincent *et al.*, 1982). Once again only the last two types of techniques are capable of providing the high rate of measurement which is required to compare the transport processes with the oscillatory in beach work. Optical techniques involving either the extinction of a light path (Hardisty, 1986) or the back scattering of light from a small source (Hanes and Huntley, 1986) are used for work on the transport of sediment is suspension above the bed. The only principle which appears to offer the potential for high frequency (> 1Hz) measurements of bedload processes on the beach is based upon the idea that, particularly for coarser sediment, the moving grains undergo inter-granular collisions and generate noise which can be detected by hydrophones.

2.7 The Measurement of Orthogonal Form

The final set of parameters in Table 2.1 which are measured in the orthogonal system represent the beach form itself. Measurements are made using standard surveying techniques on the exposed beach surface and by hydrographic surveying along the orthogonal profile offshore. Details of beach and nearshore profiling are to be found in Inman and Rusnak (1956), Nordstrom and Inman (1975) and Aubrey (1978) and recent reviews can be found in papers in Horikawa (1988) and Seymour (1989). The results of such survey work are discussed in Chapter Fourteen.

chapter three

PHYSICAL PROPERTIES OF BEACH SEDIMENTS

3.1 Introduction

Although the beach forms the interface between the three phases of matter, that is between the sediment, the seawater and the air, we are here concerned only with the first two, the effects of onshore and offshore winds being locally negligible, though relevant to the processes of wave generation which are discussed in Section B. Before proceeding to consider the properties of moving water and sediment, it is firstly necessary to define the physical properties of the stationary phases. The present chapter deals with the properties of beach sediments and the following chapter then deals with the properties of sea water. The presentation will be restricted to those properties which are appropriate to the work in later chapters. The following is based on the treatments by Leeder (1982) and Allen (1985) along with more specific references which are cited in the body of the text.

3.2 The Size of Beach Sediments

The grain size distribution of natural detritus has long been of interest to sedimentologists and geologists because it is one of the few diagnostic properties which can be identified in the field and which can help the palaeoenvironmental intepretation of sedimentary deposits. A great many samples of modern sediments have been collected in the field and analysed in the laboratory in the hope that characteristic sizes and distributions would be representative of particular environments. Although taken to extremes in many statistical papers, the hope proved to be forlorn and the last decade has seen a successful return to process-oriented grain size intepretations based upon the physics of sediment transport. Initially however a description of the size and size distribution of beach sediments is required. The techniques involve sampling, size measurement, and the analysis and representation of the results.

The size or grade of natural sediments is defined by the conventional terms shown in Figure 3.1. The use of Coulter Counters, pipettes, burettes, hydrometers and the various optical fall instruments is most suited to clays and silts, which are finer than most beach sediments. Instead, sieves are used to determine the percentage by weight of the total sample falling between particular size bands as shown in Figure 3.2. Careful choice of the sieve size interval allows the distribution to be well represented, and it has been found that a logarithmic or Wentworth (Wentworth, 1922) size scale is preferable to a linear one. It has also been found that the distribution of the logarithm of the grain diameters of natural sediment follows an approximately Gaussian normal form. The ϕ (phi) scale is preferred by sedimentologists, where the ϕ grain size D_ϕ is related to the millimetre size D_{mm} by:

$$D_\phi = -\log_2(D_{mm})$$

Eq.3.1

	PEBBLES	phi	Size mm	microns		
		-2	4	4000	Medium Gravel	
	GRANULE			2000		
		-1	2		Fine Gravel	
	V. Coarse	0	1	1000		
	Coarse	+1	1/2	500	Coarse	
	Medium	+2	1/4	250	Medium	
	Fine	+3	1/8	125	Fine 1/10 mm	
	V. Fine	+4	1/16	62.5	V. Fine	
	Coarse	+5	1/32	31.3	SILT	
	Medium	+6	1/64	15.6	1/200 mm	
	Fine	+7	1/128	7.8	CLAY	
	V. Fine	+8	1/256	3.9		
	CLAY					

Left label: Wentworth (1922) after Udden (1898). *SAND / SILT* columns as shown. *Right label:* U.S. Bureau of Soils, SAND column.

Figure 3.1 Standard terms for sediment grain sizes

The ϕ value increases with reduction in the grain diameter so that the most common sand sized sediments have reasonable values in the range 0ϕ to 4ϕ as shown in Figure 3.1. It has been found that the size distribution of natural sediments approximates to a Gaussian normal curve for which:

$$P(D) = \frac{1}{\sqrt{2\pi}\,\sigma_\phi} e^{-(D_\phi - M_\phi)^2/2\sigma_\phi^2}$$

Eq.3.2

where σ_ϕ is the standard deviation of the distribution about a mean value of $M\phi$. The integral of this curve is the cumulative size distribution (Figure 3.3). The conventional percentage greater than the stated size is plotted. Simple descriptive statistical measures can then be determined by reading off percentiles at the 16, 50 and 84% values signified by D_{16}, D_{50} and D_{84}. These are chosen because the D_{16} and D_{84} values represent two standard deviations from the modal value in a normal distribution. The methods used are discussed in a variety of papers of which Inman (1952), McCammon (1962) and Folk (1966) are commonly cited. Four statistical measures are normally derived:

$$M_{d\phi} = D_{50}$$
modal grain size
Eq.3.3

$$M_\phi = \frac{(D_{16}+D_{50}+D_{84})}{3}$$
mean grain size
Eq.3.4

$$\sigma_\phi = \frac{(D_{84}-D_{16})}{4}$$
standard deviation
Eq.3.5

$$sk_\phi = \frac{D_{16} + D_{84} - 2D_{50}}{2(D_{84} - D_{16})} \qquad \text{skewness} \qquad \text{Eq.3.6}$$

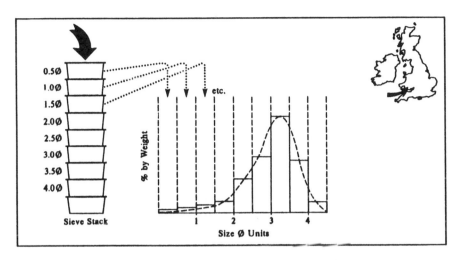

Figure 3.2 Standard percentage grain size distribution for beach sediment from Pendine Sands, S.W.Wales.

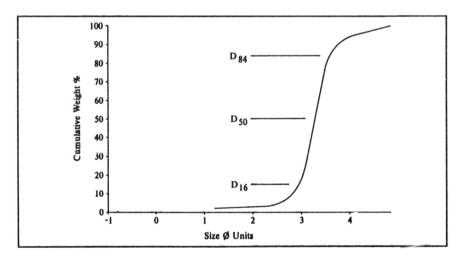

Figure 3.3 Cumulative grain size distribution of the sediment shown in Figure 3.2.

The first two represent the average size of the sediment grains, the third represents the sorting or peakedness of the distribution and the fourth indicates an asymmetry towards coarse (negative skewness) or fine material (positive skewness) in the sample. The sample from Pendine Sands shown in Figures 3.2 and 3.3 has $D_{50} = 2.1\phi$ and the cross beach variations are shown in Figure 3.4. In addition to these purely descriptive parameters, the similarity to normal distributions has prompted the use of a Gaussian probability axis, on which a normal curve plots as a straight line, as shown in Figure 3.5. It is apparent that

27

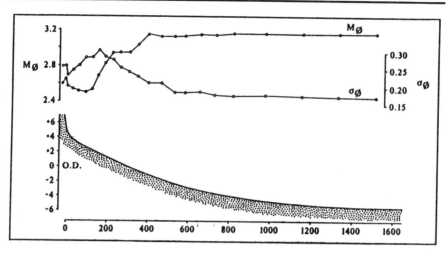

Figure 3.4 Cross beach variation in grain size at Pendine Sands, SW Wales (After Jago and Hardisty, 1984)

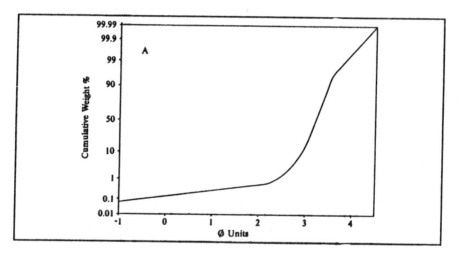

Figure 3.5 Probability axes size distribution

not one but three straight line sections are often present. This observation is very commonly the case with beach sediments, and has led to two contrasting though not mutually exclusive hypotheses. Authors such as Visher (1969) have suggested that the three sub-populations are caused by the three different modes in which loose sediment moves. The coarsest sediment known as bedload, rolls and slides along the bed, the finest known as suspended load moves within the body of the water whilst the middle population is called intermittant suspension. These modes of sediment transport will be rigorously defined and explained in Chapter 9.

The alternative hypothesis is due to Moss (1972) and co-workers who suggest that the sub-populations are due to packing controls on the grain matrix. The median size is seen as a regular framework with the finer size resting in the spaces between the larger grains and the coarsest being a lag deposit. A fuller appreciation of this idea requires an understanding of the geometrical packing considerations which follow.

3.3 The Packing of Beach Sediments

The controls on the packing of natural sediments have been determined in general and are described by Allen (1985), though detailed analyses are rare. Consider an accumulation formed of equally sized, spherical grains. The packing can be defined by the concentration, C, such that:

$$C = \frac{\text{Total volumes of grains}}{\text{Total volume of grains plus voids}} \qquad \text{Eq.3.7}$$

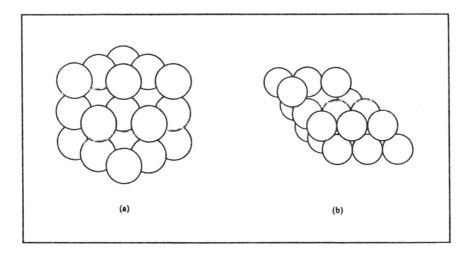

Figure 3.6 The arrangement of spherical particles in (a) cubic packing and (b) rhombohedral packing.

The loosest and most open packing of these grains occurs when the centres are at the corners of a cube as shown in Figure 3.6(a). If the grain diameter is D, then the total volume is $8D^3$ in which are eight grains each of volume $4/3 \pi (D/2)^3$. Thus this packing, known as cubic, has, regardless of the grain size, a grain concentration :

$$C = \frac{8 \, 4/3 \, \pi \, (D/2)^3}{8D^3} \qquad = 0.523 \qquad \text{Eq.3.8}$$

Alternatively Figure 3.6(b) shows an extremely close arrangement of the spheres which is known as rhombohedral packing and for which C = 0.740. In practice, natural sediments are not spherical and have a haphazard rather than regular packing. The actual concentration is determined as shown in Figure 3.7 by measuring the total volume V_1 of a sample in a graduated cylinder. Water from a second cylinder is carefully added until the sediment is saturated and the concentration is $(V_1 - V_2)/V_1$ where V_2 is the volume of water required to saturate the sediment. Tests, with a range of natural beach sands, show that they exhibit a narrow range of C varying from 0.60 to 0.64.

An alternative description of the packing of loose spheres was introduced by Bagnold

(1954), and again begins by considering the rhombohedral case. Now, suppose that the mass is dispersed uniformly so that the distance D between centres is increased to bD. If the resulting free distance between the spheres is s, then we have:

$$b = \left(\frac{s}{D}\right) + 1 = \left(\frac{1}{\lambda}\right) + 1 \qquad\qquad \text{Eq.3.9}$$

where $\lambda = D/s$ is defined as the linear concentration of the array. The volume concentration is then:

$$C = \frac{C_o}{b^3} = \frac{C_o}{\left(\frac{1}{\lambda} + 1\right)^3} \qquad\qquad \text{Eq.3.10}$$

where C_o is the maximum possible concentration when $\lambda = \infty$ (s=0) and is equal to 0.740 as above. Bagnold (1954) continues by noting that, for cubic packing, horizontal grain layers must be vertically displaced (dilatated) by a distance $(1/2)\sqrt{2}bD$ for one to slide over another, and thus the maximum value of λ is 8.3. Conversely, for rhombohedral packing, the dilatation distance is $\sqrt{(2/3)}bD$ and thus λ cannot exceed 22.5. In practice he suggests that an intermediate value of λ of 12 to 14 will apply to natural sediments. This definition of linear concentration will be utilised when consideration is given to grain shear in bedload transport in later chapters.

3.4 The Porosity of Beach Sediments

An alternative, and less common, representation of particle packing involves the porosity, P, of the sediment which is similar to the concentration for:

$$P = \frac{\text{Total volume of voids}}{\text{Total volume of grains and voids}} \qquad\qquad \text{Eq.3.11}$$

Thus, P = 1-C and therefore the porosity of beach sediments lies within the range 0.36 to 0.40.

3.5 The Density of Beach Sediments

The density, which is the mass per unit volume, of beach sediments is expressed in two

Table 3.1:		
Composition and density of common beach sediments		
Name	Formula	Density (kg m^{-3})
Quartz	SiO_2	2650
Muscovite	$KAl_2(AlSi_3O_{10})(OH)_2$	2800-2900
Orthoclase	$KAlSi_3O_8$	2550
Plagioclase	$Na,Ca(Al,Si)AlSi_2O_8$	2620-2760
Microcline	$KAlSi\ O$	2560
Garnets	$(Fe,Al,Mg,Mn,Ca,Cr)_5(SiO_4)_3$	3560-4320
Hornblende	$NaCa_2(Mg,Fe,Al)_5(Si,Al)_8O_{22}(OH)_2$	3000-3470

forms. Firstly there is the solid density of the particles themselves which depends only on their mineralogical composition, and secondly there is the bulk density of the sediment sample including the void spaces which depends not only on particle mineralogy but also on the packing which was discussed earlier. Table 3.1 lists the composition and the solid density of the chief detrital minerals found in beach sediments. Of these the commonest are quartz and the feldspars, though black layers of heavy minerals (largely hornblends and the garnets) are sometimes concentrated at particular localities. Their density is roughly 2.5 to 3.5 times that of water, but about 2,000 times the density of air at normal temperatures and pressures.

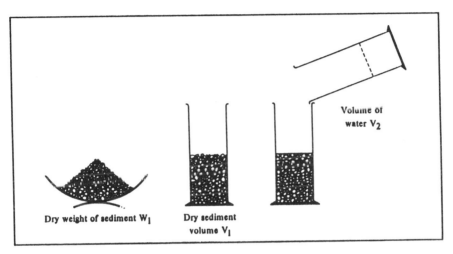

Figure 3.7 The practical determination of sediment packing and density

The symbol ρ_s is commonly used for the solid density of sediments and is distinguished from the bulk density ρ_b which includes the void spaces:

$$\rho_b = C \rho_s \qquad\qquad \text{Eq.3.12}$$

In practice both densities are determined by weighing the volumes V_1 and V_2 shown in Figure 3.7. Referring to the weights as W_1 and W_2 respectively it is apparent that:

$$\rho_s = \frac{W_1}{V_1 - V_2} \qquad\qquad \text{Eq.3.13}$$

$$\rho_b = \frac{W_1}{V_2} \qquad\qquad \text{Eq.3.14}$$

Typical beach sediments are composed of quartz sand thus have a bulk density given by Eq.3.12 of about $0.6 \times 2650 = 1590$ kg m^{-3}. This value is frequently quoted in the literature and will be used throughout the remainder of the book. Conversely, it will later be important to translate from a weight of sand which is eroded or accreted to a volume and surface height change. Thus 1 kg of sand occupies 0.00063 m^3 which is 630 cm^3. Erosion of 1 kg of quartz sand from a unit area of 1 m^2 will result in a surface erosion of 0.63 cm.

3.6 The Shape of Beach Sediments

Although essentially spherical, individual grains of beach sediments can evidence a wide range of shapes from plate-like to needle-like. There is presently little agreement on nomenclature, less still upon actual results, but this is not too important. Qualitatively grains are ascribed either a roundness which refers to the ratio of the radius of curvature at the corners to the radius of curvature of the maximum subcribed sphere (Waddell, 1932) or a sphericity, which refers to the degree to which the particle approaches a sphere. Thus, a cube has a high sphericity, whereas the corners are sharp, giving it a low roundness.

The quantitative analysis of roundness requires reference to some such diagram as that of Powers (1953), produced here in Figure 3.8. One hundred grains are each ascribed to one of the six classes. The number in each class is then multiplied by a factor of 0.14 for very angular, 0.21 for angular, 0.30 for subangular, 0.41 for subrounded, 0.59 for rounded, and 0.84 for well rounded, and the class results are summed to produce an overall value. Beach sediments which have been subjected to considerable abrasion in the surf, tend to exhibit values of 0.7 or more. Particle sphericity is also quantified through the detailed and very tedious measurements of the length of the three mutually perpendicular grain axes, with a caliper gauge or with a microscope graticule. Conventionally, the longest intermediate and shortest axes are designated a, b and c respectively. A sphere would have $a = b = c$, a needle would have $a \gg b$ and $a \gg c$ and $b = c$, and a plate would have $a = b$ and $a \gg c$ and $a \gg b$. Choosing two of these, the longest and the shortest, sphericity can be defined by the ratio c/a. Again, well worn beach sediments tend to have values close to unity whilst plates and needles are closer to zero.

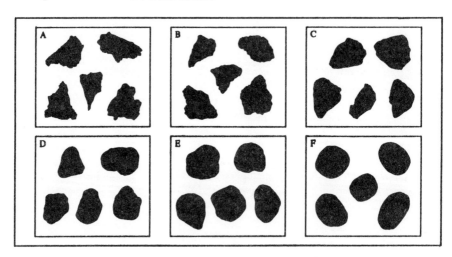

Figure 3.8 Roundness table (After Shepard, 1974).

3.7 The Yield of Beach Sediments

Although lacking in cohesion, which term is, in sedimentological parlance, reserved for the electrostatic attraction which exists between the finer particles of silts and clays, the non-cohesive beach sands and gravels do possess an inate strength. This is exhibited, most obviously, in their ability to form quite steep slopes and failure results when these slopes oversteepen and avalanche. The terminology and relevant symbols are diverse, but we shall refer to this important property with the following definitions based on Allen (1970).

The maximum surface gradient of a loose sediment surface is the yield angle. The yield angle is also symbolised by ϕ and is elsewhere also called the slump angle, the angle of repose or the angle of internal friction. The value of the surface gradient immediately after avalanching is the residual angle, symbolised by ϕ_r and, for a given sediment, the difference between these, which is the change in gradient due to slumping, is called the dilatation angle, symbolised by Λ_ϕ, The various angles are measured by rotating a horizontal cylinder half filled with sediment.

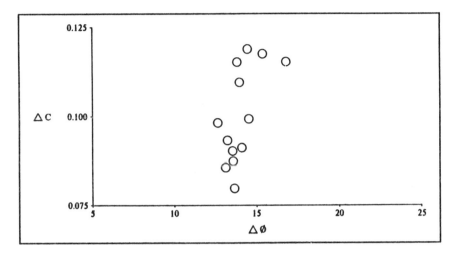

Figure 3.9 Experimental results on the change in grain concentration during slumping as a function of the dilatation angle (After Allen, 1970).

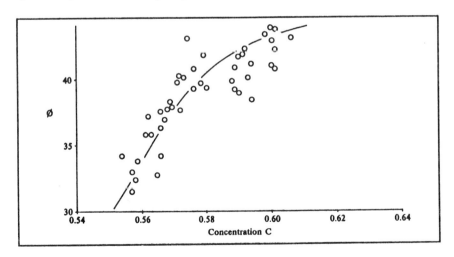

Figure 3.10 Variation in yield with grain concentration (After Allen, 1985).

In general the residual angle after shearing is constant and corresponds to loose (cubic) packing. The yield angle increases above ϕ_r as the grain concentration increases to a

33

maximum in dense (rhombohedral) packing as shown in Figure 3.9 and 3.10. The dilatation angle ranges from about 8° for round, spherical grains to about l4° for angular natural particles.

3.8 The Permeability of Beach Sediments

The resistance of the beach to the discharge of liquid is of particular importance to the present work because it determines, in part, the dissipation of energy from waves moving over the bed. The discharge rate of a liquid through a porous granular bed depends on grain and liquid properties as well as on the impelling force of the pressure head producing the flow. The complete expression for the discharge is known as Darcy's Law, and is valid for all directions of flow and for all velocities small enough that forces of inertia are negligible compared to those of viscosity. The volume discharge, q, of fluid crossing a unit area in a unit time is given by

$$q = GD^2 \frac{\rho}{\mu} \text{-g} \frac{dh}{dz}$$

Eq.3.15

where G is a dimensionless factor of proportionality depending upon the geometry of the pore space, D is the mean grain diameter, ρ and μ are the density and dynamic viscosity of water, g is the acceleration due to gravity, and dh/dz is the pressure gradient in the direction of flow. The grain property term GD^2 is often expressed by the symbol K and is called the permeability of the sediment. It has units of length squared and is often quoted in Darcies, where one Darcy equals approximately 10^{-8} cm^2 or 10^{-12} m^2.

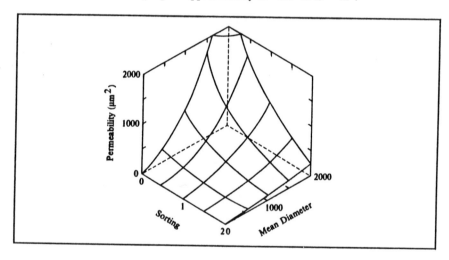

Figure 3.11 Beach sand permeability (After Krumbein and Monk, 1942).

Krumbein and Monk (1942), experimenting with water flow through sands of differing sizes and standard deviations, obtained the following empirical expression for K in darcies:

$$K = 760D^2 \, e^{1.31 \, \sigma_\phi}$$

Eq.3.16

where D is the mean grain diameter in millimeters, e is the base of natural logarithms and σ_ϕ is again the standard deviation of the sediment expressed, confusingly, in phi units.

Thus, well sorted sands are less permeable than poorly sorted sands and coarser sediments are more permeable than finer sediments. The results are shown in Figure 3.11 for a unity pressure head and a standard deviation of 0.25 ϕ which is typical of beach sands.

It is interesting to place these results in perspective with a practical example. Consider a wedge shaped swash which lies stationery at the point of maximum uprush for one second, and is at a point 0.1 m deep and the sand beneath is dry to a depth of 0.1 m, that is the conditions of Figure 3.11 apply because $dz/dh = 1$. If the beach sediment has a mean diameter of 1 mm, q is about 1 cm s^{-1}, that is about one tenth of the water is lost. If the sediment is coarser, say 10mm diameter gravel, then q is greater than 10 cm s^{-1} and all of the water is lost through percolation and the backwash is negligible, whilst a fine sand of about 0.1 mm loses less than one millimetre of the depth.

chapter four

PHYSICAL PROPERTIES OF SEA WATER

4.1 Introduction

Sea water fulfills a variety of roles within the beach system. It is a conductor of energy, by way of surface wave activity, from the offshore meteorological systems to the beach sediment. It is also the pervading fluid which tends to lift the sediments and transports and suspends the solid particles. The important properties of sea water are dealt with in this chapter.

The following sections are based largely upon the presentation of Neumann and Pierson (1966). Additional material is taken from introductory textbooks in oceanography by Anicouchine and Sternberg (1973), and from the more specialist texts on sea-water chemistry by Harvey (1955) and Riley and Skirrow (1965).

4.2 The Molecular Structure of Sea Water

Water consists of covalently bonded oxygen and hydrogen atoms with the single electron from each of two hydrogen atoms sharing one of the six electrons in the outer, third, shell of the oxygen atom; so that completed shells are formed on all three atoms as shown in Figure 4.1. The water molecule is therefore H_2O.

The chemical composition of pure water does not, however, offer any explanation for the dissolving power of this liquid. The dissolving power is a direct consequence of the ability of the water molecules to disassociate solute ions, and is due to water's uniquely high dielectric constant.

The dielectric constant, ε_e, of any material is a number which expresses how much smaller the electric intensity is in the space filled by the material than in the vacuum if the same electric field is applied. This ratio is expressed by considering a condenser with capacitance C_o in a vacuum which changes to C_m with a particular material between the plates (Figure 4.2). Then:

$$\varepsilon_e = \frac{C_m}{C_o}$$

Eq.4.1

ε_e has a value of one for a vacuum by definition, but this increases to a value of 1.0006 for air, 2.0 for petroleum, 5 to 7 for glass, 6 to 8 for mica and up to 81 for water. This is the highest dielectric value of all the liquids. The reason for the large value of ε_e for water lies in a fortuitous abnormality in the structure of the H_2O molecule. The peculiarity of the water molecule is that the two valence electrons of oxygen join the valence electrons of the hydrogen atom not at 180° as might intuitively be thought, due to the mutual repulsion of the hydrogens, but at an angle of 105° to 110° as shown in Figure

37

4.1. The result is a molecule with a positively charged hydrogen region and a negatively charged oxygen region, which results in a so-called dipole moment within the molecule. This, in turn, accounts for two of water's fundamental properties.

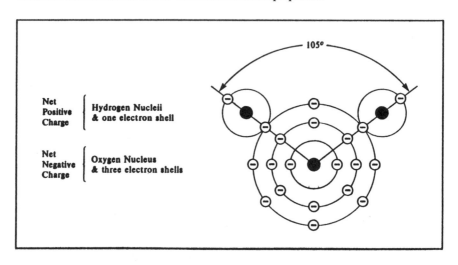

Figure 4.1 Molecular structure of pure water

Firstly, Coulomb's Law states that the force of the electrical attraction between two charges is inversely proportional to the dielectric constant of the surrounding media. Thus the force of attraction between the positively charged sodium atom and the negatively charged chlorine atom in common salt, NaCl, is considerably reduced when in water because of the liquid's very high dielectric constant. That is why water is a very powerful solvent, and why the sea is saline.

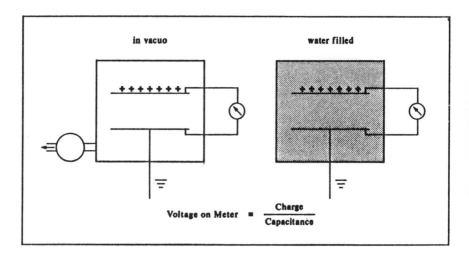

Figure 4.2 Parallel plate capacitor experiment.

Secondly, the dipole moment is the reason for the existence of strong forces acting

between the water molecules themselves. The negatively charged hydrogen regions of one molecule attract the positively charged oxygen regions of another so that the two, three or more H_2O molecules are bonded together and form multiple groups known as polymers.

Groups containing one, two and three molecules are known respectively as monohydral, dihydral and trihydral; while still larger groups can occur depending largely on the temperature of the liquid. At a temperature of $0°C$ the groups correspond to $(H_2O)_6$, but the degree of polymerization decreases with increasing temperature. The physical properties of water, including its high surface tension, high melting and boiling points and its peculiar change in density with temperature, are all due to this polymerizing behaviour, and thus to the dipolar nature of the molecule.

4.3 The Composition of Sea Water

The relatively high salt content of sea water is of special importance in oceanography and is quantified through the measurement of salinity, which is defined as the weight of dissolved solids in g per 1000 g of water, usually expressed in parts per thousand (ppt) and symbolised by $°/oo$.

In the open ocean the salinity varies from about $34°/oo$ to about $38°/oo$, with an average close to $35°/oo$. Land locked seas in humid regions with a large fresh water river output and little evaporation often have a much lower salinity, and values of less than $1°/oo$ are reported from open fiords and from the Baltic Sea. Conversely landlocked seas in arid regions, where evaporation concentrates the brine, can have much higher values with more than $45°/oo$ being recorded in the Red Sea and the Dead Sea.

Careful chemical titrations were traditionally undertaken to determine the salinity and the chemical composition of sea water. Work since Dittmar (1884) who analysed samples from the world's oceans collected on the famous Challenger expedition has shown that, regardless of the absolute concentration, the relative proportions of the different salts in sea water are constant to the second decimal place of the salinity. The proportions are shown in Table 4.1. Since chloride ions are one of the major constituents, and the relative proportions are so constant, it is only necessary to determine the chlorinity ($Cl°/oo$) of a sample. The oldest, and still often applied, method is based upon Mohr's Cl^- titration, where the chlorine content is determined with siver nitrate using a potassium chromate solution as the indicator. The salinity is then given by the internationally agreed empirical relationship:

$$S°/oo = 0.030 + 1.8050 \ Cl°/oo \hspace{3cm} Eq.4.2$$

Salinity is now determined on a routine basis with electrical instruments which measure the conductivity of the water sample and are calibrated to directly calculate the value of either the chlorinity or salinity.

Table 4.1.
Major Constituents of Seawater

Cations:			Anions		
Sodium	Na^+	30.62	Chloride	Cl^-	55.07
Magnesium	Mg^{++}	3.68	Sulphate	SO_4^{--}	7.72
Calcium	Ca^{++}	1.18	Bicarbonate	HCO_3^-	0.40
Potasium	K^+	1.10	Bromide	Br^-	0.19
Strontium	Sr^{++}	0.02	Borate	$H_2BO_3^-$	0.01

Figures are percentages by weight of the total major constituents

4.4 The Density of Sea Water

Water, and specifically the density of water, was used as a primary property in organising the metric system of units so that by definition the density of pure water at four degrees centigrade and one atmosphere of pressure is $1000\ kg\ m^{-3}$. The standard atmospheric pressure is equivalent to 760 mm of mercury and liquid density is symbolised by ρ.

This definition recognised the fact that water, unlike all other liquids, attains a maximum density at a temperature above its freezing point. This is due not only to the polymerization of molecules, but also to the interatomic angle of 109° 28' which allows the polymers to stack most closely at 4°C and to open out, resulting in a lower density, at temperature below and above this point.

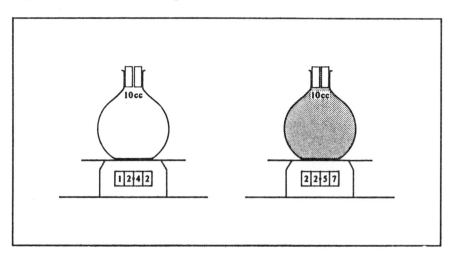

Figure 4.3 Density bottle determination for water.

The density of pure water and of solutions such as sea water which contain many dissolved substances is most readily determined by using a density bottle (Figure 4.3). The bottle is weighed and then filled with the liquid. The stopper is replaced, any excess is removed, and the bottle is reweighed. If V is the volume of the bottle, W_1 is its empty weight and W_2 its weight when full then:

$$\rho = \frac{W_2 - W_1}{V}$$

Eq.4.3

Sea water has been found by this method to have density of $1034.78\ kg\ m^{-3}$ with a salinity of 35°/oo at 20°C and atmospheric pressure. This density increase with salinity and Knudsen (1901) showed empirically that, at 0°C, the following applies:

$$\rho = 1 - 0.000093 + 0.0008149S°/oo$$

Eq.4.4

The effect of temperature on the density of sea water again demonstrates that water behaves somewhat uniquely on freezing in two respects. Firstly, the salts dissolved in sea water depress the freezing point, T_{fp}, to temperatures below the value of 0°C for pure water. Neumann and Pierson (1966) suggest that an empirical relationship for the freezing point as a function of the salinity is:

$$T_{fp} = -0.003 - 0.0527 \ S^o/oo - 0.00004 \ S^o/oo^2 \qquad \text{Eq.4.5}$$

Hence seawater of $10^o/oo$ salinity freezes at $-0.53°C$ whilst that of $30^o/oo$ salinity freezes at $-1.63°C$.

Secondly, water is once again different from any other liquid because it expands on freezing to solid ice, resulting in a decrease in density from the liquid to the solid phases. The actual density of the ice depends on the air and salt content of the crystals. Typically fresh water ice at a temperature of $0°C$ has a density of $916.76 \ kgm^{-3}$ of the pure water in the liquid state at $0° C$. Although ice forming on the sea surface tends to exclude salts and concentrate them on the remainig water, various results show that the density of ice increases with salt content. Ice with no air has a density of about $942 \ kg \ m^{-3}$ at $S^o/oo = 30$, though the usual inclusion of air pockets reduces this value *pro rata*, that is by 10% for the same percentage of air.

The effect of temperature on the density of fresh water is shown by the tabulated results given in data books, and a second order polynomial yields the following functional formula for temperatures between $5°C$ and $25°C$:

$$\rho = 0.99996 + 3.3223219 \times 10^{-5} \ T - 6.0275827 \times 10^{-6} \ T^2 \qquad \text{Eq.4.7}$$

4.5 The Surface Tension of Sea Water

Surface tension is the phenomenon which causes a drop from a tap to remain intact and almost spherical as if surrounded by some form of film. Consider the short line AB drawn in a water surface (Figure 4.4(a)), and suppose that the state of tension of the surface is such that it exerts a pull of F Newtons on both sides of the line. The surface tension γ is defined as the force exerted per unit length of the surface, $\gamma=F/AB$. The explanation of the phenomenon of surface tension lies in the inter-molecular attractions in the water. In Figure 4.4(a), the molecule at C is attracted in all directions by the surrounding molecules. The surface molecule, D, however is only attracted by those below it and thus experiences a net force into the water. This force is manifested as the surface tension. It is only when the thermal energy of the molecule becomes large enough to overcome this net force that the molecule can evaporate out of the body of the water. This reduction in surface tension is what happens when water boils, and the resulting variation in γ with temperature is shown in Figure 4.5.

Surface tension can be measured directly using the apparatus shown in Figure 4.4(b). The wire frame is suspended from the left hand arm of the balance and counterpoised when the lower ends of its vertical members are lowered in a soap solution. The balance is then clamped. The dish of the solution is then raised to cover the frame completely and then lowered to its original position, thus leaving a plane film in the frame above the surface of the solution. Weights are added to the right hand side of the balance so as to counterpoise it again. These additional weights (m kg) balance the downward force exerted on the top of the frame by the film:

$$m \ g = 2 \ \gamma \ l \qquad \text{Eq.4.7}$$

Surface tension is of particular importance because of its effect on capillary wave formation. Fleming and Revelle (1939) found that the surface tension of sea water decreases with increasing temperature and decreasing chlorinity, and fits the empirical relationship:

$$\gamma = 75.64 - 0.144T°C + 0.0399 \ Cl^o/oo \qquad \text{Eq.4.8}$$

where γ is measured in dynes cm^{-1}.

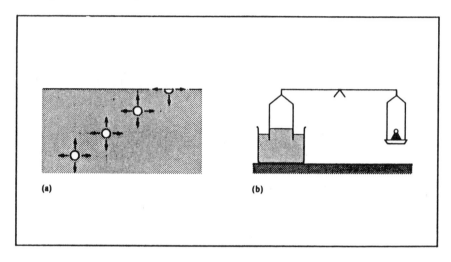

Figure 4.4 The surface tension of water showing (a) definition diagram and molecular attractions, (b) apparatus for the experimental determination of surface tension.

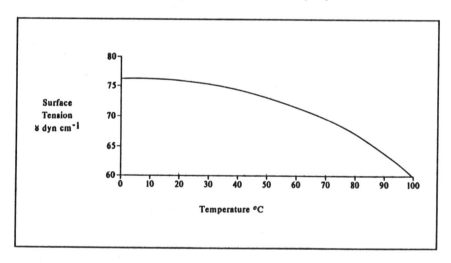

Figure 4.5 Experimental variation in surface tension with temperature.

4.6 The Hydrostatic Pressure of Sea Water

Pressure is defined as *the force acting per unit area* so that *hydrostatic pressure* is defined as the pressure *acting beneath the surface due to the weight of overlying water*. Strictly speaking hydrostatic pressure refers to pressures due to stationary water, but we shall see later that the results are also applicable to pressures beneath surface waves, when the fluid is itself in motion. Consider Figure 4.6, a unit area at depth h is overlain by a volume of h m^3 and therefore supports a mass of ρh kg of water. The pressure, P$_H$, is therefore:

$$P_H = \rho\ g\ h \qquad\qquad Eq.4.9$$

and will have units of N m^{-2}.

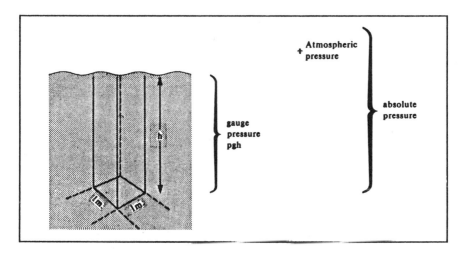

Figure 4.6 Hydrostatic pressure definition sketch.

In practice the total pressure at a given depth is equal to the hydrostatic pressure plus the atmospheric pressure acting on the water surface. For this reason hydrostatic pressure is sometimes called gauge pressure, because most instruments (except of course barometers) read zero at the surface. Atmospheric pressure is equal to about 10 m depth of water that is 98.1 x 10^3 N m^{-2}. Table 4.2 lists the most commonly used of the alternative units and gives conversions between them. The mks units will be used here.

Table 4.2				
Conversions between alternative pressure units				
Depth(m)	10	20	50	100
Pressure				
N m^{-2} (x10^3)	98.1	186	490	981
Atmospheres	1	2	5	10
p.s.i.	15	30	75	150

4.7 The Buoyancy of Particles in Sea Water
The buoyancy of an object is defined as its tendency to float in a fluid. The existence of a buoyancy force was recognised by the Greek mathematician Archimedes who deduced that any object when floating or submerged in a fluid experiences an upthrust or loss in weight. This is because the upwards pressure of the surrounding fluid on its base (Figure 4.7) is larger than the downwards pressure on its top, due to the increase in pressure with depth.

Taking V as the submerged volume in Figure 4.7 and ρ as the density of the liquid then the upthrust F will be:

$$F = \rho V g$$ Eq.4.10

Archimedes principle is very easily verified by using the hydrostatic balance shown in Figure 4.7. The solid object is weighed in air, W_1, and again when suspended in a partially filled measuring cylinder, W_2. The volume V of the solid is noted from the increased reading on the cylinder. We verify Archimedes principle by finding:

$$W_1 - W_2 = \rho \, V \, g \qquad\qquad Eq.4.11$$

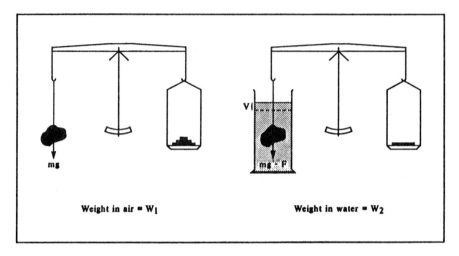

Figure 4.7 Buoyancy force definition diagram.

The buoyancy force acting upon a submerged grain of sediment causes a weight loss as above so that the submerged grain weight, W_g, is:

$$W_g = (\rho - \rho_s)V \qquad\qquad Eq.4.12$$

The behaviour of solid particles in the fluid medium is the *sediment transport problem* which forms the subject of Section C of this book.

SECTION B

HYDRODYNAMICS

chapter five

THE SHOALING TRANSFORMATIONS

5.1 Introduction

The objective of this chapter is shown in Figure 5.1. Beach morphology is controlled by sediment transport which is, in turn, controlled by currents generated due to the passage of a surface wave. We wish, therefore, to be able to take a deep water wave of known characteristics, bring it into shallow water and predict the surface profiles which are generated at each depth. In general, the wave becomes shorter and steeper in profile, whilst the seabed experiences a current which is onshore beneath the crest and offshore beneath the trough. Changes in the surface profile of the wave, which are known as the *shoaling transformations*, are dealt with in the present chapter, and the resulting nearbed flows are dealt with in Chapter Six. Wave theory is traditionally based upon the assumption that the process is *monochromatic*, that is the surface profile can be adequately described by a simple, cyclic function. This assumption has many attractions, but is clearly inappropriate for real ocean waves and therefore the development of a mixed frequency or *spectral* approach is given in Sections 5.8 and 5.9.

This chapter is concerned with the hydrodynamic processes which operate on a vertical plane lying along the direction of wave advance. This direction is normal to the wave crests and is known as the *wave orthogonal*. Earlier texts have referred to these as shorenormal processes but, as will become evident in Section 5.6(b), orthogonals are rarely shorenormal and therefore such terminology is misleading. We shall here refer to these as *orthogonal processes* in order to distinguish them from the separate group of longshore processes which are described in the Chapter Seven. The presentation is based upon Wiegel (1964), Komar (1976), the Shore Protection Manual (1984), Dyer (1986) and Sleath (1984) along with more specific references which are cited in the text. Kinsman (1984) also presents a rigorous, yet readable, description of the mathematics involved in the wave equations.

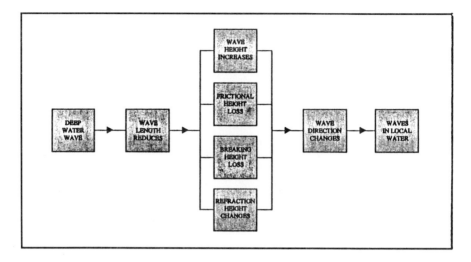

Figure 5.1 Schematic overview of the shoaling transformations detailed in the present Chapter

5.2 Wave Parameters

Wind, blowing over the surface of the world ocean, transfers energy to the surface water. The energy is manifested as a series of more or less regular crests which develop in lines normal to the wind and travel in a downwind direction. These are known properly as *wind waves* but more commonly as waves. They are of higher frequency than the other periodic oceanic water motions (Figure 5.2) which are generated by gravity and are known correctly as *tidal waves*, but colloquially as tides. We are here concerned with the equations of motion for wind waves, and the relevant sections of tidal theory will be presented in Chapter 8.

Typical monochromatic wave trains are shown in Figure 5.3, and from this four parameters are defined:

The **WAVE HEIGHT** is defined as the vertical distance between the crest and the trough. It is symbolised by H, has units m, and the subscripted H_∞ refers to the deep water value far from the coastline. The precise meaning of deep water is explained in Section 5.4.

The **WAVE PERIOD** is defined as the time interval between the occurence of successive, corresponding points on the wave profile, at a fixed reference point. It is symbolised by T, has units s, and is unaltered as the wave enters shallow water and is not therefore subscripted.

The deepwater values of H and T, together with the wave direction signified by α can completely describe a regular wave train and all other parameters can then be calculated for a particular water depth. The effects of mixed, so called random wave trains having distributions of height, period and direction are dealt with in Section 5.8 and 5.9.

The **WAVELENGTH** is defined as the horizontal distance between corresponding points on successive wave profiles. It is symbolised by L, has units m, and the subscripted L_∞

refers to the deep water value.

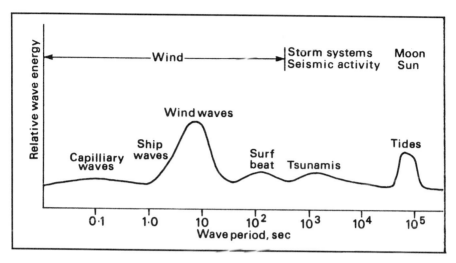

Figure 5.2 Spectrum of ocean movements.

The **WAVE CELERITY** is defined as the horizontal distance traversed by the wave in unit time. Strictly speaking the speed of the wave refers to the speed of the wave profile, which is called the *phase celerity* and is symbolised by C, with units m s^{-1}. C_∞ refers to the deep water value and, by definition, the phase celerity is:

$$C = \frac{L}{T}$$

Eq.5.1

In practice the phase celerity is usually faster than the speed of transmission of energy within the wave form, and the speed of transmission of energy is called the *group celerity*, C_g. The phase and group celerities are related by the definition $C_g = Cn$ where the coefficient of proportionality varies from n=0.5 in deep water to n=1 at the waters edge. The value of n for all water depths is given by Eq.5.12 and is also shown by the results plotted in Figure 5.8 in a later section of this chapter.

5.3 Wave Profiles

The equations for wave motions were developed, with certain approximations, by a number of 19th century mathematicians, and have been found to agree remarkably well with certain aspects of field and laboratory measurements of the phenomena. The simplest theoretical analysis is due to Airy (1845), and considers only first order powers of H. It is therefore often referred to as Airy, first-order or linear theory and predicts a wave profile:

$$\eta(x,t) = \frac{H}{2} \cos(kx - \omega t)$$

Eq.5.2

where η (the lower case greek letter "h" pronounced eta) is the elevation of the water surface, and the distance and time axes, x and t, have origins at the wave crest as shown in Figure 5.3. This concise form includes two commonly used abbreviations: k=2π/L which

49

is the *wave number* with units m^{-1} and $\omega=2\pi/T$ which is the *wave frequency* expressed in s^{-1}.

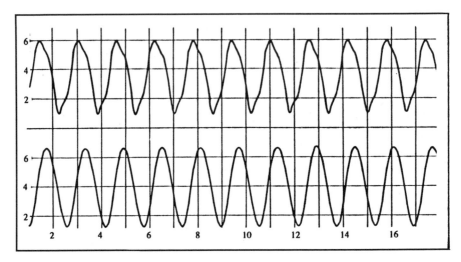

Figure 5.3 Monochromatic wave profiles shown normal to the crest line in a laboratory flume channel. The upper trace was obtained with a Druck 1 bar pressure transducer and the lower trace with a capacitative wave staff. The waves were generated with simple harmonic oscillations of a vertical plate about 7 m from the sensors. Water depth 20cm, wave height 5 cm and period of about 1.5s.

First order theory provides an adequate description of wave height and length changes during the shoaling transformations discussed below, but more sophisticated wave theories will be required to model the nearbed wave induced currents, and these are detailed in Chapter 6. Second-order wave theory developed by Stokes (1847) predicts a surface profile given by:

$$\eta(x,t) = \frac{H}{2}\cos(kx-\omega t) + \frac{\pi H^2}{8L}\frac{\cosh(kh)[2+\cosh(2kh)]}{\sinh^3(kh)}\cos[2(kx-\omega t)] \qquad \text{Eq.5.3}$$

Details of the hyperbolic functions sinh and cosh are given in Appendix I, and the Stokes wave theory is discussed in greater detail in Section 6.2(b). The trochoidal wave profile is predicted by Gerstner's (1802) theory:

$$\eta(x,t) = \frac{H}{2}\,[1-\cos(kx-\omega t)] \qquad \text{Eq.5.4}$$

In order to compare Eq.5.4 with Eqs.5.2 and 5.3, the trochoidal solution must be written as H/2[1-cos(kx - ωt +π/2)] - H/2 which is the same as the first-order profile phase shifted by 90° and displaced above rather than about the still water level. The solitary wave theory predicts an infinite wave length, and the wave profile is given by Longuet-Higgins (1981a) as:

$$\eta(x,t) = H\ \text{sech}^2\left(\sqrt{\frac{3H}{4h}}\ \frac{[x - t\sqrt{(hg+H)}]}{h}\right) \qquad \text{Eq.5.5}$$

where h is the water depth below the still water level and the hyperbolic secant (sech) is defined equal to the inverse of the hyperbolic cosine given in Appendix I. The routine in Hardisty (1990a) calculates these surface profiles and the results are shown in Figure 5.4. A fifth approach which is of use is the cnoidal wave theory developed by Korteweg and de Vries (1895). The surface profile is given by Skovgaard *et al.* (1974) as:

$$\eta(x,t) = h_{min} + H\ \text{cn}^2\left(K\left(\frac{t}{T} - \frac{x}{l}\right)K\right) \qquad \text{Eq.5.6}$$

The constant, h_{min}, is the water depth beneath the trough and cn denotes one of the cnoidal Jacobian elliptic functions which gives its name to the wave theory. The variable argument, K(t/T-x/L) is called a complete elliptic function of the first kind and K^2 is a parameter which defines the wave form (Wiegel, 1964, p43). Details of this theory are given in Chapter 6, and it is sufficient here to show typical surface profiles (Figure 5.5) and to note that, in shallow water, cnoidal waves become identical to solitary waves whilst in deep water they become sinusoidal and are adequately described by the Airy and Stokes theories given above.

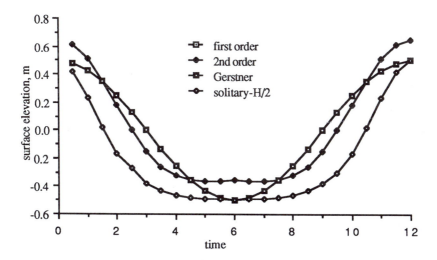

Figure 5.4 Surface profiles predicted by the wave theories using Eq.5.2 to 5.5 for T=5, h=4 and H=1. Note that, for these parameters, the first order and Gerstner profiles overlie one another. (From Hardisty, 1990a).

Comparison with experimental results

Measurements by Wiegel (1964) have confirmed that the surface profile predicted by Airy wave theory compares with the time history of a laboratory wave (Figure 5.6), whilst field observation (e.g. Bagnold, 1947) suggests that wave crests become sharper, and troughs relatively longer close to the beach as predicted by the other theories. It is, however, the changes in wave height and length rather than the overall profile which are of importance here, and for these Airy wave theory is routinely employed.

5.4 Reduction in Wave Length due to Shoaling

First-order wave theory is based upon the wave dispersion equation:

$$\omega^2 = gk \tanh(kh) \qquad\qquad \text{Eq. 5.7}$$

The hyperbolic tangent is one of the hyperbolic functions defined in Appendix I.

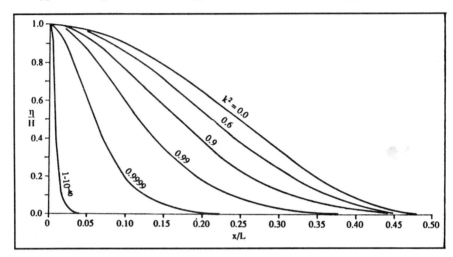

Figure 5.5 Cnoidal wave profiles (after Wiegel, 1964)

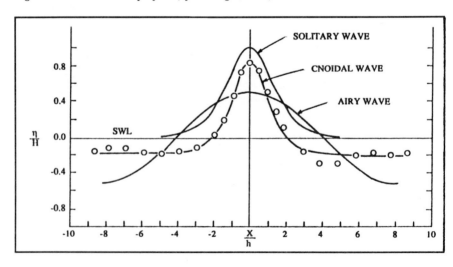

Figure 5.6 Experimental assessment of wave profiles (after Wiegel, 1964)

52

The hyperbolic functions are calculated by the program given in Hardisty (1990a) and the results are graphed in Figure 5.7. The wave celerity decreases as the water shoals, and therefore reduces the wavelength (Eq.5.1) since the period must remain constant. A full solution for the wave length in any water depth is obtained by substituting the identity for k into the dispersion equation and rearranging the resulting expression to:

$$L = \frac{g}{2\pi} T^2 \tanh\frac{2\pi h}{L}$$ Eq. 5.8

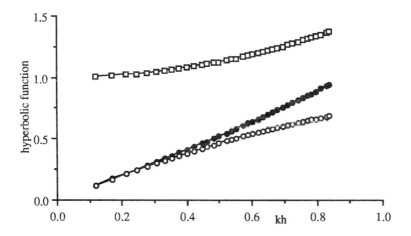

Figure 5.7 The hyperbolic tangent (open circles) sine (closed circles) and cosine (closed squares) functions with kh argument from Eq.5.7 (from Hardisty, 1990a).

Eq.5.8 is not directly soluble for the wave length because L occurs on both the left hand side and within the hyperbolic tangent, and values must be calculated by iterative techniques. Eq.5.8 was therefore rarely used to obtain values for the wave length, and recourse was frequently made to the approximations which are detailed by, for example, Komar (1976, p41). These give the wave length as $L = gT^2/2\pi$ ($\approx 1.56T^2$ in SI units) with an error of <5% in water for which $h>L_\infty/4$ (because here $\tanh(2kh)\approx1$) and, with the same degree of accuracy, as $L = T\sqrt{(gh)}$ in water for which $h<L_\infty/20$ (because here $\tanh(kh)\approx kh$). These two limits have been used to define a *deep water* and a *shallow water* regime. There is no such simple approximation for intermediate depths between these two ranges. Nowadays, however, the advent of micro-computers has greatly facilitated the use of iterative techniques, and Eq.5.8 is used for all water depths. The general result is expressed as the non-dimensional ratio L/L_∞ and is graphed in Figure 5.8. It is a simple task to calculate the wave celerity from this equation since $C = L/T$ (Eq.5.1) and the general result is again expressed as the ratio C/C_∞ which is, of course, identical to the wavelength curve in Figure 5.8.

Although the SLOPES model which is detailed in later chapters utilises second-order Stokes wave theory for which the wave celerity and wave length are as above (Komar, 1976), third-order theory predicts a longer wave length and is given by Horikawa (1978) as:

$$L = \frac{gT^2}{2\pi} \tanh \frac{2\pi h}{L} \left(1 + \left(\frac{\pi H}{L} \right)^2 \frac{\cosh(4kh) + 8}{8 \sinh^4(kh)} \right) \qquad \text{Eq.5.9}$$

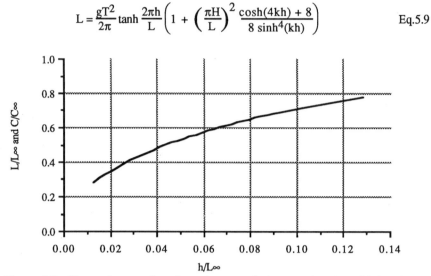

Figure 5.8 Change in wave length and wave celerity as a function of h/L$_\infty$ during shoaling. (from Hardisty, 1990b).

Comparison with experimental results

It is difficult to measure the wavelength directly but the theory can be tested by determining the wave celerity from measurements of the time taken for crests to progress between fixed points in a laboratory flume or on a beach profile. A computer routine which iteratively solves the wave length equation is given in Hardisty (1990b) and shown in Figure 5.8.

5.5 Increase in Waveheight due to Shoaling

Since a certain amount of wave energy per unit width is contained within each wave length, and the preceeding section showed that the wavelength reduces as the wave enters shallow water, then the wave height must increase to maintain sufficient mass above the still water level to conserve the potential energy of the wave. In more detail the average rate at which wave energy is transmitted in the direction of wave propogation is called the *wave power*, P, and is given by Reynolds (1877) and Rayleigh (1877) and by Komar (1976, Eq.3-22) as:

$$P = \frac{1}{8}\rho g H^2 C \frac{1}{2} \left(1 + \frac{2kh}{\sinh(2kh)} \right) \qquad \text{Eq.5.10}$$

The wave energy, E, is calculated by integrating this expression over a full wave length:

$$E = \frac{1}{8}\rho g H^2 \qquad \text{Eq.5.11}$$

This energy is equally divided between the kinetic energy due to the wave induced orthogonal currents and the potential energy due to the displacement of the surface profile of the wave from the still water level. Another common substitution (which was introduced in Section 5.2 as $C_g=Cn$) is defined by writing:

$$n = \frac{1}{2}\left(1 + \frac{2kh}{sinh(2kh)}\right)$$

Eq.5.12

so that the relationship for the wave power (Eq.5.10) becomes:

$$P = E C n$$

Eq.5.13

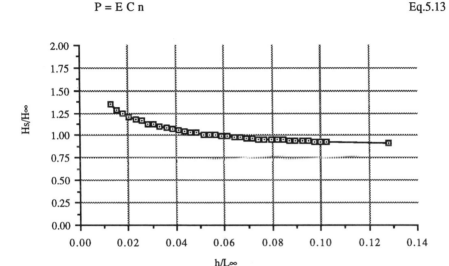

Figure 5.9 Shoaling height changes as a function of h/L∞ (from Hardisty, 1990b).

Eq.5.12 shows that n=0.5 in deep water because sinh(2kh) is then very large compared with 2kh and that n=1 in shallow water because sinh(2kh) is then equal to 2kh (Figure 5.7). The increase in wave height can now be calculated by equating the wave power at a depth h to the deep water value $(ECn)_\infty$. Combining Eq.5.12 with this equation then gives the shoaling wave height, H_s, as:

$$H_s = H_\infty \sqrt{\frac{1}{2n}\frac{C_\infty}{C}}$$

Eq.5.14

The shoaling coefficient, $K_S = \sqrt{(C_\infty/C)}$ is often substituted into Eq.5.14 to give a simpler expression for the local wave height:

$$H_s = H_\infty K_S \sqrt{\frac{1}{2n}}$$

Eq.5.15

$C_\infty=L_\infty T$ and $C=LT$ are calculated from Eq.5.1 using the wavelength given by Eq.5.8 and are combined with n from Eq.5.12 to compute the wave height for any given water depth. The resulting theoretical expression is usefully presented as the ratio H/H_∞ and is shown in Figure 5.9.

Comparison with experimental results

Snyder *et al.* (1958) have shown that Eq.5.13 is essentially correct by generating waves at one end of a channel and measuring the transmitted power at the other end. The theory which leads to Eq.5.14 and 5.15 compares well with the experimental data of Iverson (1951). The small decrease in wave height which is followed by a marked increase is due to n increasing faster than C at the commencement of the transformations as was shown in Figure 5.8. A routine in Hardisty (1990b) calculates the shoaling height change in accordance with the above theory.

5.6 Secondary Changes in Wave Height

The shoaling transformations described above result in the wave height increasing and the wave length decreasing as the wave approaches the shoreline. Additionally there are three main processes which act to change the wave height during shoaling. Firstly *seabed friction* causes energy dissipation. Secondly the wave may change direction in the process called *refraction* which can cause the wave energy per unit crest length to alter and therefore wave height can increase or decrease. Thirdly, the increasing wave height and decreasing wave length will ultimately result in the *wave breaking* and this dissipates energy, reducing the wave height. Each of these processes will be examined in this section so that a complete solution for wave height changes during shoaling can be constructed. In general the solution is of the form:

$$H = H_s - \Delta H_f - \Delta H_r - \Delta H_b \qquad \text{Eq.5.16}$$

where H_s is the wave height due to the shoaling transform, ΔH_f is the wave height loss due to seabed friction, ΔH_r is the wave height loss due to wave refraction and ΔH_b is the wave height loss due to the breaking process. The following sections present the derivation of the loss terms, and the full equation is assembled as Eq.5.29.

5.6(a) Seabed Friction

The calculation of wave energy loss due to seabed friction is usually based upon Putnam and Johnson's (1949) application of the quadratic stress law (see also Chapter Nine):

$$\tau = \frac{1}{2} f_w \, \rho \, u^2 \qquad \text{Eq.5.17}$$

where τ is the shear stress on the seabed, f_e is the wave friction factor and u is the horizontal component of the nearbed flow (Chapter Six). Sleath (1984, p195) gives the rate of change of wave energy by combining this expression with first-order theory as:

$$\left(\frac{dH}{dx}\right) = -\frac{4 \, f_e \, k^2 \, H^2}{3\pi \, \sinh(kh)(\sinh 2kh + 2kh)} \qquad \text{Eq.5.18}$$

where f_e is Sleath's "energy dissipation factor". The wave height attenuation given by Eq.5.18 is a function not only of the water depth, but also of the distance which the wave has travelled. The wave height loss, ΔH_f, between deep water (defined above as $h=L_\infty/4$) and the local depth h is then:

$$\Delta H_f = \int_{h=L_\infty/4}^{h} \frac{4 \, f_e \, k^2 \, H^2}{3\pi \, \sinh(kh)(\sinh 2kh + 2kh)} \, dx \qquad \text{Eq.5.19}$$

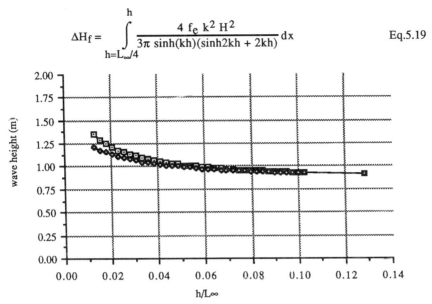

Figure 5.10 Wave height transformations predicted by Eq.5.19 with upper line energy conservation transform and lower line the effect of seabed friction with $f_e=0.01$, $T=5s$ and $H_\infty=1$ and the seabed profile given in Chapter Seventeen (from Hardisty, 1990c).

TABLE 5.1
Field Measurements of f_e

Reference	Mean f_e	Range of f_e
Bretschneider (1954) U.S.A.	0.106	0.060-1.934
Iwagaki and		
Kakinuma (1967):		
Akita coastline	0.116	0.066-0.190
Izumisano coastline	0.410	
Hiezo coastline (1963)	0.176	0.054-0.380
Nishikinomaha coastline	1.100	0.560-2.460
Hiezu coastline (1964)	0.094	0.020-0.150
Takahama coastline	0.100	0.060-0.170

Sleath (1984) notes that a number of other, related friction factors have been devised and used by Bagnold (1946), Zhukovets (1963), Kajiura (1968), Carsten's *et al.* (1969), Sleath (1976), Lofquist (1980) and Longuet Higgins (1981b). However since all rely on some empirical coefficients, and the majority of data appears to relate to f_e, we shall follow Sleath and use Johnson's friction factor here. Laboratory studies have investigated the range of f_e for smooth surfaces, but since natural seabeds are generally composed of rippled sand, it is likely that the friction factor is related to the nature of these bedforms. The existing laboratory results are rather scattered (Sleath, 1984) though they compare favourably with the site values in Table 5.1. A value of about 0.1 appears to be not unreasonable.

Comparison with experimental results

It is difficult to test the theory for wave height reduction due to seabed friction because the process cannot practically be seperated from the other shoaling transformations, and because data is, in effect, used to calibrate f_e rather than to independently test Eq.5.19. However, the height reduction due to seabed friction can be obtained by integrating Eq.5.19 across the depth range from deep water to the specified depth. A routine in Hardisty (1990c) performs these calculations and typical results are shown in Figure 5.10.

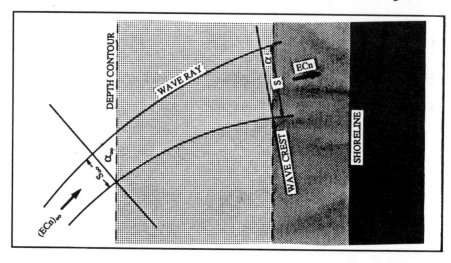

Figure 5.11   Definition diagram of refraction geometry for the case of shoreparallel, straight bathymetric contours.

5.6(b) Refraction

Upon entering shallow water the wave celerity is reduced with the result that whenever a wave crest crosses bathymetric contours at other than a normal angle of incidence, the shoaler parts of the crest will travel more slowly than the deeper parts and the crest will therefore bend to lie more parallel with the contours. This is an example of Snell's Law which is more usually applied to light waves, and both the electromagnetic and hydrodynamic processes are known as *wave refraction*. Lines which run normal to the wave crest are called *wave orthogonal* lines, and wave refraction can cause wave orthogonals to converge or to diverge as the wave crosses the contour. Since orthogonal lines may be constructed with spacings of one metre in deep water, then the energy contained between these orthogonals is, by definition, the deep water wave energy per metre of wave crest which was given by Eq.5.11. Therefore when inter-orthogonal spacings change the amount of energy per metre of wave crest changes, and the wave height will also change. However, neglecting other shoaling effects and assuming that no energy "leaks" laterally along the wave crest, the inter-orthogonal wave power must remain constant (Figure 5.11):

$$P = (E\ C\ n\ s)_h = (E\ C\ n\ s)_\infty = \text{constant} \qquad\qquad \text{Eq.5.20}$$

where s_h is the inter-orthogonal spacing of the wave in shallow water compared with the deep water value s_∞. Substituting for E from Eq.5.11 and writing $n_\infty=1$ from Eq.5.12

gives the wave height during refraction but including the shoaling from Eq.5.14 , H_r, as:

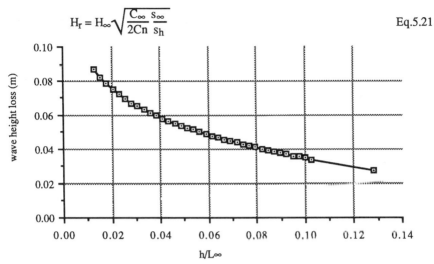

$$H_r = H_\infty \sqrt{\frac{C_\infty}{2Cn} \frac{S_\infty}{S_h}} \qquad \text{Eq.5.21}$$

Figure 5.12 Height changes due to wave refraction for angles of incidence of 30° and a shore parallel, though not necessarily planar bathymetry. Deep water wave height of 1 m. (from Hardisty, 1990d).

The ratio $\sqrt{(S_\infty/S_h)}$ is called the *refraction coefficient* signified by K_r which can be determined from Snells Law for simple bathymetries. More complex problems require full wave refraction diagrams. For the case of waves approaching a straight shoreline and for offshore contours which are straight and parallel to the shoreline (though not necessarily regularly spaced) refraction may be treated analytically by utilising Snell's Law directly:

$$\frac{\sin\alpha}{\sin\alpha_\infty} = \frac{C}{C_\infty} = \text{constant} \qquad \text{Eq.5.22}$$

where α is the local angle between the bathymetric contours and the wave crest and α_∞ is the angle between the deep water wave and the shoreline (Figure 5.11). C is again the local wave phase celerity and C_∞ the deepwater value. From the geometry of the wave rays:

$$\frac{S_\infty}{s} = \frac{\cos\alpha_\infty}{\cos\alpha} \qquad \text{Eq.5.23}$$

Combining Eq.5.21 to 5.23 we obtain:

$$H_r = H_\infty \sqrt{\frac{C_\infty}{2nC} \frac{\cos\alpha_\infty}{\cos\alpha}} \qquad \text{Eq.5.24}$$

where α_∞ is given and, from Eq.5.22, $\alpha = \sin^{-1}(C\sin\alpha_\infty/C_\infty)$, and the wave celerity expressions were given in Section 5.4. The resulting wave heights for a range of deep water angles are shown in Figure 5.12. Eq.5.24, of course, reduces to Eq.5.14 when

$\alpha_\infty = \alpha = 0$, normal incidence and no refraction.The expression can more usefully be written as:

$$\Delta H_r = H_s - H_\infty \left(\sqrt{\frac{C_\infty}{2nC}} \, \frac{\cos\alpha_\infty}{\cos(\sin^{-1}(C\sin\alpha_\infty/C_\infty))} \right)$$ Eq.5.25

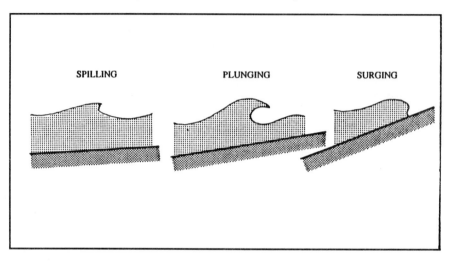

Figure 5.13 Breaker types (after Komar, 1976).

5.6(c) Breaking

When waves reach the beach and enter water which is approximately as deep as the waves are high, they break, the crest throwing itself forward and disintegrating into bubbles and foam. Breaking occurs because the wave becomes shorter and higher and therefore steeper until the water particle motion in the crest exceeds the celerity of the waveform and so surges ahead. Three types of breakers are commonly recognised and are described by Galvin (1972) as:

> *Spilling Breakers*: Foam, bubbles and turbulent water appear at the wave crest and eventually cover the front face of the wave. Spilling starts at the crest when a small tongue of water moves forward faster than the waveform as a whole. In its final stages, the spilling wave evolves into an undular bore.

> *Plunging Breakers*: The whole front face of the wave steepens until vertical; the crest curls over the front face and falls into the base of the wabe; and a large sheet-like splash arises from the point where the crest touches down.

> *Surging Breakers*: The front face and crest of the wave remains relatively smooth and the wave slides up the beach with only minor production of foam and bubbles. Resembles a standing wave.

Diagrams of the breaker types are shown in Figure 5.13, but it is important to appreciate that the classification describes parts of a continuum which ranges between the extreme types. Various authors have made this point and used a range of non—dimensional numbers to attempt to quantify the breaker type. The most common of these are summarised in Table 5.2, together with the source references and typical values for the transitions between breaker types. The symbols in this table have been edited to confirm with the usage elsewhere in the book except that t_{up} is defined as the time which elapses between the wave first breaking, and the same crest arriving at the waters edge (Kemp, 1960, 1975).

Table 5.2
Comparison of Breaker Type Parameters

Parameter	Spill/Plunge	Plunge/Surge	Source
Breaker Coefficient, $(B_0 = H_\infty/L_\infty\beta^2)$	10^2	10^{-3}	Galvin, 1968, 1972
Iribaren's Number $I_r = \beta/\sqrt{H_\infty/L_\infty}$	10^{-1}	10^1	Iribarren & Nogales 1949, Battjes, 1974a,b Losada et al., 1981
Phase Difference $P_t = t_{up}/T$	1.3	0.7	Kemp 1960, 1975 Kemp & Plinston, 1968, 1974
Surf Scaling Parameter $\varepsilon = 4\pi^2 H_b/gT^2\beta^2$	high	low	Guza & Bowen 1975, Guza & Inman, 1975
Reflectivity $\varepsilon_R = 1/\varepsilon$	low	high	Huntley & Bowen, 1975, Guza & Bowen, 1977
Bore classification $B_{hy} = H_b/h$	undular<0.28 transitional full>0.75		Suhayda & Pettigrew, 1977

Since the object is to relate the wave energy loss to the breakers it is necessary to determine exactly where on the shoaling wave profile the wave begins to break and, when it does so, how much energy is lost during its continued shorewards progress. The first of these two questions is relatively simple to answer because a number of studies have demonstrated a relationship between the water depth and the wave height at the break point. Most theoretical analyses have been based upon the work of McCowan (1894) who demonstrated that the ratio of the wave height to the water depth at the break point, γ_b, will be 0.78. Other theoretical studies predict a range of values:

McCowan (1894)	0.78
Boussinesq (1872)	0.73
Rayleigh (1876)	1.00
Gwyther (1900)	0.83
Davies (1951)	0.83
Packham (1952)	1.03
Yamada (1957)	0.83
Laitone (1959)	0.73
Lenau (1966)	0.83

Although it is surprising that so many theoretical solutions exist, it is apparent that the problem is even more complicated, because empirical work has shown that the critical ratio is also dependent upon the seabed gradient. Sverdrup & Munk (1946) report that field measurements on ocean beaches support McCowans results for very low beach gradients. However Ippen and Kulin (1955) found that values range from 2.8 for a beach slope of 0.065 down to 1.2 for a beach slope of 0.023. Kishi and Saeki (1967) further conclude that the value does not drop to the theoretical result until the beach slope becomes gentler than 0.007. Alternatively Tucker *et al.* (1983), for example, find that a value as low as 0.5 is appropriate. Galvin (1972) finds empirically that:

$$\gamma_b = 0.72(1 + 6.4\tan\beta) \qquad\qquad \text{Eq.5.26}$$

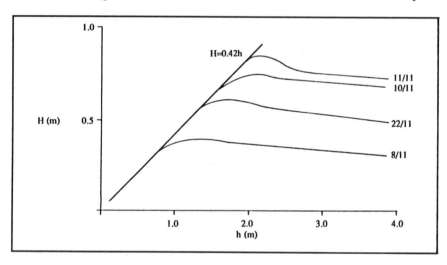

Figure 5.14 Saturation of wave height in the breaker zone

Such results allow the depth at which breaking commences to be determined and it is now necessary to compute the energy loss at each point shorewards of that depth in order to correct the wave height transformations. Unfortunately there is again a diversity of opinion, but some general observations can be made. Firstly at one end of the breaker type spectrum, that is under surging breakers, there is clearly very little energy loss actually at the break-point. At the other extreme, that is under spilling breakers, it seems that energy loss continues from the breakpoint right up to the waters edge. Explicit arguments have not been presented which quantify this energy loss but it may be possible to answer the question implicitly using a very interesting observation by Thornton and Guza (1982). They found that, to a very good approximation, waves shorewards of the breakpoint are *saturated*, that is the wave height at any point is limited by the local depth as shown in Figure 5.14. The results may then be interpreted as suggesting that the wave height is kept at a point of incipient breaking by the breaking process itself. This is what might be expected on dimensional grounds (Longuet-Higgins, 1972) and is what is found in both laboratory experiments using monochromatic waves and in detailed field experiments in conditions of both broad and narrow spectral distributions of wave energy (Thornton and Guza, 1982; Bowen and Huntley, 1984). The wave height in any water depth shorewards of the breakpoint is then given by $H = \gamma_b h$, where the coefficient depends upon beach slope as in Eq.5.26 so that, shorewards of the breakpoint, wave height decreases linearly with the water depth:

$$H = 0.72h(1 + 6.4 \tan\beta) \qquad\qquad Eq.5.27$$

Eq.5.27 can be rewritten in the required form:

$$\Delta H_b = H_s - 0.72h(1 + 6.4 \tan\beta) \qquad\qquad Eq.5.28$$

Such an analysis does not produce any sudden change in beach processes at the break point, and is therefore unsuitable in attempting to account for the intermediate breaker types, that is those of the plunging type. Here, clearly, there is a sudden, quite catastrophic loss of wave energy in a region which is very close to the breakpoint, and shorewards from that point a new wave develops and continues shorewards, perhaps repeating the process a number of times before finally reaching the shoreline. Again the details are unclear but a surmise could be that as the beach gradient increases, the celerity of the wave brings the wave into shallower water too quickly for the wave energy to be gently dissipated in spilling breakers, but instead a lot of energy is rapidly dissipated.

5.7 Summary of Wave Height Changes

Figure 5.15 depicts the total energy lost by a typical shoaling wave due to each of the processes discussed in the preceeding sections. The local wave height is written by combining the various equations:

$$H = H_\infty \sqrt{\frac{1}{2nC}\frac{C_\infty}{}} \qquad\qquad \textit{Shoaling transformations}$$

$$- \int_{h=L_\infty/4}^{h} \frac{4\, f_e\, k^2\, H^2}{3\pi\, \sinh(kh)(\sinh 2kh + 2kh)}\, dx \qquad\qquad \textit{Frictional Losses}$$

$$- \left(H_s - H_\infty \left(\sqrt{\frac{C_\infty}{2nC}\, \frac{\cos\alpha_\infty}{\cos(\sin^{-1}(C\sin\alpha_\infty/C_\infty))}} \right) \right) \qquad\qquad \textit{Refraction Losses}$$

$$OR = 0.72h(1 + 6.4 \tan\beta) \qquad \textit{Breaking Losses} \qquad Eq.5.29$$

where the last term is only invoked if H/h<0.78, and in this case the others are set to zero. It is clear that the shoaling transform and any refraction changes are the most important processes seawards of the breakers, and that within the surf zone it is the breaking characteristics which control the wave height.

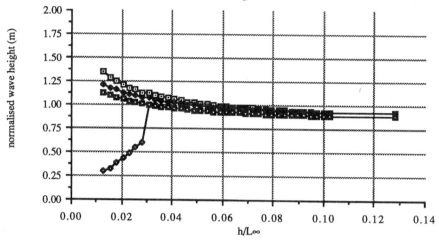

Figure 5.15 Comparison of normalised wave height changes due to the different processes. The example shows a wave T=5s, H∞=1m and a∞=30° shoaling over a typical seabed profile (see Chapter Seventeen for details of the profile). The open squares represent the shoaling height transforms only, the closed diamonds represent the addition of seabed friction with f_e=0.01, the closed squares represent the addition of refraction and the open diamonds represent the addition of breaking (from Hardisty, 1990e).

Additional processes which have not been considered here include percolation currents beneath the seabed, and Horikawa (1978) presents an analysis which is similar to that above for frictional losses with :

$$\left(\frac{dP}{dx}\right) = -\frac{\pi K \rho g H^2}{4L \cosh^2(kh)} \qquad Eq.5.30$$

where K is the coefficient of permeability (Chapter Four). However Putnam (1949) showed that such losses were negligible except in the case of a very wide and very coarse nearshore bathymetry. Bathymetric change also results in some reflection of the incident wave. Battjes (1974a) found that the surf similarity parameter given by:

$$\xi = \frac{\tan\beta}{\sqrt{H_o/L_\infty}} \qquad Eq.5.31$$

is important for determining the amount of reflection of waves approaching a beach at right angles. The total amount of reflection is given by the reflection coefficient, K_{rf}, defined by:

$$K_{rf} = \frac{H_r}{H_i}$$ Eq.5.32

where H_r and H_i are the reflected and incident wave heights respectively. Seelig and Ahrens (1981) developed curves from the results of several authors and Battjes (1974a) presents an empirical formulae for K_{rf} in terms of ξ as:

$$K_{rf} = 0.1\,\xi^2$$ Eq.5.33

5.8 Random Waves

Waves at sea do not generally have the regular and precise properties of the simple harmonic waves described in the preceeding sections. Nevertheless a given sea state can be expressed mathematically as the sum of simple harmonic waves each with a specific T, H and α. The result is a sea state which can only be adequately described by the distribution of periods (which are called frequency spectra because $\omega=2\pi/T$), the distribution of wave heights (called the height probability distribution), and the distribution of wave direction (called the wave directional spectrum). Considerable success is being achieved in, at least, describing a real sea in these terms and the following sections are based on Earle and Bishop (1984), together with other more specific references in the body of the text. It should be noted that our present state of understanding of irregular waves concentrates more on the height and period distributions, at the expense of directional spectra. Thornton and Guza (1983) note that shallow water wave direction is obtained by applying Snell's Law to the individual monochromatics.

5.8(a) The Distribution of Wave Periods

If the monochromatic deep water wave train shown in Figure 5.3 is subjected to spectral analysis, that is a graph is drawn of energy on the y axis versus frequency (or its inverse, period) on the x axis, then a line spectrum as shown in Figure 5.16(a) results. These waves have all of their energy at a single frequency, and the scale on the vertical axis can be plotted simply as wave height, or as the root mean squared of the wave height. More commonly though the square of the wave height is plotted, and the spectrum then represents wave power and is called a power spectrum. The water depth was reduced in the experimental wave channel which had been used to generate the profiles in Figure 5.3, so that the waves changed from deep water, Airy waves to shallower water waves. A spectrum of the surface elevations in these conditions is shown in Figure 5.16(b) which reveals the generation of the so called harmonics of the fundamental frequency, and the spectrum now consists of a number of lines. Finally surface elevations were measured with pressure transducers on an English Channel beach (Hardisty, 1989), and the power spectrum was obtained and is shown in Figure 5.16(c). Now a large number of frequencies are revealed and the graph is no longer a line spectrum, but a continuous function. The spread of wave energy in the frequency domain is commonly given for a fully developed sea state in the form derived by Pierson and Moskowitz (1954):

$$E(\omega) = \frac{2\pi A g^2}{\omega^5}\exp\left(-B(\omega/\omega_0)^4\right)$$ Eq.5.34

where $E(\omega)$ is known as the spectral density function, ω is the wave frequency (Section 5.3), A is an empirical constant ≈ 0.0081, B is a second empirical constant ≈ 0.74, and ω_0 is related to the wind speed 19.5 m above the sea surface, $U_{19.5}$, by $\omega_0 = g/U_{19.5}$. These are known as P-M spectra and an example is shown in Figure 5.17.

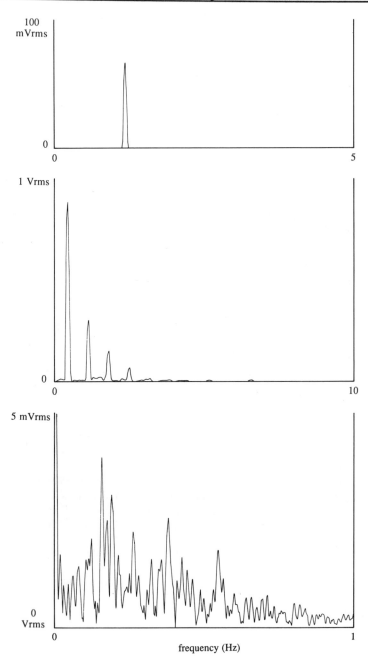

Figure 5.16(a) line spectrum of a sinusoidal wave in deep water, (b) the presence of harmonics in shallow water and (c) beach spectrum. (Author's data).

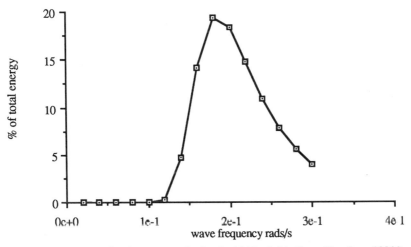

Figure 5.17 Pierson-Moskowitz spectra obtained with Eq. 5.34. (from Hardisty, 1990f)

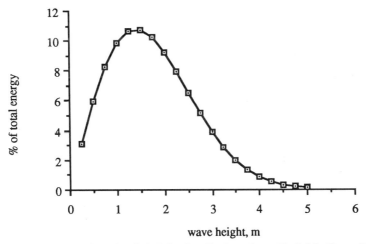

Figure 5.18 Range of Rayleigh height distributions from Eq.5.36. (from Hardisty, 1990h).

An alternative spectrum which is used in engineering work resulted from wave measurements in the southern North Sea (Ewing, 1986) and is known by the acronym JONSWAP (Joint North Sea Waves Analysis Project) and is:

$$E(\omega) = \alpha \, H_{1/3} \, \frac{1}{\omega_o^4 \omega^5} \, \exp\left(-\frac{1.25\omega_o^4}{\omega^4} \, \gamma_b \, \exp\left(\frac{(\omega/\omega_0-1)^2}{2\sigma_b^2}\right)\right) \quad \text{Eq.5.35}$$

where $\gamma_J = 1 \sim 7$ (mean 3.3) and $\sigma_J = 0.07$ if $\omega < \omega_o$ or $\sigma_J = 0.09$ if $\omega > \omega_o$. Solutions for this equations are given by Hardisty (1990g).

5.8(b) The Distribution of Wave Heights

In addition to spectral representations, it is often important to know the probability of a given wave height occuring in a given wave record. Longuet-Higgins (1952) showed that the wave height distribution can be approximated by a Rayleigh distribution:

$$P(H) = \frac{2H}{H_{rms}^2} \exp -\left(\frac{H}{H_{rms}}\right)^2 \qquad\qquad Eq.5.36$$

where H_{rms} is the root mean square wave height (the root of the mean of the square of all of the wave heights in the record) as shown in Figure 5.18. The Rayleigh distribution is used to examine the SLOPES model in Chapter Twenty.

5.9 Random Wave Transformations

The monochromatic solutions given above do not directly account for the changes in wave heights and frequency spectra which occur when a random wave field enters shallow water. The following sections extend the monochromatic analysis to account for the additional changes.

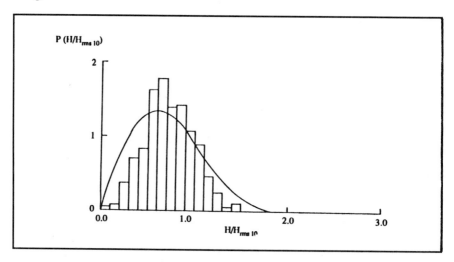

Figure 5.19 Empirical probability density functions in 102 cm water depth plotted against a Rayleigh distribution. $H_{rms}10$ is the root mean squared wave height in 10m of water. (Data from Guza and Thornton, 1983).

5.9(a) Transformations of Wave Height Distributions

There are two generic classes of random wave shoaling and breaking models. The earlier types are called local models in which shoaling is only dependent on the local depth, whilst the more recent type are called integral path models and are based on integrating the energy flux balance equation from deep water shorewards.

Thornton and Guza (1983) recommend a refinement of the integral path approach and show that a Rayleigh distribution of the same form as Eq.5.34 can adequately describe the distribution of the shoaling waves as shown in Figure 5.22. Furthermore they suggest that the linear shoaling transformations discussed earlier are adequate for the wave height distribution up to the break point and used their results on the saturation of wave height in the surf zone (Figure 5.17) to suggest that the r.m.s. value of the wave height can, for simple profiles, be predicted by:

$$H_{rms} = 0.42h \qquad \text{Eq.5.37}$$

More complex models simulate wave breaking by truncating the tail of the Rayleigh distribution based upon various breaker criteria as shown in Figure 5.20. Collins (1970) and Battjes (1972) used a sharp cut off and one of the various forms of the breaking criterion discussed in Section 5.6.

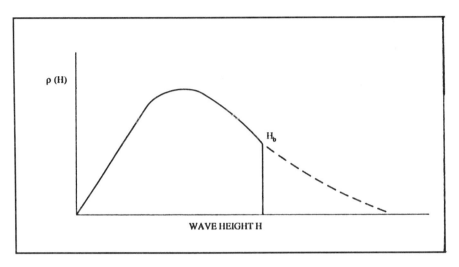

Figure 5.20 Random wave shoaling model of Collins (1970). The original Rayleigh distribution (dotted line) is truncated according to the wave and beach characteristics.

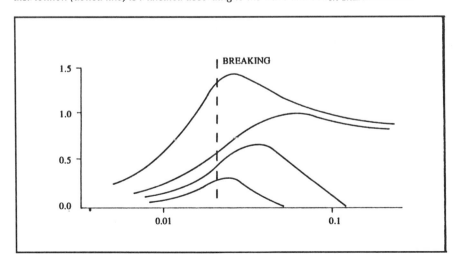

Figure 5.21 Growth of harmonics during shoaling (after Flick, 1981)

Kuo and Kuo (1974) used a similar cutoff, but redistributed the energy of the broken waves across the range of heights in proportion to the probability of unbroken waves at each height. Goda (1975) follows a similar but more complex breaking criterion and also generates a more gradual cutoff. The spectral SLOPES program described in the last chapter of this book simulate the processes by combining Eq.5.35 and Eq.5.36.

5.9(b) Transformations of Wave Period Distributions

We have already seen (Figure 5.16(a) and (b)) that the so called non-linear effects result in the relative growth of higher harmonics in the wave spectrum during shoaling. Flick *et al.* (1981) give a graphic demonstration of this effect by analysing the wave spectrum in laboratory experiments as shown in Figure 5.21. Guza and Thornton (1980) report similar results from field measurements at Torres Pine Beach, San Diego, California. These effects are predicted by the higher order wave equations, and the overall height transformation requires the application of the monochromatic results to each frequency within the wave period distribution. The generation of harmonics can be predicted by non-linear shoaling models such as described by Freilich and Guza (1984) and Thornton and Guza (1989) but are too complex for inclusion in the beach models detailed here at the present time. Furthermore, various types of shallow water resonance and edge waves appears to be responsible for the generation of long waves in the nearshore. It is undoubtedly true that, in the future, the investigation and analysis of these effects may lead to completely new developments in the modelling of the orthogonal system. However, at the present time, we must confine this book to less complex solutions.

chapter six

ORTHOGONAL CURRENTS

6.1 Introduction

The objective of this chapter is shown in Figure 6.1. Beach morphology is controlled by sediment transport which is in turn controlled by currents generated due to the passage of a surface wave. Here, therefore, we wish to be able to take a local surface wave of known characteristics and predict the near bed currents which are generated in a particular water depth.

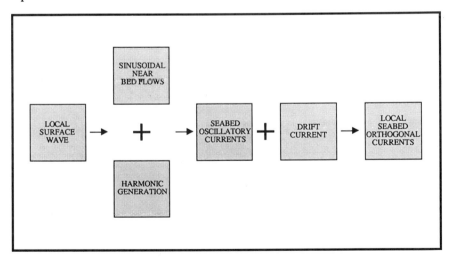

Figure 6.1 General overview of the chapter

The near bed currents are divided into two groups. The first group includes the oscillatory currents which are generated along a vertical plane normal to the wave crest line and these are known as orthogonal currents and are detailed in the present chapter (Figure 6.2). The second group includes those currents which are generated in shallow water as a result of waves breaking at an angle to the shoreline and are known as longshore currents. The longshore currents are detailed in the following chapter.

The presentation is based upon Wiegel (1964), Komar (1976), the Shore Protection Manual (1984), Sleath (1984) and Dyer (1986) along with the more specific references which are cited in the body of the text.

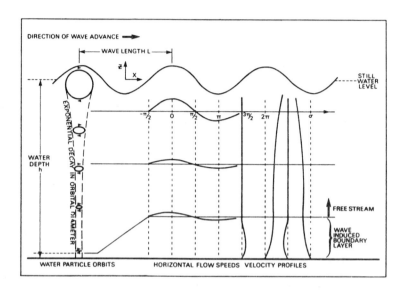

Figure 6.2 Definition diagram of orthogonal currents (After Hardisty, 1989b)

6.2 Nearbed Wave Induced Currents

It will become apparent in Chapter Nine that, although seabed sediment is transported as a result of forces acting upon individual grains, it is impossible to quantify those forces from either a theoretical or a practical standpoint. The theoretical problems of correctly describing the wave induced boundary layer are outlined in, for example, Davies (1983). Practical problems are associated with the laboratory or field measurement of either the hydrodynamic or sediment transport variables involved in the process. Instead, and in order to make progress here, suitable results are obtained by relating the mass transport of sediment to the wave induced currents immediately outside the boundary layer. In this section the wave equations which predict such current velocities will be analysed and tested to provide expressions for the currents in terms of the local wave parameters which were derived in the preceding chapter. The Airy, Stokes, solitary, cnoidal and Gerstner wave theories each predict the required parameters and are detailed in the following sections.

6.2(a) Airy Wave Theory

Airy (1845) assumed that the flow is two dimensional, incompressible and irrotational, and that the wave height is small compared with the water depth and the wave length. The theory approximates the non-linear boundary condition by using Taylor's theorem but includes only first order terms in H. The full expression for the horizontal flow velocity is given by Komar (1976) as:

$$u(x,z,t) = \frac{\pi\ H\ \sinh\ [k(z + h)]}{T\ \sinh(kh)}\cos\ (kx - \omega t) \qquad\qquad \text{Eq.6.1}$$

where z is measured upwards from the still water level. We solve for the nearbed flows (z=-h), and for the time history of the flow at a fixed point (x=0):

$$u(t) = \frac{\pi\ H}{T\ \sinh(kh)}\cos\ (\omega t) \qquad\qquad \text{Eq.6.2}$$

It is useful to assess the geomorphological significance of the various wave theories from the degree of asymmetry which each predicts (Section 6.3). The various parameters which have been used to describe the flow asymmetry are detailed in section 6.3 where it will be seen that each parameter depends upon a measure of the peak onshore, u_{in}, and peak offshore, u_{ex}, flow velocity. These occur beneath the crest and the trough and we therefore solve Eq.6.1 for x=0 and x=L/2 at t=0, and follow Clifton and Dingler (1984, p190):

$$u_{in} = u_{ex} = \frac{\pi\ H}{T\ \sinh(kh)} \qquad\qquad \text{Eq.6.3}$$

Airy wave theory therefore predicts symmetrical onshore and offshore flow velocities. The approximations for the hyperbolic function which were detailed in Section 5.4 have often been used to define *deep* and *shallow* water conditions giving:

Airy Deep Water ($h>L_\infty/2$)

$$u_{in} = u_{ex} = \frac{\pi\ H\ e^{-kh}}{T} \qquad\qquad \text{Eq.6.4}$$

Airy Shallow Water ($h<L_\infty/20$)

$$u_{in} = u_{ex} = \frac{H}{2}\sqrt{\frac{g}{h}} \qquad\qquad \text{Eq.6.5}$$

However these approximations are unnecessary since it is a simple task to calculate the flow speeds from Eq.6.1. The program given by Hardisty (1990i) provides solutions for Eq.6.1 in all water depths and the results are shown, along with those for the other wave theories considered here, in Figure 6.3 for a typical wave of 1 m height and 8 s period between depths of 40 m and the shoreline. For the sake of clarity this diagram has been constructed without utilising any of the shoaling wave transformations discussed in the previous chapter.

6.2(b) Stokes Wave Theory

Stokes (1847) used the same assumptions but included second order terms for the wave height. The full expression for the horizontal orbital velocity is given by Komar (1976) as:

$$u(x,z,t) = \frac{\pi H}{T}\ \frac{\cosh[k(z+h)]}{\sinh(kh)}\ \cos\ (kx - \omega t)$$

$$+ \frac{3}{4} \left(\frac{\pi H}{L}\right)^2 C \, \frac{\cosh[2k(z+h)]}{\sinh^4(kh)} \cos[2(kx - \omega t)] \qquad \text{Eq.6.6}$$

Solving for near bed flows (z=-h) the hyperbolic cosine becomes equal to unity and the time history of the flow at x=0 is:

$$u(t) = \frac{\pi H}{T\sinh(kh)} \cos(\omega t) + \frac{3}{4} \left(\frac{\pi H}{L}\right)^2 C \, \frac{1}{\sinh^4(kh)} \cos(2\omega t) \qquad \text{Eq.6.7}$$

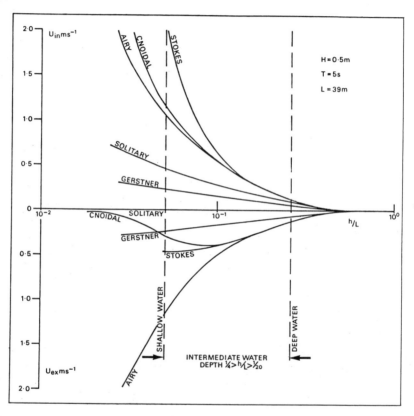

Figure 6.3 Predictions of the wave equations (after Hardisty, 1989b)

Solving for peak flows beneath the crest and trough we follow Clifton (1984, p191) to give:

$$u_{in} = \frac{\pi \, H}{T \sinh(kh)} \left(1 + \frac{3}{4} \frac{(\pi H)^2}{LT \sinh^4(kh)} \right) \qquad \text{Eq.6.8}$$

and

$$u_{ex} = \frac{\pi\ H}{T\ \sinh(kh)}\left(-1 + \frac{3}{4}\ \frac{(\pi\ H)^2}{LT\ \sinh^4(kh)}\right)$$

Eq.6.9

where u_{ex} is, of course, negative signifying offshore flow.

The first term in these expressions is the Airy Wave theory discussed above and is known as the fundamental. The second term has twice the frequency of the Airy wave component and is known as the first harmonic. It is positive under the crest and the trough and negative 1/4 and 3/4 wavelengths from the crest as is shown in Figure 6.4. The Stokes Wave nearbed velocity which results from the addition of these two terms is also shown in the diagram. Stokes Wave theory predicts a symmetrical orbital flow which is identical to first-order theory in deep water but which becomes increasingly asymmetric as the higher frequency term becomes larger when the wave shoals. It is noted that theories have been formulated which include higher order wave height terms (3rd order by Skjelbreia, 1959; 5th order by Skjelbreia and Henderson, 1962) but that workable solutions are not easily derived and the differences are small.

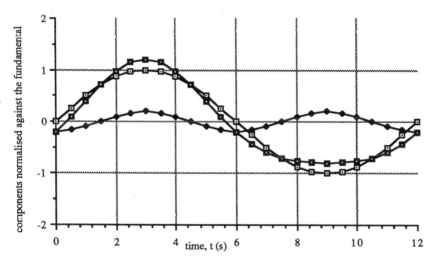

Figure 6.4 Construction of the Stokes wave second order solution, showing the addition of the first harmonic (closed diamond) to the fundamental (open square) giving the second-order flow speed (closed square).

6.2(c) Solitary Wave Theory

The solitary wave theory of Scott-Russell (1838, 1844) was recently reviewed by Miles (1980) and a reasonable and commonly used formulation is due to Korteweg and de Vries (1895). Longuet-Higgins (1981a) shows that, to lowest order, the wave profile is given by Eq.5.5 and that the horizontal flow component is:

$$u(x) = \frac{g\eta(x)}{C}$$

Eq.6.10

where $\eta(x)$ is again the surface elevation of the wave profile and C is the wave celerity which in solitary wave theory is given by:

$$C = \sqrt{gh\left(1 + \frac{H}{h}\right)} \qquad\qquad \text{Eq.6.11}$$

where h is again the undisturbed water depth and H is the wave height. At the crest therefore, where x=0, $\eta(x)$=H and the hyperbolic secant becomes unity:

$$u_{in} = \frac{gh}{\sqrt{\{gh(1 + H/h)\}}} \qquad\qquad \text{Eq.6.12}$$

The solitary wave is a wave of translation (rather than a progressive wave as is the case with those so far covered) and, *senso stricto*, there is no return flow so that u_{ex}=0 in all water depths. The results are shown in Figure 6.3.

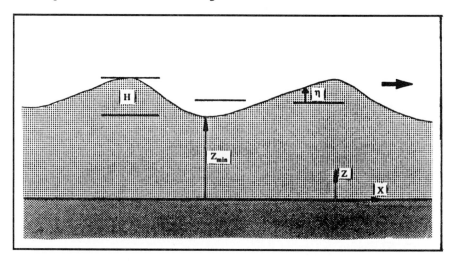

Figure 6.5 Notation for cnoidal wave theory.

6.2(d) Cnoidal Wave Theory

The horizontal velocity component in cnoidal wave theory is given in the notation of Figure 6.5 by, for example, Sleath (1984, p 12) as:

$$u = \sqrt{(gh)}\left[-\frac{5}{4} + \frac{3h_1^2}{4h^2} + \frac{3H}{2h} - \frac{h_1 H}{2h^2}cn^2(\Omega) - \frac{H^2}{4h^2}cn^4(\Omega) - 8HK^2(K)\frac{d}{3} - \frac{z^2}{2h}\right.$$

$$\left. + \frac{-K^2 sn^2(\Omega) + cn^2(\Omega)dn^2(\Omega) - sn^2(\Omega)dn^2(\Omega)}{L^2}\right] \qquad \text{Eq.6.13}$$

Even Sleath remarks that 'equally unpleasant' expressions are available for other quantities of interest (e.g. Wiegel, 1964). Workable expressions are however available for the peak flow speeds, and Skovgaard *et al.*(1974) quotes the following equations for the peak onshore and offshore flow speeds which for z=-h are given by

76

$$u_{in} = C \left(\frac{z_{max}}{h} - \left(\frac{z_{max}}{h} \right)^2 - 4F \right) \qquad \text{Eq.6.14}$$

and

$$u_{ex} = C \left(\frac{z_{min}}{h} - \left(\frac{z_{min}}{h} \right)^2 + 4mF \right) \qquad \text{Eq.6.15}$$

where:

$$z_{max} = z_{min} + H \qquad \text{Eq.6.16}$$

$$z_{min} = H \left(\frac{1}{m} \left(1 - \frac{E}{K} \right) - 1 \right) \qquad \text{Eq.6.17}$$

$$F = \left(\frac{1}{3} - \frac{(z+h)^2}{h^2} \right) \frac{H}{3gT^2} K^2 \qquad \text{Eq.6.18}$$

$$A = \frac{2}{m} - 1 - \frac{3E}{mK} \qquad \text{Eq.6.19}$$

$$m_1 = 1 - m \qquad \text{Eq.6.20}$$

$$C = \sqrt{gh \left(1 + A\frac{H}{h} \right)} \qquad \text{Eq.6.21}$$

The three cnoidal functions, m_1 (which is a parameter of the elliptic functions and integrals), K (which is a complete elliptic integral of the first kind) and E (which is a complete elliptic integral of the second kind) and depend upon a non-dimensional depth ratio known as the Ursell Number HL^2/h^3. Tables of these functions are included in Skovgaard et al. (1974), and the appropriate sections are reproduced in Table 6.1 below. Clearly even these equations do not provide an amenable form for cnoidal wave theory in geomorphic models. Nevertheless a sub-routine is included in Hardisty (1990i) which interpolates between the tabulated values to obtain nearbed flows as a function of the water depth and the wave parameters. The results are also shown in Figure 6.3 for a wave in various water depths. It is apparent that the cnoidal wave is the same as solitary wave theory when the wave length is very large, and at the other extreme, reduces to Stokes and Airy wave theories.

6.2(e) Gerstner Wave Theory

The fifth wave theory which we consider here is that developed by Gerstner (1802) for waves of finite height, and produces a trochoidal surface wave profile (cf Section 5.3). Froude (1862) and Rankine (1863), on the other hand, started with an assumed trochoidal profile and developed their equations from this curve, which is generated by the motion of a point on a circle, as the circle rolls along the underside of a line. The particle orbits are the same as for deep water Airy Wave theory so that:

Table 6.1 Jacobian Elliptic Functions for use in Cnoidal Wave Theory

$\dfrac{HL^2}{h^3}$	m	K	E	$\dfrac{HL^2}{h^3}$	m	K	E
1	.073	1.601	1.542	55	.975	3.252	1.034
2	.141	1.631	1.514	60	.981	3.386	1.027
3	.204	1.662	1.487	65	.985	3.516	1.022
4	.262	1.692	1.462	70	.989	3.643	1.017
5	.315	1.723	1.438	75	.991	3.766	1.014
6	.365	1.754	1.416	80	.993	3.886	1.012
7	.412	1.786	1.394	85	.995	4.003	1.009
8	.454	1.817	1.373	90	.957	4.117	1.008
9	.494	1.849	1.354	95	.966	4.228	1.006
10	.530	1.881	1.335	100	.997	4.336	1.005
11	.564	1.912	1.318	150	.999	5.304	1.001
12	.595	1.944	1.301	200	.999	6.124	1.000
13	.624	1.976	1.285	250	1.00	6.847	1.000
14	.651	2.008	1.270	300	1.00	7.500	1.000
15	.675	2.041	1.256	350	1.00	8.101	1.000
16	.698	2.073	1.243	400	1.00	8.660	1.000
17	.719	2.105	1.230	450	1.00	9.186	1.000
18	.739	2.137	1.218	500	1.00	9.682	1.000
19	.757	2.169	1.207	550	1.00	10.15	1.000
20	.774	2.201	1.196	600	1.00	10.60	1.000
				650	1.00	11.04	1.000
22	.804	2.266	1.176	700	1.00	11.45	1.000
24	.829	2.329	1.158	750	1.00	11.85	1.000
26	.851	2.393	1.142	800	1.00	12.24	1.000
28	.870	2.456	1.128	850	1.00	12.62	1.000
30	.886	2.519	1.116	900	1.00	12.99	1.000
32	.901	2.581	1.104	950	1.00	13.34	1.000
34	.913	2.643	1.094				
36	.923	2.704	1.085	1000	1.00	13.69	1.000
38	.932	2.764	1.077	2000	1.00	19.36	1.000
40	.940	2.824	1.070	3000	1.00	23.71	1.000
42	.947	2.883	1.063	4000	1.00	27.38	1.000
44	.953	2.942	1.057	5000	1.00	30.61	1.000
46	.958	3.000	1.052	6000	1.00	33.54	1.000
48	.963	3.057	1.047	7000	1.00	36.22	1.000
50	.967	3.113	1.043	8000	1.00	38.73	1.000

$$u_{in} = u_{ex} = \frac{\pi \, H \, e^{kh}}{T}$$ Eq.6.22

Typical results are shown in Figure 6.3. We shall see in the following section that the Gerstner Wave theory is less suitable for modelling nearbed flows than others, but it is worthwhile noting that the surface profile of a trochoidal wave (as well as that of a Stokes wave) compares well with those observed in experimental studies (Komar, 1976). However, mass transport is not predicted and the velocity field is rotational, whereas waves formed by conservative forces must be irrotational. Even worse, in a trochoidal wave the particles rotate in the opposite sense to the rotation that would be produced in a wave generated by wind stress on the water surface. That is particles beneath the crest would move seawards whereas the reverse is in fact the case.

6.3 Comparison of Wave Theories

The foregoing presentation suggests that up to five wave theories may be appropriate for beach models, and evidently some reasoned choice must be made here. There are three criteria on which the choice can be made. Firstly, it will become apparent in later sections that a suitable description of the flow must be asymmetric and must in general describe stronger onshore and weaker offshore flow velocities. Secondly, the chosen theory must be appropriate to the complexity of the other process formulae in the model, there is little to be gained by, for example, including a sophisticated but untested theory for the nearbed currents, when the sediment transport produced by those currents is a simplistic estimate. Finally, the wave theory should be supported by laboratory and field measurements, which is in itself a stiff criterion because it is only in the last decade that flow measurement technology has become sufficiently advanced to permit an adequate measurement of the rapidly reversing flows on natural beaches. The following sections address the wave theories with these criteria.

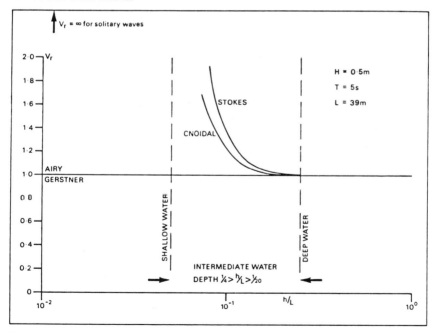

Figure 6.6 Flow asymmetries predicted by the five wave theories (after Hardisty, 1989b).

Morphodynamic comparison of wave theories

Bowen (1980), Hardisty (1986) and others (cf Chapter Fifteen) argued that the downslope component of the gravity force on seabed sediment will inevitably result in continuous offshore transport unless an onshore asymmetry is present in the flow field. There have been various descriptions of flow asymmetry by different authors. Clifton (1976) proposed the use of a velocity magnitude asymmetry equal to the algebraic difference between the moduli of u_{in} and u_{ex}, though that author is less happy with the idea in a later paper (Clifton and Dingler, 1984). Alternatively, Hardisty (1984) followed Kemp (1975) and used the velocity ratio, V_r, which is the non-dimensional ratio of the peak onshore to the peak offshore flow velocities ($V_r=u_{in}/u_{ex}$). Hardisty (1989b) argued that morphodynamic stability of the orthogonal seabed profile (Chapter Fifteen) required a flow with $V_r\approx1$ in deep water, and increasing so that $V_r>1$ in intermediate and shallow water depths. The velocity ratio predicted by the various theories is shown in Figure 6.6, from which it is apparent that neither the Airy wave theory, nor the rotational wave theory fulfill these criteria, because the predicted flow is symmetrical in all water depths. It would appear then that Stokes, cnoidal or solitary wave theory is preferred, but that cnoidal theory is difficult to apply (Eq.6.11) and solitary is a wave of translation with no return flows. The chosen course here is to utilise Stokes theory over the broadest range to which it applies.

Comparison of Stokes wave theory with experimental results

Hardisty (1989b) conducted experiments in a wave channel and measured u_{in} and u_{ex} for a variety of waves in intermediate water depths with an ultrasonic current meter.. The results for u_{in} are shown in Figure 6.7 and for u_{ex} are shown in Figure 6.8. In addition Morison and Crooke (1953) conducted experiments at the University of California and measured flow speeds with neutrally buoyant droplets and their results are indicated by "1" in Figures 6.7 and 6.8. Le Mehaute et al. (1968) also measured the positions of neutrally buoyant particles in a wave tank, but only reported the results for the peak onshore flow as indicated by "2" in Figure 6.7. Iwagaki and Sakai (1970) used hydrogen bubble tracers and hot film anemometers to determine peak onshore flow speeds and their results are indicated by "3" in Figure 6.7. Grace and Rocheleau (1973) measured near bottom crest and trough velocities with a small, ducted impellor located about 0.5m above the seabed near the Hawaiian Island of Oahu in a water depth of about 11m. Their results are reported by Dean and Perlin (1986) and although finding that linear theory agreed reasonably well with the measured trough velocities (u_{ex}), the values of u_{in} are underpredicted by linear theory by approximately 23%. They therefore find that the wave induced near bed flow is asymmetric with $V_r>1$.

These data all confirm the asymmetry of the nearbed flow, but regression analysis of the results suggests that the peak flows are less than predicted by theory by between 63 and 89%:

$$u_{in\ obs} = 0.63\ u_{in} \qquad\qquad\qquad Eq.6.23$$

$$u_{ex\ obs} = 0.89\ u_{ex} \qquad\qquad\qquad Eq.6.24$$

where the 'obs' subscript represents the corresponding flow measurement. The correction is not, however, suprising because it is well known that seabed pressures are attenuated to an extent which exceeds theory by a similar amount (Draper, 1957, Driver, 1980), with $H_{obs} = 0.84\ H_{pred}$ (Draper, 1957). It is apparent then that, wanting better answers, we must attenuate the nearbed flows predicted by second-order wave theory by a value of about 0.8.

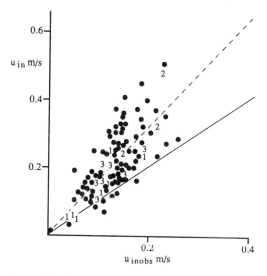

Figure 6.7 Comparison of the theoretical prediction for the peak onshore velocity with the experimental results of Hardisty (1989b) and others (see text for details).

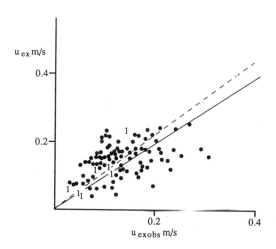

Figure 6.8 Comparison of the theoretical prediction for the peak offshore velocity with the experimental results of Hardisty (1989b) and others (see text for details).

Shallow Water Limit

Stokes (1847) recognised that small-amplitude theory, on which the second order solution is based, is inappropriate in shallow water and suggested that the ratio (here signified by R_h) of the first to the fundamental harmonics of the surface profile must be small if the equations were to remain valid. Thus, from Eq.5.3, this ratio is given by:

$$R_h = \frac{H/2}{\dfrac{\pi H^2 \cosh(kh)[2 + \cosh(2kh)]}{8L \sinh^3(kh)}}$$

Eq.6.25

Substituting for $\sinh(kh) \approx kh$ and $\cosh(kh) \approx 1$ in shallow water (cf Section 5.4) this yields:

$$R_h = \frac{3}{32\pi^2} \frac{HL^2}{h^3}$$

Eq.6.26

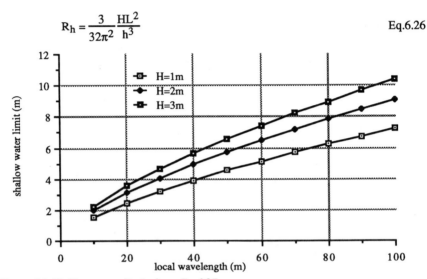

Figure 6.9 Shallow water limits from Eq.6.27.

The shoreward limit of the second-order equations is usually taken to be given by a value of the dimensionless grouping on the right hand side of Eq.6.26, HL^2/h^3, which is known as the Ursell Number, Ur (cf section 6.2(d) and Ursell, 1953). Various versions of this and other parameters have been utilised in the literature (De, 1955; Longuet-Higgins, 1956; Laitone, 1962). Most usefully, Madsen (1971) observed that when $R_h>0.25$ the secondary wave becomes visible in the trough of the fundamental wave profile. Substituting this criteria into Eq.6.26 suggests that the shallow water limit of second-order theory should be taken as $Ur<8\pi^2/3$, which is equivalent to:

$$\frac{HL^2}{h^3} < 26.3$$

Eq.6.27

6.4 Wave Induced Drift Currents

In addition to the oscillatory, orthogonal currents described in the preceding sections, surface waves generate steady, mass transport currents in a direction normal to the wave advance. In terms of the second-order theory utilised here these involve an onshore drift current because an interesting departure of the Stokes wave from the Airy wave is that the particle orbits are not closed. This leads to a non-periodic drift or mass transport in the direction of wave advance, the associated velocity, U_s, being given by Komar (1976) as:

$$U_s = \frac{1}{2}\left(\frac{\pi H}{L}\right)^2 C \frac{\cosh[2k(z+h)]}{\sinh 2(kh)}$$

Eq.6.28

which, in deep water, reduces to:

$$U_{s\infty} = \left(\frac{\pi H_\infty}{L_\infty}\right)^2 C_\infty\, e^{2kz}$$

Eq.6.29

Integration of Eq.6.24 with depth yields the discharge q:

$$q = \frac{\pi}{4}\frac{H_\infty{}^2}{T}$$

Eq.6.30

which is the volume of water transported towards the shore per unit crest length and per unit time. For narrow channels of finite length Longuet-Higgins (1953) formulated the solutions shown in Figure 6.10:

$$U(h) = \frac{1}{4}\left(\frac{\pi H}{T}\right)\left(\frac{\pi H}{L}\right)\frac{1}{\sinh 2(kh)}$$

$$\left\{2\cosh\left(2kh\left(-\frac{z}{h}-1\right)\right) + 3 + 2kh\left(3\,\frac{z^2}{h^2} + 4\,\frac{z}{h} + 1\right)\sinh 2kh\right.$$

$$\left. + 3\left(\frac{\sinh 2kh}{kh} + \frac{3}{2}\right)\left(\frac{z^2}{h^2} - 1\right)\right\}$$

Eq.6.31

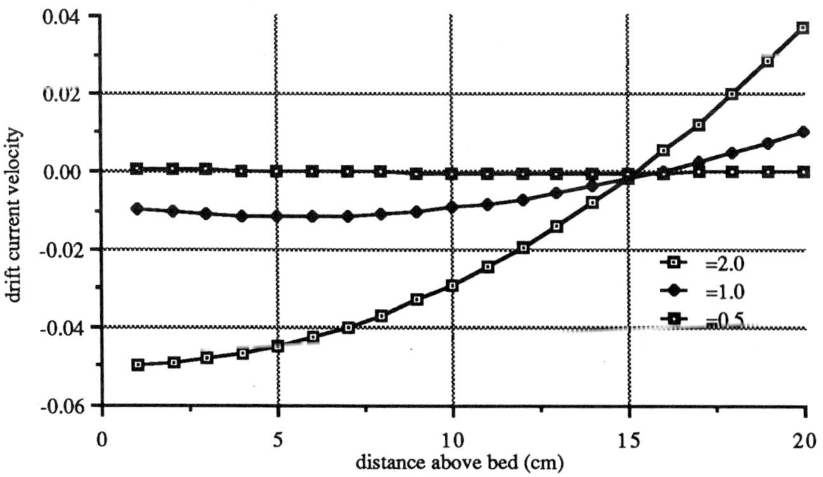

Figure 6.10 Vertical distribution of the Stokes drift current for various wave periods and for 0.10m waves in 0.20m water depth (after Hardisty, 1990j).

The results show a nett shoreward flow near the surface and near the bottom balanced by a return flow at mid depth. It is not known if the distribution applies on natural beaches. Russell and Osario (1958) note that when the channel width is not constrained there is a tendency to develop a horizontal circulation and continuity need not then be confined in two dimensions.

The net shoreward velocity near the bottom, U_{so}, is given by Longuet-Higgins (1953) as:

$$U_{so} = \frac{5}{4}\left(\frac{\pi H}{L}\right)^2 C \frac{1}{\sinh^2(kh)}$$

Eq.6.32

from Eq.6.28 where z=-h and therefore the hyperbolic cosine becomes equal to unity (Figure 5.7). Komar (1976) notes that this may be responsible for producing a net shorewards transport of sediment close to the seabed.

6.5 Currents due to Random Waves

In the presence of random waves, Guza and Thornton (1980) suggest that the equations for the horizontal velocity is given by the summation of linear theory for the n constituent sinusoids:

$$u(t) = \sum_{n=1}^{n=\infty} \frac{a_n \, \omega_n \, \cosh|k_n|(h+z)}{\sinh|k_n|h} \cos(k_n x + \omega_n t + \varepsilon_n)$$

$$= \sum_{n=1}^{n=\infty} \left(\frac{\omega_n \cosh |k_n| (h+z)}{\sinh |k_n| h}\right) \eta_n$$

Eq.6.33

where a_n, ω_n, k_n and ε_n are the wave amplitudes, frequencies, numbers and phase angles respectively and η_n is the surface elevation given by:

$$\eta_n = a_n \cos (k_n x + \omega_n t + \varepsilon_n)$$

Eq.6.34

Specifically this theory states that the spectrum of the near bed flow is adequately described by a linear transformation of the surface elevation spectrum. In other words, if the shoaling transformations discussed in the previous chapter are correct, then the near bed flows follow from transforming the surface elevation at each frequency using Airy theory.

Comparison with experimental results

Comparisons between spectra of sea surface elevation and velocity, measured at the same horizontal location in relatively shallow water, have been made by various authors. Bowden and White (1966), and Simpson (1969) made observations in 4 to 6 m of water and the spectral peaks were about 0.2 Hz, that is in a non-dimensional depth, h/L_∞ of $0.7<h/L_\infty<0.1$. Thornton and Krapohl (1974) studied long Pacific swell (0.06 Hz) in 19 m of water ($h/L_\infty=0.04$) and Cavaleri et al. (1978) took data in 16m of water with peak spectral energy at 0.2 Hz ($h/L_\infty=0.4$). Reporting these results, Guza and Thornton (1980) note that these measurements are of weakly non-linear waves at some distance offshore in depths where Airy theory can be expected to apply. The measured horizontal velocity spectra generally agree with predicted spectra with about 10% error in the vicinity of the spectral peak, although Simpson's data showed a larger disagreement of about 30%.

Guza and Thornton (1980) extended the analysis into shallow water with an Ursell number from Ur<0.05 to Ur>>1 with Ur=a/h(kh)^2 (with the amplitude a taken as $a_{1/3}=H_s/2$ and k is taken as the wave number at the spectral peak (see Chapter Five for details of the significant wave height notation). The measurements were made at Torrey Pines Beach, San Diego which is a gently sloping, moderately sorted, fine grained sand beach. The surface elevation was measured with dual resistance wave staffs and the orthogonal currents were measured with Marsh-McBirney electromagnetic current meters.

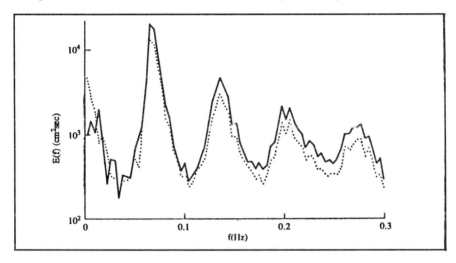

Figure 6.11 Comparison of sea surface elevation (solid line) and horizontal current (broken line) spectra for 1.76 m water depth at Torrey Pines (source: Guza and Thornton, 1980).

Figure 6.11 shows the results for one of their comparisons of the predicted surface elevation from the transformed horizontal current spectrum, with the results from a wave staff. Guza and Thornton (1980) conclude that the data show considerable agreement with the linear theory from well outside to inside the surf zone despite the fact that non-linearities might be anticipated close to the shoreline. There was also evidence of growth in the long wave energy as discussed in Section 5.9(b).

6.6 Wave Induced Boundary Layers

Although a complete description of the wave induced boundary layer has still to be achieved, there are some useful results which should be discussed. The velocity profile within the wave induced boundary layer for a flat impermeable bed is shown in Figure 6.2, and it can be seen that both the phase and the amplitude of the velocity oscillation vary through the layer, with the phase lead increasing with distance from the bed. For laminar flow, the maximum bed stress leads the maximum flow by about $\pi/4$.The thickness of the boundary layer, δ, can be defined in a number of ways (Dyer, 1986), but can be given by:

$$\delta = \sqrt{4\pi\upsilon T}$$ Eq.6.35

where υ is the kinematic viscosity of the fluid defined by μ/ρ, and μ is the dynamic viscosity of the fluid. The definition of δ is rather arbitrary since the free stream velocity is approached only slowly at the top of the boundary layer. Nevertheless the boundary layer is very thin, with $\delta \approx 0.7$ cm for a 4s wave. However as the roughness of bed increases, or as the maximum orbital velocity increases, the laminar boundary layer thins and a transition to a turbulent boundary layer will eventually occur. The first turbulent effects that can be seen are small vortices in the lee of individual bed grains. On flow reversal these vortices tend to be ejected upwards into the flow and to be carried back over the grain to interact with the vortex forming on the other side. This leads to an increase in thickness of the boundary layer and, when fully turbulent, the boundary layer is a few centimetres thick.

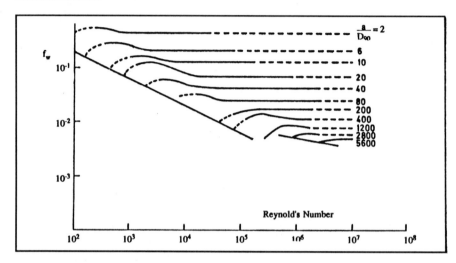

Figure 6.12 Wave friction factor variation with wave Reynolds number (after Jonsson, 1966)

As a measure of the structure of the wave induced boundary layer, a wave Reynolds Number can be formulated (see also Chapter Twelve):

$$Re_w = \frac{u_{max} \, d_o}{\upsilon} \qquad \qquad Eq.6.36$$

where u_{max} is the maximum orbital velocity, given in first-order theory by Eq.6.3 and d_o is the amplitude of the nearbed flow particle orbits given, also in first-order theory, by:

$$d_o = \frac{H}{2 \sinh(kh)} \qquad \qquad Eq.6.37$$

Jonsson (1966) stated that the boundary layer remains laminar provided that $Re_w < 1.26 \times 10^4$. The maximum bottom stress, τ_{omax}, can be defined from Eq.5.17 as:

$$\tau_{omax} = \frac{1}{2} f_w \, \rho \, u_{max}^2 \qquad \qquad Eq.6.38$$

where, for a laminar boundary layer:

$$f_w = \frac{2}{\sqrt{Re_w}}$$ 　　　　　　　　　　　　　　　　　　 Eq.6.39

For a turbulent boundary layer, the friction factor depends upon the roughness of the boundary and the wave Reynolds number as shown by Jonsson's diagram for f_w against Re_w (Figure 6.12), but it is still not apparent exactly when, in the wave cycle, the peak stress is achieved.

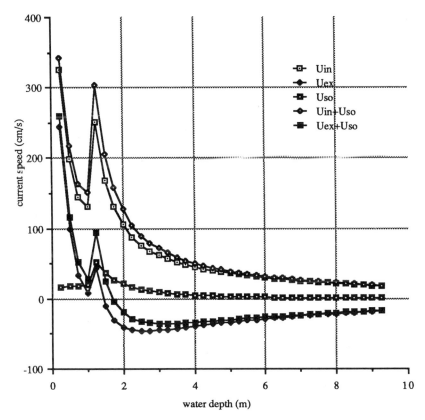

Figure 6.13 Example of orthogonal flows generated by Eq. 6.38 for a deep water wave of height 1m and period 5s, and including the shoaling transformations detailed in the preceding chapter. In this diagram only, Uin, Uex and Uso represent the peak onshore, peak offshore and Stokes drift currents. Note that the wave shoals and breaks in about 1.5 m of water, but that the second-order solutions become unstable (Uex>0) in about the same depth.

6.7 Conclusions

During shoaling from deep water, second-order Stokes wave theory will be utilised in the model presented in later chapters. The orthogonal currents are given by the sum of the attenuated fundamental and first harmonic terms (Eq.6.7) and the Stokes drift current (Eq.6.32) at the bed:

$$u(t) = A_a \left(\frac{\pi H}{T \sinh(kh)} \cos(\omega t) + \frac{3}{4} \left(\frac{\pi H}{L} \right)^2 C \frac{1}{\sinh^4(kh)} \cos(2\omega t) \right)$$

$$+ \frac{5}{4} \left(\frac{\pi H}{L} \right)^2 C \frac{1}{\sinh^2(kh)} \qquad \text{Eq.6.40}$$

where the attenuation coefficient, A_a, has a value of 0.8, L is given by the dispersion relationship (Eq.5.8) and H by the full shoaling form of Eq.5.29. C will be taken to be defined as L/T since the second order solutions for C are identical to linear theory on which the dispersion equation is based. Stokes theory will be assumed to apply into water as shallow as defined by the Madsen criteria (Ur<26.3). The resulting flows are modelled by Hardisty (1990e) and typical results are shown in Figure 6.13. It can be seen that, strictly, the second order solution becomes unstable and therefore a correction is introduced in Chapter Seventeen which permits the wave modelling to be continued into shallower water for the SLOPES work.

chapter seven

LONGSHORE CURRENTS

7.1 Introduction

The theories of wave transformation and shoaling dealt with in the preceding chapters lead to expressions for orthogonal sediment transport and will later be used to model the orthogonal system in a vertical plane. The theories of wave refraction and longshore current generation are briefly introduced here. These lead to expressions for shore parallel sediment transport and thus provide both a source and a sink for the orthogonal transport in the region shorewards of the breakpoint. Thus the present chapter, and Chapter Eleven, Longshore Sediment Transport, are included in this book so that profile response due to longshore movement of sediment in the surf and swash zones can be assessed. Additionally the chapters form an appropriate introduction to the more complex problems of nearshore resonance and edge wave effects.

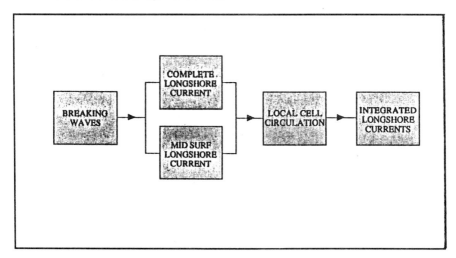

Figure 7.1 Definition sketch of chapter objectives

The objective of this chapter is shown in Figure 7.1, and it is designed to provide functional formulae for the longshore current velocities in terms of the wave parameters at the break point. The chapter is based on the review by Komar (1983) with other more specific references detailed in the text.

Currents which flow parallel to the shoreline are generated by both oblique wave approach (Sections 7.2 and 7.3) and by longshore variations in the breaker height (Section 7.4). Each type is dealt with separately and a combined solution is then presented in Section 7.5. The former type will, for convenience, be referred to here as longshore currents, whereas the latter type will be referred to as local currents.

7.2 Wave Set-up and Set-down

It is convenient to introduce the theoretical solution for longshore currents through consideration of the concept of radiation stress. Radiation stress is familiar from electromagnetic field theory and has been used to analyse the change in the mean water level or set-up of water at the shoreline due to the presence of waves (Longuet-Higgins and Stewart, 1964; Bowen et al., 1968). The radiation stress is defined as "the excess flow of momentum due to the presence of the waves" (Komar, 1976). If the x-axis is placed in the direction of wave advance (Chapter One) and the y-axis parallel to the wave crests, then there are two components to the radiation stress. The radiation stress (x momentum flux) across the plane x=constant (i.e. across a plane parallel to the shore in the direction of wave advance) is given by:

$$S_{xx} = E\left(\frac{2kh}{\sinh(2kh)} + \frac{1}{2}\right) = E\left(2n - \frac{1}{2}\right)$$ Eq.7.1

where n was given by Eq.5.12. The radiation stress (y momentum flux) across the plane y=constant is given by:

$$S_{yy} = E\left(\frac{kh}{\sinh(2kh)}\right)$$ Eq.7.2

In both of these equations the wave energy E is taken to be given by Eq.5.11: $E=\rho gH^2/8$. The onshore momentum flux must be balanced by an opposing force (this effectively was Newton's definition of a force as a rate of change of momentum), which is manifested as a water slope so that the pressure gradient of the sloping water surface balances the change (spatial gradient) in the incoming momentum:

$$\frac{dS_{xx}}{dx} + \rho g(h+\eta)\frac{d\eta}{dx} = 0$$ Eq.7.3

where h is the difference between the still water level and the mean water level in the presence of waves. This difference is known as the set-up or set-down due to the waves. Longuet-Higgins and Stewart (1964) integrated Eq.7.3 along with Eq.7.1 to obtain:

$$\eta = -\frac{1}{8}\frac{H^2k}{\sinh(2kh)}$$ Eq.7.4

outside the breaker zone, whilst inside the breaker zone:

$$\frac{d\eta}{dx} = \frac{1}{1+8/3\gamma_b^2}\tan\beta$$ Eq.7.5

where the ratio of the wave height to the water depth at the breakers ($\gamma_b=H_b/h_b$) was assumed constant at 0.8 (cf Section 5.6(c)) and β is again the beach slope angle.

Comparison with experimental results

The laboratory experiments of Bowen *et al*. (1968) are shown in Figure 7.2 and found general agreement with the theory. Hardisty (1990l) provides routines for the calculation of wave set up from the radiation stress, and typical results are shown in Figure 7.3.

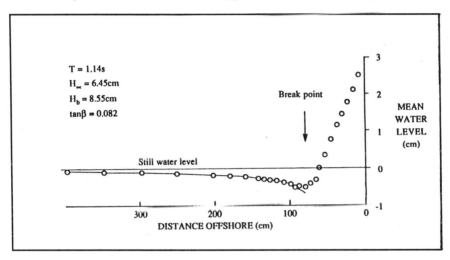

Figure 7.2 Wave set-up and set-down in laboratory experiments compared with Eq.7.4 [after Bowen et al (1968)]

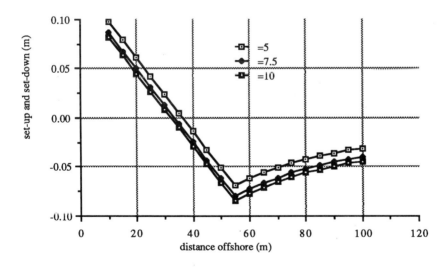

Figure 7.3 Wave set-up and set-down for a constant H=2m, an orthogonal gradient of 0.05 and a range of wave periods (from Hardisty, 1990l).

7.3 Longshore Currents: A Complete Solution

A large number of equations (Table 7.1) have been developed which purport to predict the longshore current velocity. One approach (Inman and Bagnold, 1963; Eagleson, 1965;

Table 7.1

Mean Longshore Current Equations

Putman, Munk and Traylor (1949)

$$V = \left(0.871 \, \frac{H_b^2}{T} \, \frac{gs_L}{c_f} \, \tan\beta \, \sin2\alpha_b \right)^{1/3} \qquad \text{Energy flux}$$

where s_L is the fraction of the total energy flux that is responsible for the longshore current.

Inman and Quinn (1952)

$$V = \left(\left(\frac{1}{4\varepsilon_L^2} + C_b \, \sin\alpha_b \right)^{1/2} - \frac{1}{2\varepsilon_L} \right)^2 \qquad \text{Momentum flux}$$

where $\varepsilon_L = 108.3 \, \frac{H_b}{T} \, \tan\beta \, \cos\alpha_b$ and the units are fps. Reduces to $V \approx C_b \sin\alpha_b$ where C_b is the wave celerity at the break point.

Inman and Bagnold (1963)

$$V = 4 \, \sqrt{\frac{\gamma_b}{3}} \, \frac{1}{T} \, \tan\beta \, \sin\alpha_b \, \cos\alpha_b \qquad \text{Mass continuity}$$

Galvin and Eagleson (1965)

$$V \approx 2gT \, \tan\beta \, \sin\alpha_b \, \cos\alpha_b \qquad \text{Mass continuity}$$

Longuet-Higgins (1970a and b)

$$V = \frac{5\pi}{8} \, \frac{\tan\beta}{c_f} \, u_m \, \sin\alpha_b \qquad \text{Radiation stress}$$

Komar (1975, 1976)

$$V = 2.7 \, u_m \, \sin\alpha_b \, \cos\alpha_b \qquad \text{Empiricised radiation stress}$$

Galvin and Eagleson, 1965) is based upon discharge relationships produced by the shorewards mass transfer of water at the breakers. These are possibly unfounded because they do not account for the return of this water to the breakers and therefore confuse longshore currents with the rip current systems described in Section 7.4.

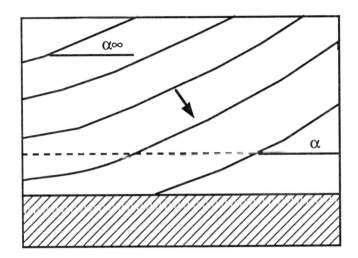

Figure 7.4 Definition sketch of refraction geometries

The subject was, however, placed on a sounder theoretical footing with the publication of papers by Bowen (1969a and b), Thornton (1971) and Longuet-Higgins (1970 a and b) making use of the concept of radiation stress. Longuet-Higgins assumed that the longshore currents are steady and constant in the longshore direction, so that the y-component of the equations of motion reduce to:

$$\frac{\partial S_{xy}}{\partial x} - [R_y] + \frac{\partial}{\partial x}\left(\mu_e(\eta+h)\frac{\partial v}{\partial x}\right) = 0 \qquad \text{Eq.7.6}$$

which states that the gradient of the onshore flux of longshore directed momentum $(\partial S_{xy}/\partial x)$ and the gradient of the longshore current are balanced by a time averaged frictional drag $[R_y]$ and h is the water depth, η the wave set-up or set-down (cf Section 7.2) and μ_e is an eddy coefficient for the horizontal mixing across the surf zone. The solution depends upon the horizontal mixing term and upon the form given to the frictional drag term $[R_y]$. Considering that the stress is that due to the combined motion of the waves and the longshore current, Longuet-Higgins used a quadratic stress formulations:

$$[R_y] = \frac{2}{\pi} c_f \rho u_m v \qquad \text{Eq.7.7}$$

where u_m is the maximum orbital velocity of the waves given by first-order theory (cf section 6.2(a)):

$$u_m = \frac{\gamma_b}{2}\sqrt{g(\eta+h)} \qquad \text{Eq.7.8}$$

and η is again the wave set-up or set-down. Longuet-Higgins (1970b) took the eddy dimensions to scale with the width of the surf zone and obtained two solutions: shorewards of the breakers:

$$v = \frac{v_0(B_1 x^{P_1} + Ax)}{x_b}$$

Eq.7.9

and seawards of the breakers:

$$v = \frac{v_0 B_2 x^{P_2}}{x_b}$$

Eq.7.10

where x is the distance from the shoreline, x_b is the distance to the breaker zone and the reference velocity at the break point is given by linear theory as:

$$v_0 = \frac{5\pi}{16} \gamma_b \zeta^2 \frac{\tan\beta}{c_f} \sqrt{(gh_b)} \sin\alpha_b$$

Eq.7.11

where c_f is a friction factor (cf Section 7.4), α_b is the angle of the breakers to the shoreline (Figure 7.3), h_b is the depth at the breaker zone, $\tan\beta$ is the beach gradient, ζ is a constant factor which results from inclusion of wave set up:

$$\zeta = \left(1 + \frac{3\gamma_b^2}{8}\right)^{-1}$$

Eq.7.12

and γ_b (Chapter Five) is the ratio of the wave height to the water depth (including set-up or set-down) at the breakpoint:

$$\gamma_b = \frac{H}{\eta+h}$$

Eq.7.13

The other variables in Eq.7.9 and 7.10 are:

$$A = \left(1 - \frac{5}{2}\zeta P\right)^{-1}$$

Eq.7.14

$$B_1 = \frac{P_2 - 1}{P_1 - P_2} A$$

Eq.7.15

$$B_2 = \frac{P_1 - 1}{P_1 - P_2} A$$

Eq.7.16

$$P_1 = \frac{3}{4} + \left(\frac{9}{16} + \frac{1}{\zeta P_e}\right)^{1/2}$$

Eq.7.17

$$P_2 = \frac{3}{4} - \left(\frac{9}{16} + \frac{1}{\zeta P_e}\right)^{1/2}$$

Eq.7.18

The dimensionless parameter P_e reflects the significance of the horizontal eddy transfer: the larger the value of P_e the more important the effect.

Inside the breaker zone, certain approximations can be utilised to simplify these equations. Consider $\gamma_b \approx 0.78$ (Chapter Five) then ζ (Eq.7.12) becomes about equal to 0.81. Replacing these parameters in Eq.7.11 yields, with a little algebra and in SI units because of the replacement of the gravitational constant:

$$v_o = 253 \tan\beta \sqrt{H_b} \sin\alpha_b \qquad\qquad Eq.7.19$$

Comparison with experimental results

Hardisty (1990m) presents a program for the distribution of the longshore current across the surf zone based upon the theory given above and typical results are shown in Figure 7.5. A family of longshore current profiles is obtained which depend upon the choice of P in the equations. Longuet-Higgins (1970b) compares his solution with the laboratory measurements of Galvin and Eagleson (1965), concluding that P_e lies in the range 0.1-0.4, though Komar (1983) questions the validity of the data set and thus of the empirical conclusions. The horizontal mixing couples adjacent water columns and it can be seen that this lateral diffusion also enables the momentum flux inside the surf zone to drive currents in the region beyond the breakers. A straight line profile results when there is no horizontal eddy mixing (i.e. $P_e = 0$) and the maximum current becomes more evenly distributed and moves seawards with greater mixing.

7.4 Longshore Currents: Mid Surf Solution

The general solution was simplified by Komar (1975) from Longuet-Higgins (1970 a) to give the longshore current at the mid surf position, v, as:

$$v = \frac{5\pi}{8} \frac{\tan\beta}{c_f} u_m \sin\alpha_b \qquad\qquad Eq.7.20$$

In this expression u_m is the maximum value of the horizontal orbital velocity at the breakers given by Longuet-Higgins (1970 a) as:

$$u_m = \sqrt{\frac{2E_b}{\rho h_b}} \qquad\qquad Eq.7.21$$

Eq.7.21 is, of course, identical to the shallow water approximation for first-order theory (Eq.6.5) with the substitution for H^2 from Eq.5.11. Longuet-Higgins (1970 a) reviewed the available estimates and arrived at $c_f = 0.01$ though he also related this to a turbulent factor, B_t:

$$c_f = 0.036 \, B_t \qquad\qquad Eq.7.22$$

Where B_t lies between 0.167 and 0.50. Longuet-Higgins suggested that, physically, B_t increases with increasing horizontal mixing lengths and with decreasing surf zone width. Komar (1975) notes that these both decrease with increasing beach slope and therefore suggests that $\tan\beta/c_f$ is essentially constant for natural beaches at about 0.138. Komar (1975) therefore simplifies the Eq.7.20 to:

$$v = 2.7 \, u_m \sin\alpha_b \cos\alpha_b \qquad\qquad Eq.7.23$$

where the cosine factor is included by Komar (1975) because of the support of laboratory data at larger angles of incidence for which $\cos\alpha_b$ is no longer equivalent to unity.

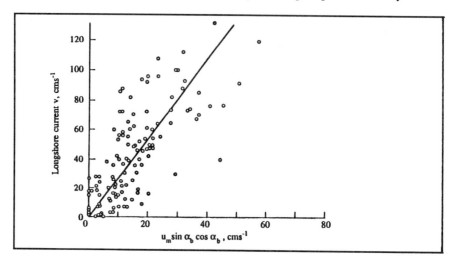

Figure 7.5 Mid surf longshore current velocities (after Komar, 1976).

Comparison with experimental results

Komar (1976) provides a comprehensive review of available data on the mean longshore current velocity, v, in comparison with Eq.7.23 and his results are shown in Figure 7.5.

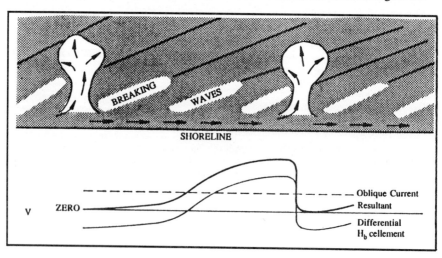

Figure 7.6 Local current generation

7.5 Local Currents

The longshore currents described in 7.2 and 7.3, are the major flows which result in large amounts of sediment movement. On a smaller, secondary scale there are however a variety of cell like flow structures which can develop at various scales. Since the net effect of any single or paired cell system is zero, these features do not normally contribute to the gross form of the beach, but their local effects are nevertheless of interest and will be detailed here.

7.5(a) Cell Circulations

The most apparent feature of the cell circulation is the rip currents: strong narrow currents that flow seawards from the surf zone. These are fed (Figure 7.6) by currents which run parallel to the beach and which increase in velocity from zero midway between two adjacent rips reaching a maximum just before turning seawards into the rip current itself. It is important to realise that both the cell circulation and the steady, longshore currents can be present simultaneously. The current pattern actually observed is the sum of the two processes. This is why it is often difficult to distinguish between the two types in field measurements.

There is no reason for the generation of cell circulations if the wave height at the breakers is constant along the shore. This, however, is frequently not the case and then a cell circulation may develop. Komar (1976) considers the problem and finds that (based on the radiation stress approach) the velocity of the cell circulation, V_c, is:

$$V_c = \frac{\pi\sqrt{2}}{c_f\gamma_b{}^3}\left(1 + \frac{3}{8}\gamma_b{}^2 - \frac{\gamma_b{}^2}{4}\cos2\alpha_b\right)\frac{\partial H_b}{\partial y} \qquad \text{Eq.7.24}$$

Where $\partial H_b/\partial y$ is the longshore change in wave height. The cell circulation therefore depends primarily on the existance of variations in the wave height along the shore. There are two principle ways in which such variations can be produced. The more obvious is by the process of wave refraction which was discussed earlier, producing a local relative concentration in wave energy and hence locally higher waves. An example of such control is the well documented case where canyon heads cause local wave shadows and rip current development. The alternative cause of longshore wave height variation is the generation of edge waves in the nearshore region, and this phenomena is described in the following section.

7.5(b) Edge Waves

Many authors (for example Shepard and Inman, 1950,1951b) have observed cell circulation on long straight beaches with regular topography and normally incident waves. These have been attributed to a form of forced resonance in the nearshore region, which was described theoretically by Lamb (reprinted in 1975). The resonant energy is manifested as a series of waves running parallel to the shore and called edge waves. Under certain circumstances these can be standing or stationary and then the interaction or summation of the incoming edge waves produces alternatively high and low breakers along the shoreline and therefore gives rise to a regular pattern of circulation cells with evenly spaced rip currents. The problem was addressed by Stokes (1847) who obtained the edge wave dispersion equation which is analogous to Eq.5.7: $\omega_e{}^2 = gk_e\tan\beta$, where ω_e is the frequency and k_e the wave number of the edge waves. Eckart (1951), Ursell (1952) and a series of papers after Bowen (1969b) and Bowen and Inman (1969) have investigated this phenomena and show that a more general solution is given by the dispersion equation:

$$\omega_e{}^2 = gk_e\sin(2n_e + 1)\tan\beta \qquad \text{Eq.7.25}$$

where ω_e is the frequency of the edge wave and n is the integer mode number. The Stokes solution clearly applies only to n=0 mode. The lines for n_e = 1, 2, 3 etc. represent the free modes which are "resonant", whereas the areas between lines do not represent solutions to the dispersion equation and any motion here must be of a forced wave. Where $(2n_e +1)\tan\beta > \pi/2$ there are no trapped solutions, and waves having these (ω_e, k_e) values may propogate to, or from, deep water and are consequently known as leaky "modes". They include the normal incident waves. Huntley *et al.* (1981) demonstrated the applicability of Eq.7.25 by measuring the energy density over a range of (ω_e, k_e) and confirmed that maxima correspond to modal number values, m. Re-arrangment of Eq.7.25 provides the edge wave length as:

$$\lambda_e = \frac{gmT^2}{2\pi} \sin(2n + 1)\tan\beta \qquad \text{Eq.7.26}$$

Typically $m = 1$, and Wright *et al.* (1979) find that edge waves can exist as subharmonics of the incident waves, so that $m = T_e/t$ having values of 2, 4, 6 etc. Wright *et al.* (1979) note that reflective beaches coincide with low m values whilst dissipative beaches coincide with higher values. It is apparent that the presence of edge waves will have an as yet unquantified and possibly dominant effect on orthogonal currents and sediment transport. It appears likely therefore that future orthogonal models will have to include predictive equations for edge wave generated transport and beach response. Such analysis will fundamentally change our models before the end of this century, but is beyond the scope of the present book.

7.6 Conclusions

Komar (1975, 1976) combines the longshore current and local current equations given above to suggest that the mean, total surf zone current, V_t, will be:

$$V_t = 2.7 \ u_m \ \sin\alpha_b \ \cos\alpha_b - \frac{\pi\sqrt{2}}{c_f\gamma_b^3}\left(1 + \frac{3\gamma_b^2}{8} - \frac{\gamma_b^2}{4}\cos2\alpha_b\right)\frac{\partial H_b}{\partial y} \qquad \text{Eq.7.27}$$

It is apparent that, if $\partial H_b/\partial y = 0$ then only the longshore current component is present, and conversely if $\alpha_b = 0$ then only the local currents are present. The SLOPES model, described in later chapters will utilise the approximation to the full longshore current solution given in Section 7.3:

$$v = 2.7 \ u_m \ \sin\alpha_b \ \cos\alpha_b \qquad \text{Eq.7.23}$$

to compute typical values of the longshore current, but will assume zero gradient in the longshore current, so that the profile evolution is purely orthogonal. The limitations of this assumption are discussed in Chapter Twenty.

chapter eight

TIDES

8.1 Introduction

The regular rise and fall of the sea at coastal sites has been of considerable concern to man since earliest times. It is caused by the gravitational attraction of, principally, the moon and the sun on the waters of the world's oceans. The theoretical analysis of these tidal motions was central to Sir Isaac Newton's *Principia Mathematica* published in the late seventeenth century and provided corroboration for his, then novel, concept of force, a concept which is in turn so central to the mechanics contained in every chapter of this book.

Earlier observations of tidal motions date back to antiquity and there is a rudimentary tide table in the British Museum, for London Bridge in the year 1213. The preparation of regular tide tables for the main ports of Great Britain was started during the seventeenth and eighteenth centuries. The first Admiralty tide tables were issued in 1833, and are now regularly produced on a world wide basis. A fuller introduction to the history of tidal measurements will be found in Hardisty (1990aa).

The analysis of tidal motions divides into two approaches and each is detailed here. Newton's equilibrium theory of tides is described below and provides the theoretical explanation of the range and period of ocean tides. Once generated, however, these tides propogate as waves in the same manner as, but on a longer time scale than, the wind waves described earlier. Entering shallow shelf waters, and being deflected, reflected and resonated around the local coastline causes the ocean tide to change considerably before reaching the beach. Therefore the equilibrium theory, although undoubtedly correct in its explanation of the nature of tides, does not provide the site specific predictions which are required in the orthogonal model.

Instead the alternative approach, known as harmonic analysis, is more useful. This involves the decomposition of the tide into a series of regular sinusoidal constituents of a given or determinable period and amplitude and then the recomposition of the result by simple addition. The constituents represent the influence of the gravitational forces postulated by Newton's theory but their individual parameterization is performed from an analysis of the local tidal measurements and the result is therefore very site specific.

Both of these approaches, which are not after all mutually exclusive, can be employed to examine two aspects of the tide. The first is the actual currents produced by the movements of the sea water from troughs (low water) to crests (high water) as the tidal wave progresses along a coastline. Given our definition of the orthogonal system as a wind wave controlled environment, these tidal currents, except where they flow strongly close to the shore, are of no direct consequence to beach morphodynamics and will not be considered here. The second aspect of the tide, the rise and fall of sea level, is however of crucial importance as will be apparent if one compares the morphology of the inter-tidal profile of a 6m tidal range in the Atlantic to that of a 20 cm tidal range in the Mediterranean. These regular changes in the position of the water's edge and of the surf zone as a whole are discussed in detail below.

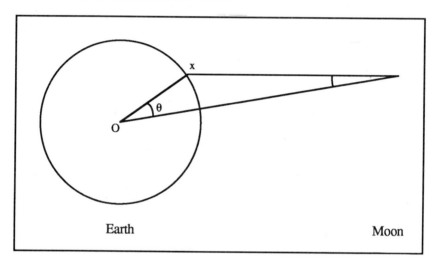

Figure 8.1 Definition sketch of the Earth-Moon system.

This Chapter is based on the now classical Doodson and Warburg (1941), Neumann and Pierson (1966), Muir-Wood (1969), the British Admiralty Tide Tables (Volume 1, 1988) and Pugh (1987), along with other more specific references cited in the body of the text. The objective of the chapter is to define a relationship for the changing height of the tide on a specific beach profile through time.

8.2 Tide Generating Forces

Newton showed that the force of attraction between two bodies m_1 and m_2 was proportional to the product of their masses and inversely proportional to the square of their separation r. Newton defined the constant of proportionality as G, the universal gravitational constant so that the force is:

$$F = G \frac{m_1 m_2}{r^2}$$
Eq.8.1

By definition, the more usual gravitational acceleration of a mass m at the earth's surface is due to a force:

$$F = m\,g \qquad\qquad \text{Eq.8.2}$$

and combining for unit mass ($m_1=m=1$) with E the radius and e the mass of the earth:

$$F = G\frac{E}{e^2} \qquad\qquad \text{Eq.8.3}$$

and rearranging:

$$G = \frac{ge^2}{E} \qquad\qquad \text{Eq.8.4}$$

Table 8.1

Dimensions of the Earth, Moon and Sun

Diameter of earth	12,753 km
Diameter of moon	3,479 km
Mass of earth	5×98^{10} kg
Mass of moon	7×34^{10} kg
Mass of sun	1×98^{10} kg
Average distance between:	
Centres of earth and moon	384,329 km
Centres of earth and sun	149×10^6 km

Using Equation 8.1, the moon's gravitational pull on a particle of mass m_x at x in Figure 8.1 which represents a plane containing the earth and the moon is thus:

$$F_x = \frac{Gm_x M}{x^2} \qquad\qquad \text{Eq.8.5}$$

Where M is the mass of the moon and x the distance to the centre of the moon. For unit mass ($m_x = 1$) and substituting for G from Eq. 8.4 we have:

$$F_x = \frac{gMe^2}{Ex^2} \qquad\qquad \text{Eq.8.6}$$

The equilibrium theory shows that the proximal and distal tidal bulges are actually raised by the departures of the magnitudes of these forces acting on particles of the sea from their mean value for the earth as a whole. This is different from the idea of a centrepetal force which is invoked in some texts to explain the second distal bulge. The mean value for the earth as a whole in the equilibrium theory is approximated to the moon's gravitational pull at the centre of the earth. That is:

$$F_o = \frac{gMe^2}{Er^2} \qquad\qquad \text{Eq.8.7}$$

Where r is the distance between the centres of the earth and the moon. With a and b as defined in Figure 8.1, the tangential component of F_x normal to OX is then:

$$\frac{g \ M \ e^2}{Ex^2}$$

Eq.8.8

The component of F_o along OX is:

$$\frac{g \ M \ e^2}{Er^2}$$

Eq.8.9

So that the tangential component of the differential attractive force is the difference between these two:

$$F_t = \frac{M \ (\ e^2 \ \sin(a + b) \ - \ e^2 \ )}{E(X^2 r^2)}$$

Eq.8.10

Which reduces to:

$$F_t = \frac{g \ M \ e^3}{Er^3} \sin(2\theta)$$

Eq.8.11

This is the equilibrium balance of forces responsible for the ocean tide. It is seen to have a minima and a maxima of about $\pm 0.84 \times 10^7$ g for E/M = 81.5 and r/e = 60.3 (Table 8.1) when θ = 45°, 135°, 225°, and 315° and is zero intermediately at θ= 0°, 90°, 180°, and 270° as shown in Figure 8.2. Similar relationships may be established for the attractive forces due to the sun, which are equivalent to the numerical values obtained from Eq. 8.11 multiplied by a factor of 0.46 to take account of the different values, for the sun, of the ratios corresponding to M/E and e/r.

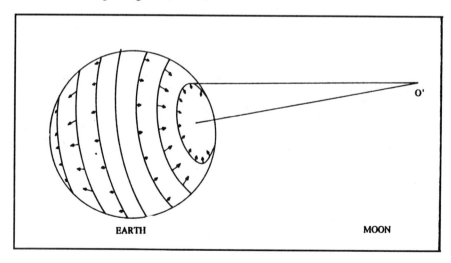

Figure 8.2 Global distribution of the tide generating force.

8.3 Equilibrium Tide

It can be shown (Muir-Wood, 1969 p.5) that, for a point on a particular latitude θ, the sea level $\eta(\theta)$ is given by:

$$\eta(\theta) = \frac{KE}{2}\left(3\left(\sin^2\theta - \frac{1}{3}\right)\left(\sin^2\theta - \frac{1}{3}\right)\right) - \sin^2\theta \, \sin^2\delta \, \cos z$$

$$+ \cos^2\theta \, \cos^2\delta \, \cos(2z) \qquad\qquad \text{Eq.8.12}$$

Where from Eq. 8.3:

$$K = \frac{3Me^2}{2Er^3} \qquad\qquad \text{Eq.8.13}$$

Table 8.2
Major Tidal Potential Constituents

Symbol	Name	Period hrs	Speed °/hr	Coefficient
Semi-Diurnal Components				
M_2	Principal lunar	12.42	28.9841	0.908
S_2	Principal solar	12.00	30.0000	0.423
N_2	Larger lunar elliptic	12.66	28.4397	0.174
K_2	Luni-solar semi diurnal	11.97	30.0821	0.115
υ_2	Larger lunar evectional	12.63	28.5126	0.033
μ_2	Variational	12.87	27.9682	0.028
L_2	Smaller lunar elliptic	12.19	29.5285	0.026
T_2	Larger solar elliptic	12.01	29.9589	0.025
$2N_2$	Lunar elliptic second order	12.91	27.8954	0.023
Diurnal Components				
K_1	Luni-solar diurnal	23.93	15.0411	0.531
O_1	Principal lunar diurnal	25.82	13.9430	0.377
P_1	Principal solar diurnal	24.07	14.9589	0.176
Q_1	Larger lunar elliptic	26.87	13.3987	0.072
M_1	Smaller lunar elliptic	24.86	14.4921	0.040
J_1	Small lunar elliptic	23.10	15.5854	0.030
Long Period Components				
M_f	Lunar fortnightly	327.8	1.0980	0.156
M_m	Lunar monthly	661.3	0.5444	0.083
S_{sa}	Solar semi-annual	≈2191	0.0821	0.073

Table 8.3
Tidal Amplitudes and Phases during March 1936

	Immingham England		San Francisco California		Do San Vietnam	
Lat.	53°38' N		37°48'N		20°43'N	
Long.	0°11'W		122°27'W		106°48'E	
	Phase	Amplitude	Phase	Amplitude	Phase	Amplitude
M_2	161°	223.2 cm	330°	54.2 cm	113°	4.4 cm
S_2	210°	78.8 cm	334°	12.3 cm	140°	3.0 cm
N_2	141°	44.9 cm	303°	11.5 cm	99°	0.8 cm
K_2	212°	18.3 cm	328°	3.7 cm	140°	1.0 cm
K_1	279°	14.6 cm	106°	37.0 cm	91°	72.0 cm
O_1	120°	16.4 cm	89°	23.0 cm	35°	70.0 cm
P_1	257°	6.4 cm	104°	11.5 cm	91°	24.0 cm

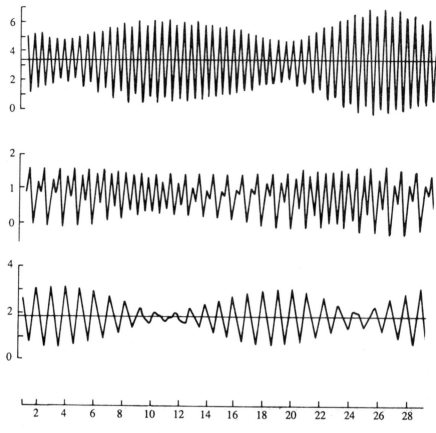

Figure 8.3 Tides at Immingham, San Francisco and Do San for March 1936 (after Defant, 1961).

Where δ is the declination of the moon and z is the easting of the point with respect to the moon's meridian, and changes as the earth rotates. This expression has a constant, long period component for given values of θ and δ, has a diurnal (once daily) component given by cos(z) and a semi-diurnal (twice daily) component given by the cos(2z) term. The declination has a maximum value of 30° but can be zero so that cosδ remains non zerobut sin(2δ) can become zero. Therefore the diurnal but not the semi-diurnal terms can become zero. Eq.8.13 represents the general equation for the equilibrium tide due to the moon. A similar expression may be derived for the equilibrium tide due to the sun and the summation of the two will give that for the combined effects of the sun and the moon.

In practice the situation is even more complicated because the earth moon distance, r, varies on a monthly basis, the solar lunar declination changes and other celestial bodies exert weaker attractive forces. Denoting the main solar components by S and lunar components by M, each subscripted by a daily frequency so that the twice daily, semi-diurnal lunar component is called M_2, Doodson and Warburg (1941), and Neumann and Pierson (1966) tabulated the most important tide generating components as shown in Table 8.2.

8.4 Harmonic Analysis of Tides

It is seldom possible to derive correctly the full theoretical expression for tidal heights on beaches because of the shallow water and coastal effects which are discussed below. It is more useful to derive, by harmonic analysis, the main constituents and then express the resultant at a sinusoidal constant:

$$\eta(t) = A_1 \cos (\omega_1 t - p_1°) + A_2 \cos (\omega_2 t - p_2°) + \ldots\ldots \qquad \text{Eq.8.14}$$

Where A_n represents the amplitude, ω_n the angular velocity (usually expressed as the speed number in degrees per solar hour) t is the time and p accounts for phase differences between the terms. The relationship between the speed number and the period T is simply $T = 360/\omega$ solar hours. Various almanacs and the Admiralty manual of tides provides details of the many constituents currently used for tidal analysis, with mean values for speed numbers and coefficients for the equilibrium tide.

The important frquencies and potential amplitudes of the the major components are given in Table 8.2, although not all of the components have to be considered at a particular location. For practical purposes the problem reduces to predicting the tides at a particular coastal site. No matter how complex the boundary conditions, and no matter how strong or weak the effects of friction, the solution for the tidal amplitude must contain these same periodic terms. Particular sites however, by reason of their depth and configuration, respond more strongly to certain periods than to others. Also the shape of the local coastline can cause the deep water tide to be reflected, and therefore it can change from being a purely progressive to a standing wave. Thus, although one is sure that a particular frequency can be present, its phase and amplitude would be difficult to derive from purely theoretical principles. Practical tide prediction involves the collection of long term records of the water level and subjecting them to analysis to determine the required amplitude and phases of the tidal terms. Table 8.3 and Figure 8.3 show the results for the major components at three sites, Immingham on the North Sea where the semi-diuirnal contribution is dominant, Do San in Vietnam, where the diurnal component is dominant, and San Francisco in California where both diurnal and semi-diurnal components are important.

At Immingham (the river front of which, incidentally, is visible from the desk at which this book was written) there are usually two high and two low tides each day. For the month shown in Figure 8.4 (March, 1936) the range of the tide decreases for the first few days and then increases until about March 12th. It then decreases again until about March 18 and then builds up to an even greater range at about March 26. The time of the highest tides follow the full and new moon by a few days. The are called the *spring tides* and occur near the times when the moon and sun are either in opposition (on opposite sides of the earth) or in conjunction (on the same sides of the earth). The low tidal ranges occur shortly after the first and third quarters of the moon when the moon is in quadrature (lines between the centre of the earth and the centres of the moon and the sun are roughly perpendicular). These low tidal ranges are called *neap tides*. These ranges from neap to spring tide are explained by the beat effect of the M_2 and S_2 tides which re-inforce at spings and are in opposition at neaps (See section 8.6). The added effect of N_2 and K_2 is also evident, as the second set of spring tides in the month has a greater range than the first set.

At San Francisco, the semi-diurnal tides (Species 2 tides) are less dominant and the diurnal tides (Species 1 tides) are greater producing a intriguing interference pattern with successive high and low tides being, at times, quite unequal in amplitude.

At Do San the Species 2 tides are very small compared with the Species 1 tides and there is usually only one tidal oscillation per day. The Species 2 tides cause only minor perturbations on the dominant diurnal oscillations, and these are visible only when the tidal range is small. Since K_1 and O_1 are nearly the same amplitude there is an almost complete cancellation of the tide at some times during the month.

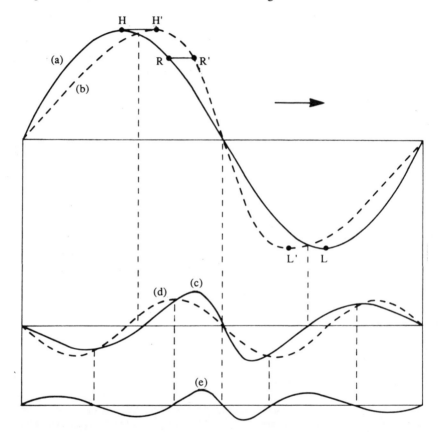

Figure 8.4 Generation of harmonics of the principal tidal component in shallow water..

8.5 Shallow water effects

The progress of the ocean tide into shallow water results in the local amplifications discussed above. In addition, it produces harmonics of the tide and local reflections. These effects are discussed in the following sections.

8.5(a) Generation of harmonics

In a process which is analogous to the generation of the second harmonic in a wind wave (Chapter Six), tides steepen as they shoal as shown in Figure 8.4. Doodson and Warburg (1941) present a mathemtical analysis of the change in the surface profile of the tide which is beyond the scope of this book, however it is apparent that, since the region closer to high water is in effectively deeper water, than the region close to low water then the sinusoidal profile, (a) in Figure 8.4, becomes distorted. The equation of points on the surface profile is given by:

$$c = \sqrt{g(Z_o + 3\eta)}$$ Eq.8.15

where Z_o is again the mean sea level and η is the elevation of the point above Z_o.

The deviation from the sinusoid is well represented by curve (c) which is itself composed of a sinusoid and a further distortion. The result is the generation of a harmonic of the primary with twice the initial frequency and a further harmonic with three times the initial frequency. Since these effects are most evident in the M_2 tide, these harmonics are known as the M_4 and M_6 tides. The phase difference between the M_2 and M_4 tides results in an asymmetry but no change in wave height, whereas for the wind waves discussed above the phase shift resultas in an increase in the crest and decrease in the trough.

8.5(b) Co-tidal and Co-range Lines

The progress of the ocean tide into shallow water, the effects of Coriolis and the reflection, or partial reflection at coastlines all contribute to produce a complex pattern which is plotted as a chart of:

Co-tidal lines which link positions of coincidental high and low water (or any other phase point) on the wave

Co-range lines which link positions of equal tidal range.

The result is a tidal wave which appears to rotate, in the northern hemisphere, in a clockwise direction about points of zero tidal range. These points are called amphidromic points.

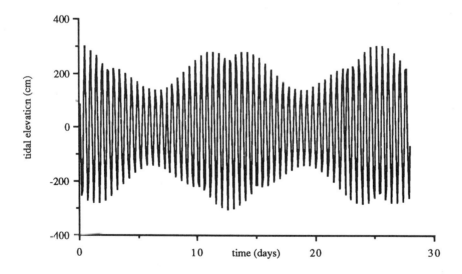

Figure 8.5 Tides at Immingham (from Hardisty, 1990n).

8.6 Beach Tides

In the orthogonal models it is necessary to predict both the horizontal and the vertical position of the water's edge throughout the tidal cycle. The resulting curves are known as beach tides and, notwithstanding the impressive theoretical basis provided by the equilibrium theory, it is rarely necessary to consider more than the M_2 and S_2 components. Neglecting the shallow water constituents, the mean tidal level and mean sea level are approximately the same. Doodson and Warburg (1941) note that, if Z_o is the mean sea level, M_2 is the amplitude of the semi-diurnal lunar component, and S_2 is the amplitude of the semi-diurnal solar component, these constants can be calculated from local tide tables:

Mean High Water Springs	$= Z_o + (M_2 + S_2)$
Mean High Water Neaps	$= Z_o + (M_2 - S_2)$
Mean High Water	$= Z_o + M_2$
Mean Low Water	$= Z_o - M_2$
Mean Low Water Neaps	$= Z_o - (M_2 - S_2)$
Mean Low Water Springs	$= Z_o - (M_2 + S_2)$

Hence a suitable model for the beach tide, which will include the Spring Neap cycle is:

$$z(t) = Z_0 + M_2 \cos(28.9841t) + S_2 \cos(30.0000t + p°) \qquad \text{Eq.8.16}$$

where the angular velocities are taken from Table 8.2, and $p°$ is the phase of the S_2 component at $t=0$ and all angles are in degrees or:

$$z(t) = Z_0 + M_2 \cos(0.5059t) + S_2 \cos(0.5236t + p) \qquad \text{Eq.8.17}$$

where all angles are in radians. The second solution is preferred here for consistency with the use of radian frequency elsewhere in this book.

Comparison with experimental results

Hardisty (1990n) provides a program for calculating the tidal elevation as given by Eq.8.17 and typical results for Immingham are shown in Figure 8.5 for a twenty eight day lunar month, which compare well with the actual results for this site shown in Figure 8.3(a).

8.7 Meteorological effects

Meteorological conditions cause differences between the predicted and the actual tide at a coastal site. Variations in tidal heights are mainly caused by strong or prolonged winds or by unusually low or high barometric pressures. Differences between predicted and actual times of high and low water are caused mainly by the wind. The two effects are discussed seperately below.

8.7(a) Barometric Effects

Tidal predictions are computed for average barometric pressure. A difference from the average of 34 millibars can cause a difference in height of about 0.3m (Admiralty Tide Tables, 1988). A low barometric pressure will tend to raise sea level and a high barometer will tend to depress it. The water level does not, however, adjust itself immediately to a change of pressure and it responds, moreover, to the average change in pressure over a considerable area.

8.7(b) The Effect of Wind

The effect of wind on sea level - and therefore on tidal height and times - is very variable and depends largely on the topography of the area in question. In general it can be said that wind will raise sea level in the direction in which it is blowing in a process which is analogous to the wave set up and set down detailed in the preceeding chapter.

8.8 Tidal Immersion Time

Beach process work requires not only an understanding of the height of the water level on the profile but also an appreciation of how long the water level remains at such heights. The duration of submergence of any point on the intertidal profile is called the tidal immersion time and has recently been discussed by Trenhaile (1978, 1980) and Trenhaile and Layzelle (1981).

8.9 Conclusions

The height of the tide on the beach, $\eta(t)$, will, in the nearshore models developed in subsequent chapters, be described by:

$$z(t) = Z_0 + M_2 \cos(0.5059t) + S_2 \cos(0.5236t + p) \qquad \text{Eq.8.18}$$

where Z_0 is the mean depth above datum, M_2 and S_2 are the amplitudes of the lunar and solar constituents and p is the phase difference between the lunar and solar constituents.

SECTION C

SEDIMENT DYNAMICS

chapter nine

BASIC SEDIMENT TRANSPORT

9.1   Introduction

The subaqueous movement of sand and gravel in the natural environment has long been of interest to hydraulic engineers as is evidenced by Graf's (1971) reference to siltation problems in aqueducts and harbours in Roman and Greek civilisations. The objective of research has been, and still is, simply to define the relationships between the hydrodynamic parameters which were derived in the preceding chapters, and the mass or volume of solid material which is moved. The relationships are called sediment mass transport formulae, and the objective is known as the two phase or sediment transport problem. The work of the late Ralph Bagnold is central to the development of solutions to the two phase problem, and it is worth noting that, at the age of ninety four, he spoke at a research conference in London University during May 1989 and suggested that the behaviour of sand grains in a fluid is a subject about which we still know very little, and that the term "conflux" could be applied to the dynamic behaviour of the mixture. Bagnold died in London on 28th May 1990.

The following treatment is based on the presentations of Sleath (1984) and Dyer (1986), though reviews are provided in many engineering textbooks such as Raudkivi (1976), Graf (1971) and Yalin (1977). The subject is complex, and presently poorly understood, so that texts have, quite rightly, sought empirical and often site specific solutions for practical purposes. The objective here is rather different in that a physically sound, but more general approach is described which will be developed for wave induced sediment transport in the following chapters. In particular it is now apparent that the effect of slope on the movement of sediment is of paramount importance in the orthogonal system. Unusually then, the present chapter seeks to define slope inclusive formulae.

9.2 Unidirectional Flow Characteristics

This chapter considers uni-directional sediment transport in preparation for the reversing, wave induced transport which is dealt with later. We must therefore begin by defining a number of characteristics which are used to describe unidirectional flows:

Bed Shear Stress is defined as the force per unit area acting on the sediment surface, is sysmbolised by τ and has units N m^{-2}.

The Shear Velocity is a useful and alternative representation of the shear stress, is symbolised by u_* and has units of velocity, m s^{-1}. It is defined by:

$$u_* = \sqrt{\frac{\tau}{\rho}}$$

Eq.9.1

Flows are said to be *laminar* when mixing only occurs on a molecular scale and *turbulent* when inertial mixing occurs through the tranfer of packets of water. The boundary layer which develops in a unidirectional flow consists of (Figure 9.1(a)) a viscous sublayer adjacent to the bed, a buffer layer wherein turbulence is generated, the fully turbulent layer, and the outer layer or free stream. If bed sediment grains do not protrude through the viscous sublayer, then the surface is said to be smooth, whereas for coarser sediments or higher flow speeds, the grains protrude through this layer and the boundary is said to be rough. Two equations for determining the bed stress within the boundary layer are in common use. The first utilises the observation that, to a good approximation, the shear stress is proportional to the square of the velocity so that:

$$\tau = \rho \, C_{Dz} \, u_z^2$$

Eq.9.2

which is known as the *quadratic stress law* with the coefficient of proportionality, C_{Dz}, being the *drag coefficient* and u_z being the flow velocity at a height z. It is apparent that it is necessary to reference both the drag coefficient and the flow velocity to a particular height because, for a given bed stress, the flow speed is dependent upon the height as shown in Figure 9.1. The values of C_{Dz} depend upon the seabed sediment and bedform characteristics and are well documented in the literature. The second, and alternatively equation for the velocity profile is given by the von-Karman semi-logarithmic law (Figure 9.1(b)):

$$u_z = u_* \frac{1}{\kappa} \ln \frac{z}{z_o}$$

Eq.9.3

where κ is the von-Karman constant which is generally taken to be about 0.4. The intercept, z_o, is known as the roughness length, and typically has a value of 0.5cm to 1.0 cm for rippled, sand sized sediment but is equal to about D/30 for plane beds where D is again the grain diameter.

The change from laminar to turbulent conditions is characterised by the ratio of the destabilising inertial forces to the stabilising viscous forces, which is known as the grain Reynolds Number, Re_g:

$$Re_g = \frac{\rho \, u_* D}{\mu} = \frac{u_* D}{\nu} \qquad\qquad\qquad Eq.9.4$$

where ν is the viscosity and μ is the dynamic viscosity of the fluid. This is but one of a number of non-dimensional Reynolds Numbers which can be formulated with different length and velocity scales (see Chapter Twelve).

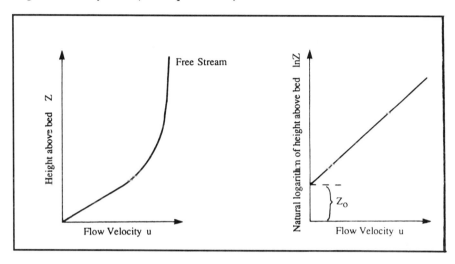

Figure 9.1 Elements of the uni-directional boundary layer (a) for a linear plot of height against flow velocity and (b) for a semi-lograithmic plot showing the von-Karman relationship.

9.3 Modes of Transport

Water, flowing over a bed of sediment, exerts forces which tend to move the individual particles. At low flow speeds the force of gravity, which opposes the movement, is greater than the fluid forces and no motion occurs. If the flow is gradually increased the more exposed particles are seen to vibrate in their niche positions (Bisal and Nielson, 1962, Lyles and Krauss, 1971 in air and Yalin, 1977 and Hammond *et al.*, 1984 in water) until firstly individual grains and then flurries and finally the whole surface grain layer is moved downstream. The flow parameters which prevail under these conditions represent an important stage in the sediment transport problem and:

"The **critical flow** condition is defined as that which, if exceeded, results in the displacement of sedimentary particles."

The critical condition is symbolised by the subscript "cr" which is variously applied to the flow speed, u_{cr}, the flow speed at a reference height, u_{zcr}, the bed shear stress, τ_{cr}, the shear velocity, u_{*cr}, or the Shields parameter, θ_{cr} (Section 9.4). Coarse sediment moving under flow conditions little greater than critical rolls and slides along the bed, and this continuing contact is used to delineate the first mode of transport:

"**Bedload transport** is defined as the movement of grains, the weight of which is wholly supported by solid transmitted forces."

115

This rather idealised description is sufficient for the present purposes but belies the complexity of the process (Leeder, 1979). In practice turbulent bursting and the inhomogeneity of the sediment surface result in a scatter of the observed values of critical conditions. If the flow speed is increased further then a stage is reached at which grains are lifted from the bed and carried along within the body of the flow, and this is used to delineate the second mode of transport:

"**Suspended sediment** transport is defined as the movement of grains, the weight of which is wholly supported by fluid transmitted forces."

These two definitions are extremely useful in simplifying the sediment transport problem, but in reality the transport process is a continuum ranging from pure bedload to pure suspended load. Between these extremes is a poorly defined region within which individual grains bounce and roll along the bed. This region has been variously termed saltation (from the Latin *saltare* to dance) or intermittent suspension with the latter term being preferred here. Intermittant suspension is more important in wind blown transport because the lower fluid density demands far more extreme conditions for true suspension. The process is explained by Bagnold (1941) for aeolian work, and will not be discussed further.

In addition to the characterisation of critical conditions for bedload and suspended load, it will be necessary to determine the quantity of sediment which is moved by a given flow:

"The **rate of sediment transport** is defined as the dry weight of solid material which passes over unit cross section in unit time."

The rate is here symbolised by i in the positive x direction (Chapter One), by j in the positive y direction and by k in the vertical z direction. The rate is subscripted i_b for the bedload, i_s for the suspended load and i_t for the total load. In geomorphological applications it is frequently necessary to obtain values for the net sediment transport in a given time interval. For steady flow, where i is constant, the net transport, symbolised by the upper case I, is given simply by the expression $I = i(t_2 - t_1)$ where $(t_2 - t_1)$ is the interval of interest. More usually however in the beach system where the magnitude and even the direction of the wave induced currents and the resulting sediment transport are constantly changing, we must evaluate:

$$I = \int_{t_1}^{t_2} i(t) \, dt \qquad\qquad\qquad Eq.9.5$$

9.4 The Threshold of Bedload Transport

A spherical particle resting on a bed of like spheres (Figure 9.2) experiences a gravitational force acting vertically downwards (F_G) which opposes the lift (F_L) and drag (F_D) forces. Theoretical expressions for these three forces and hence for the critical condition are derived in the present section and compared with experimental results.

9.4(a) Horizontal Bedload Thresholds

When the grain is at the point of movement the three forces must act through O' in Figure 9.2 with a resultant through the point of contact, and thus lying along O'P. At the critical condition the parallelogram of forces therefore gives:

$$\cos\alpha_{cr} = \frac{F_G - F_L}{O'P} \qquad\qquad \text{Eq.9.6}$$

and $\qquad \sin\alpha_{cr} = \dfrac{F_D}{O'P} \qquad\qquad$ Eq.9.7

hence $\qquad \tan\alpha_{cr} = \dfrac{F_D}{F_G - F_L} \qquad\qquad$ Eq.9.8

$$F_D = (F_G - F_L)\tan\alpha_{cr} \qquad\qquad \text{Eq.9.9}$$

Each of these three forces is derived to solve Eq.9.9 in the following sections.

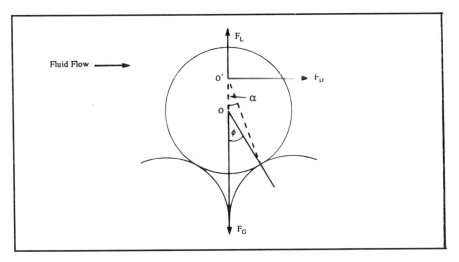

Figure 9.2 Force balance analysis of bedload threshold conditions and the parallelogram of forces at the critical condition.

Drag Force

The fluid drag will consist of the drag on the area normal to the flow which is occupied by the grain and, because it protrudes above the general bed level, a large proportion of that on the area covered by its wake. Since the drag force per unit area is τ, then the equivalent drag on the grain is τ/N where N is the ratio of the mean drag and lift forces over the whole bed to the drag and lift on the protruding grains. The value of N ranges between 0.20 and 0.30 (Dyer, 1986, p116) implying that the stress on an area between three and five times the grain's projected area in the grain's lee is imposed on the grain. Chepil (1958) has measured a value of N=0.21 for hemispherical elements placed three diameters apart in a hexagonal pattern. Thus the drag on the grain is:

$$F_D = \pi \frac{D^2}{4} \frac{\tau}{N} \qquad\qquad \text{Eq.9.10}$$

Chepil (1958) has also shown that the drag force acts some way above the centre of the sphere, and typically at a distance of 0.29 of a grain diameter below the top of the grain.

This gives a value of $\alpha_{cr}=24°$. The angle α_{cr} is the angle of dynamic friction and contrasts with the angle of static friction (Section 3.5) which is equal to the angle of repose of the grains.

Lift Force

The grain on the boundary will disturb the flow, the stream lines being deflected over the top. The distortion causes the fluid above the grain to be accelerated and, as shown by Bernoulli's equation, this results in a lowering of the pressure. Consequently there will be a difference in pressure vertically across the grain which will cause a lift force similar to that induced on an aircraft wing by an airstream. The lift force can be considered in a similar way to the drag force:

$$F_L = \frac{1}{2}\rho \; C_L \; A \; u^2 \qquad\qquad \text{Eq. 9.11}$$

The projected area of the particle, A, is set equal to $\pi D^2/4$ (Dyer, 1986) and the empirical coefficient corrects for grain protrusion and for the reference height for the velocity determinations. Einstein and El-Samni (1949) and Aksoy (1972) obtain values of between 0.1 and 0.2 for C_L at typical Reynold's numbers, and Coleman (1967) and Davies and Samad (1978) found very low, even negative lift force values at low Reynold's numbers due possibly to a significant water flow beneath the grain. Interestingly Bagnold (1972) found that the lift decreased rapidly as the particle was raised above the bed, and could be considerably increased because of the Magnus effect if the particle rotated in the sense which increases the relative velocity on its upper surface. More pragmatically Chepil (1958) has shown that, for spherical particles:

$$F_L = 0.85 \; F_D \qquad\qquad \text{Eq.9.12}$$

Gravitational Force

The immersed weight of a spherical particle was shown in Chapter Four to be:

$$F_G = \frac{\pi}{6}(\rho_s - \rho) \; g \; D^3 \qquad\qquad \text{Eq.9.13}$$

Force Balance

Substituting Eq.9.10, 9.12 and 9.13 into Eq.9.9 gives, with a little re-arrangement:

$$\tau_{cr} = \frac{0.66 D g (\rho_s - \rho) N \tan\alpha_{cr}}{1 + 0.85 \tan\alpha_{cr}} \qquad\qquad \text{Eq.9.14}$$

Where, as above, Chepil (1958) has shown that $\alpha_{cr}=24°$ and $N=0.3$. Substituting the density of quartz sand (Chapter Three) and seawater (Chapter Four) into Eq.9.14 gives the threshold bed stress equation:

$$\tau_{cr} = 103.5 \; D \qquad\qquad \text{Eq. 9.15}$$

The instantaneous bed shear stress can be much higher than the mean value and it is this maximum that will cause the sediment to move. The threshold values are then apparently lowered by a turbulence factor, T_f, which Chepil (1959) finds to be about 2.5 and Grass (1970) to be 2.2. Eq. 9.14 becomes:

$$\tau_{cr} = \frac{0.66Dg(\rho_s-\rho)N \tan\alpha_{cr}}{(1 + 0.85\tan\alpha_{cr})T_f} \qquad \text{Eq.9.16}$$

and again for quartz grains in water with T_f=2.5:

$$\tau_{cr} = 41.4 \, D \qquad \text{Eq. 9.17}$$

The two equations can be converted to the more pragmatic velocity form by utilising the quadratic stress law (Eq.9.2) so that the mean threshold formula (Eq.9.15) becomes:

$$u_{zcr} = \sqrt{\frac{103.5 \, D}{\rho \, C_{Dz}}} \qquad \text{Eq. 9.18}$$

and for the turbulent maximum Eq.9.17 becomes:

$$u_{zcr} = \sqrt{\frac{41.4 \, D}{\rho \, C_{Dz}}} \qquad \text{Eq.9.19}$$

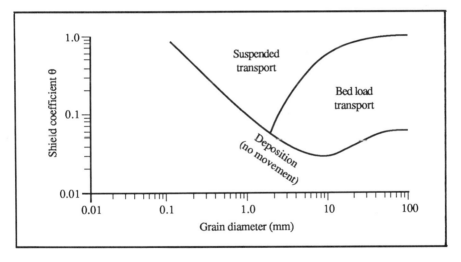

Figure 9.3 Shields threshold curve

Comparison With Experimental Results

For the high Reynolds number regime where the drag and lift coefficients, and probably N and T_f also, become independent of the grain Reynolds Number, Re_g, then both Eq.9.14 and 9.16 can be written as:

$$\frac{\tau_{cr}}{(\rho_s-\rho)gD} = \text{constant} \qquad \text{Eq.9.20}$$

or in more general terms:

$$\theta = \frac{\tau}{(\rho_s-\rho)gD} = f(Re_g) = f\left(\frac{u_*D}{\mu}\right) \qquad \text{Eq.9.21}$$

The dimensionless group on the left hand side of Eq.9.21 effectively compares the threshold shear stress with the immersed weight of a unit grain thickness layer of the bed. It is called the Shields entrainment function θ, with a threshold value θ_{cr}. The relationship was first investigated by Shields in 1936 and he produced the experimental threshold curve shown in Figure 9.3. This has three distinct zones the limits of which correspond to the three boundary layer flow regimes:

1. Up to $Re_g=3$, smooth boundary flow exists and the particles are embedded in the viscous sublayer, and θ_{cr} should be independent of the grain diameter. White (1970) has investigated the threshold of natural quartz grains and of glass ballotini in this region and finds that $\theta_{cr}=0.06Re_g^{-0.5}$ for this region, or in terms of the friction velocity:

$$u_{*cr}{}^{5/2} = 0.06\ g\left(\frac{\rho_s-\rho}{\rho}\right)u^{0.5}D^{0.5} \qquad \text{Eq.9.22}$$

2. For $3<Re_g<200$ there is a transitional region where the grain size is the same order as the thickness of the viscous sublayer. There is a minimum θ_{cr} of about 0.03 at an Re_g of about 10.

3. In the rough turbulent regime with $Re_g>200$, θ_{cr} has a constant value of between 0.03 and 0.06.

Figure 9.4 compares the empirical results of Eq.9.22 with the theoretical Eq.9.14 and 9.16, showing that there is now considerable agreement between the theory and the results in this region. On a purely empirical basis Miller et al. (1977) have compiled most of the available data for flat beds and produced curves such as Figure 9.5, which is of more direct use. Their results are summarised by best fit lines:

$$\tau_{cr} = (128.2\ D)^{0.6} \qquad \text{for } D<0.08 \text{ cm} \qquad \text{Eq.9.23}$$

$$\tau_{cr} = (43.48\ D)^{1.1} \qquad \text{for } D>0.08 \text{ cm} \qquad \text{Eq.9.24}$$

where D is in cm and τ_{cr} in dynes cm^{-1}. Alternatively in terms of the conventional flow speed at a height of 100 cm above the bed, u_{100},:

$$u_{100cr} = \sqrt{\frac{(128.2\ D)^{-1.69}}{\rho C_{D100}}} \qquad \text{Eq.9.25}$$

$$u_{100cr} = \sqrt{\frac{(43.48\ D)^{-0.9}}{\rho C_{D100}}} \qquad \text{Eq.9.26}$$

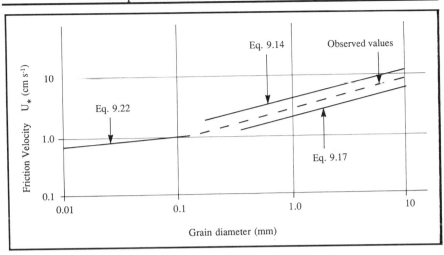

Figure 9.4 Threshold friction velocity for grain movement. Dashed line, observed values according to Miller et al. (1977).

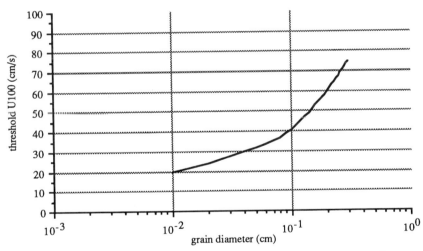

Figure 9.5 Curves of observed threshold values for the movement of quartz sediment on a flat bed for a current measured at a height of 100cm with an assumed C_{D100} of 0.003. (from Hardisty, 1990o).

9.4(b)   Slope Inclusive Bedload Thresholds

If a grain is resting on a slope, as is invariably the case on beaches, the force balance becomes rather more complicated because there will now be a component of the grain weight acting down the slope which increases u_{cr} in the upslope direction and decreases u_{cr} in the downslope direction. Dyer (1986) and Allen (1982 a, b, c) present analyses which are essentially similar and give a correction for the effect of slope on the threshold of sediment transport:

$$B_b = \frac{u_{cr\beta}}{u_{cro}} = \sqrt{\frac{\tan\phi - \tan\beta}{\tan\phi}} \cos\beta$$

Eq.9.27

where ϕ is the angle of dynamic friction of sediment, u_{cro} is the horizontal bed threshold and $u_{cr\beta}$ is the corresponding threshold on a bedslope β.

Comparison With Experimental Results

Whitehouse and Hardisty (1988) performed laboratory experiments which confirmed the applicability of Eq.9.27, and the results are given in Figure 9.6 showing that for down slope flow and $\tan\beta = \tan\phi$ the shear stress for movement will be zero and the slope will naturally avalanche. Conversely for up slope flow when $\tan\beta = \tan\phi$ a greater threshold stress is required to initiate the transport of sediment.

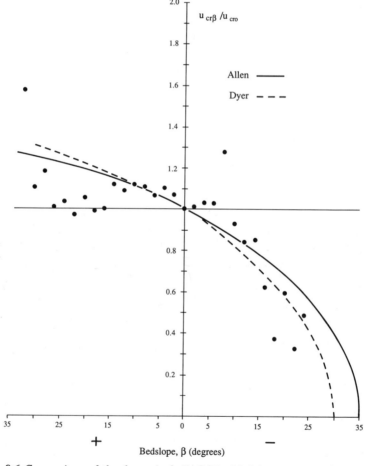

Figure 9.6 Comparison of the theoretical Eq.9.27 with laboratory experiments on the initiation of coarse grained quartz sand (After Whitehouse and Hardisty, 1988).

TABLE 9.1
SOME FORMULAE FOR THE BEDLOAD TRANSPORT RATE

Reference	Formula	Comments
Du Boys (1879)	$i = A (\tau - \tau_{cr}) \tau$	
Donat (1929)	$i = A(u^2 - u_{cr}^2)u^2$	u = mean flow
Rubey (1933)	$i = A (Q^{0.6} - Q_{cr}^{0.6})\theta^{0.6} S^{1.4}$	Q = discharge S = bedslope
MacDougall (1933)	$i = A (Q - Q_{cr}) S^B$	as above
Schoklitsch (1934)	$i = A (Q - Q_{cr}) S^{1.5}$	as above
O'Brien and Rindlaub (1934)	$i = A (\tau - \tau_{cr})^m$	
Shields (1936)	$i = A (\tau - \tau_{cr}) Q S$	as above
Kalinske (1942)	$i = A f(\tau/\tau_{cr}) u_*$	
Meyer-Peter and Muller (1948)	$i = A (\tau - \tau_{cr})^{1.5}$	
Brown (1950)	$i = A \tau^3$	Einstein-Brown
Einstein (1950)	$i = A f(\tau/D)$	

The coefficient A is different in each of these formulae, and in some it is a constant, whilst in others it depends on the characteristics of the sediment. Note that since $\tau \propto u^2$ and $Q \propto u$ then most of these formulae have a similar structure to Eq.9.43. The table is based, in part, on references in Sleath, 1984.

9.5 The Rate of Bedload Transport

There is, of course, great interest in quantifying sediment movement so that a predictive formula can be developed. Many sediment transport formulae have been proposed and Sleath (1984) lists those shown in Table 9.1. Further background is available in textbooks such as Raudkivi (1967), Graf (1971) and Yalin (1971). The formulae fall into four groups:

1) Empirical Formulae
These use a large number of flume, and occasionally field, results to obtain a completely empirical relationship between the sediment transport rate and a characteristic flow variable.

2) Dimensionless formulae
The sediment and flow variables are grouped together in nondimensional numbers and the

various constants and coefficients are determined from experimental data.

3) Probabilistic Formulae
A probability function is used to describe the relationship between the flow and the particle movement to determine the sediment transport rate.

4) Deterministic Formulae
Starting from the basic physics of the movement of individual grains, the transport relationships are developed. However, averaging these movements in space and time inevitably involves constants which are determined experimentally.

9.5(a) Horizontal Bedload Rates
It is apparent from Table 9.1 that the structure of most of these formulae is essentially similar, and that this, coupled with the presently very limited availability of reliable field measurements of the sediment transport rate in the marine environment make it difficult to choose between the different approaches. The theoretical basis of the bedload sediment transport equations is reviewed in Chapter 12, but until better analytical and experimental techniques become available we must choose one of the existing approaches for the present purposes. It appears that one of the most commonly used formulae in marine work is based on sound physical principles and is represented by the analysis introduced by R.A.Bagnold (1956). Bagnold's work so pervades the field of sediment transport that the reader is advised to obtain and thoroughly read a copy of *The Physics of Sediment Transport by Wind and Water : A Collection of Hallmark Papers by R.A.Bagnold* which was edited by C.R.Thorne, R.C.MacArthur and J.B.Bradley and published by the American Society of Civil Engineers in 1988. The evolution of his sediment transport formulae can be traced through this literature and is most fully presented in the 1956 and 1966 papers. Although these papers provide values for certain of the empirical parameters, they make difficult reading and the newer student to beach work is well advised to approach the theory through the 1963 paper on which the following is based.

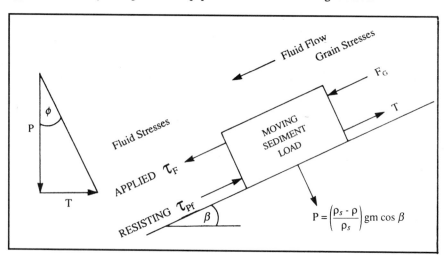

Figure 9.7 (a) Definition sketch for tanϕ=T/P, (b) force balance for the bedload layer.

By way of introduction, it is interesting to note that possibly the most important single contribution made to science by Isaac Newton was his formalisation of the proportional relationship between the force which is applied to an object and the resulting acceleration. The constant of proportionality is called the mass of the object and the formula becomes F=ma. All of Newtonian mechanics stems, by simple algebra, from this definition. Bagnold's theory, three centuries later, is based on Newtonian mechanics and begins with an equally simplistic definition. Bagnold takes the problem which is familiar to all high school students of a block moving at a constant speed down a surface and reminds us that, in this example of steady state motion, the force parallel to the surface (T in Figure 9.7(a)) is proportional to the perpendicular force (P in Figure 9.7(a)). Bagnold argues and demonstrates experimentally that the same analysis applies to a mass of sediment moving over a stationary bed so that:

$$T \propto P \qquad\qquad Eq.9.28$$

$$\text{or } T = \tan\phi \, P \qquad\qquad Eq.9.29$$

$$\text{or } \tan\phi = \frac{T}{P} \qquad\qquad Eq.9.30$$

where this constant of proportionality, $\tan\phi$, is called the dynamic friction coefficient. Since the determination of the sediment transport rate requires only that we know the mass and velocity of the mobile grains in terms of the fluid parameters, and the former is related to P whilst the latter is related to T, then Bagnold simply evaluates the terms in Eq.9.30 to derive a bedload formula. His derivation can be explained as follows.

Consider a mass m_b of sediment of density ρ_s overlying unit area of the bed at an angle β to the horizontal then (Figure 9.7(b)), if the density of the fluid is ρ, the normal force is given by the trigonometry of $\triangle ABC$ as:

$$P = \left(\frac{\rho_s - \rho}{\rho_s}\right) gm_b \cos\beta \qquad\qquad Eq.9.31$$

This force will be balanced by an equal dispersive stress supporting the submerged weight of the grains. Combining Eq.9.29 and Eq.9.31 the resisting solid frictional stress is given by:

$$T = \tan\phi \left(\frac{\rho_s - \rho}{\rho_s}\right) gm_b \cos\beta \qquad\qquad Eq.9.32$$

The grain weight will also have a downslope component, F_G, given by the trigonometry of $\triangle ACD$ as:

$$F_G = \left(\frac{\rho_s - \rho}{\rho_s}\right) gm_b \sin\beta \qquad\qquad Eq.9.33$$

The downslope forces are therefore the sum of the applied fluid stress τ_F per unit area and the downslope component of the grain weight given by Eq.9.33, as shown in Eq.9.34.

$$\text{Downslope Forces} = \tau_F + F_G = \tau_F + \left(\frac{\rho_s - \rho}{\rho_s}\right) gm_b \sin\beta \qquad\qquad Eq.9.34$$

Substituting $\sin\beta=\cos\beta\tan\beta$ this becomes:

$$\text{Downslope Forces} = \tau_F + \left(\frac{\rho_s - \rho}{\rho_s}\right) gm_b \cos\beta \tan\beta \qquad \text{Eq.9.35}$$

The upslope forces are the sum of the frictional stress, T, given by Eq.9.32 and any pore fluid friction, τ_{pf}, maintained by the shearing of the pervading fluid within the mobile grain layer:

$$\text{Upslope Forces} = T + \tau_{pf} = \left(\frac{\rho_s - \rho}{\rho_s}\right) gm_b \cos\beta \tan\phi + \tau_{pf} \qquad \text{Eq.9.36}$$

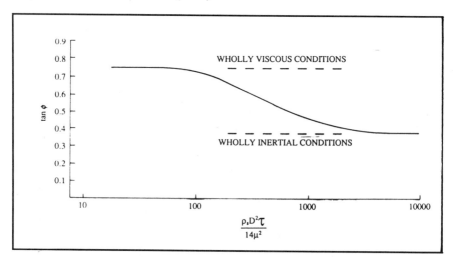

Figure 9.8 The dynamic friction angle, tanϕ, as a function of the grain Reynolds number (After Bagnold, 1966).

Bagnold (1954) showed experimentally that, at high grain concentrations such as occur near the base of the bedload layer, the value of the solid phase element T may exceed 100 times that of the fluid element τ_{pf} in Eq.9.36, even when the densities of the solid and fluid are the same. The fluid element did not begin to dominate until the concentration had been reduced to less than 9% and the whole of the resistance to motion may therefore be considered to be exerted by the shearing of the solid phase in bedload transport. Equating 9.35 and 9.36 at steady state equilibrium (i.e. when the bedload is moving at a constant speed):

$$\tau_F + \left(\frac{\rho_s - \rho}{\rho_s}\right) gm_b \cos\beta \tan\beta = \left(\frac{\rho_s - \rho}{\rho_s}\right) gm_b \cos\beta\tan\phi + \tau_{pf} \qquad \text{Eq.9.37}$$

which re-arranges to:

$$\tau_F = \left(\frac{\rho_s - \rho}{\rho_s}\right) gm_b \cos\beta \ (\tan\phi - \tan\beta) + \tau_{pf} \qquad\qquad \text{Eq.9.38}$$

where τ_{pf} is taken to be small. Finally the sediment dry mass transport rate, i_b, is by definition the product of the bedload mass m_b and the mean bedload velocity u_b:

$$i_b = m_b \ u_b \qquad\qquad \text{Eq.9.39}$$

However, since the tangential thrust stress required to maintain the bedload in motion is, neglecting τ_{pf}, given by Eq.9.38 at $[(\rho_s-\rho)/\rho_s]gm_b\cos\beta(\tan\phi-\tan\beta)$, then multiplication by u_b gives the product of the force and the distance moved per unit time which is the work done by the fluid in unit time, i.e. the fluid "power" expended per unit of the bed area in transporting the bedload. Whence:

$$\text{Fluid power expended} = \left(\frac{\rho_s - \rho}{\rho_s}\right) g \ i_b \ \cos\beta(\tan\phi - \tan\beta) \qquad\qquad \text{Eq.9.40}$$

The total available fluid power is written as the product of the excess stress $(\tau - \tau_{cr})$ and the mean flow speed; of which only a fraction e_b is consumed in transporting the bedload, thus:

$$\text{Available power} = e_b \ (\tau - \tau_{cr}) \ u \qquad\qquad \text{Eq.9.41}$$

Equating Eq.9.40 and 9.41 and re-arranging gives:

$$i_b = \frac{\rho_s \ e_b(\tau - \tau_{cr}) \ u}{(\rho_s - \rho) \ g \cos\beta \ (\tan\phi - \tan\beta)} \qquad\qquad \text{Eq.9.42}$$

Eq.9.42 is the Bagnold bedload formula which, in various guises, has proven popular in marine sediment transport work, and will be used in the morphodynamic modelling sections later in this book.

Bagnold shows $\tan\phi$ (Figure 9.8) as a function of the grain Reynolds number, suggesting that the transport rate for given flow conditions increases if either the grain size or the bed stress increases. Various presentations of Eq.9.42 have been produced (Gadd et al, 1978, Hardisty, 1983) and most usefully we replace $(\tau - \tau_{cr})$ by $\rho C_D(u^2 - u_{cr}^2)$ (Eq.9.2), and subsume ρC_D, $\rho_s/(\rho_s - \rho)$, e_b and $1/(\cos\beta(\tan\phi - \tan\beta)$ into a calibration coefficient k (although the last named group re-appears in Eq.9.46) to yield:

$$i_b = k_b \ (u^2 - u_{cr}^2) \ u \qquad\qquad \text{Eq.9.43}$$

which is, of course, similar to many of the formulae given in Table 9.1. This type of function has been successfully incorporated into beach models by Inman and Bagnold (1963), Bowen (1980), Bailard and Inman (1980), and Hardisty (1984, 1986, 1990ag) and some empirical values for k_b are given in the following section. Alternatively, Bowen (1980) suggests values for the constants of $e_b = 0.1$, and $\tan\phi = 0.6$. The choice of C_D depends upon the reference height of the flow measurements (Section 9.2), but, for the conventional u_{100}, a typical value of 0.03 appears appropriate.

				Table 9.2.			
			Experimental Sediment Transport Rates				
Run	u	f(u)	i_b	Run	u	f(u)	i_b
	cm/s		gm/cm/s		cm/s	$\times 10^3$	gm/cm/s
1	23.0	6279	0.008	42	63.7	240	2.281
2	25.0	9225	0.023	43	64.9	254	2.086
3	22.5	5630	0.024	44	79.9	487	2.086
4	22.0	5016	0.100	45	99.1	944	3.278
5	28.0	14784	0.100	46	112.2	1380	4.172
6	35.0	33915	0.170	47	131.7	2246	12.07
7	35.0	33915	0.270	50D	25.6	9.4	0.001
8	40.0	53760	0.350	51	37.8	43	0.014
9	38.0	45144	0.450	52	49.7	108	0.140
10	42.0	63336	0.450	53	58.2	180	0.656
11	40.0	53760	0.550	54	55.8	158	0.313
12	48.0	98304	0.700	55	62.8	229	1.341
13	46.0	85560	0.770	56	56.4	163	0.715
14	51.0	119595	0.780	57	40.5	54	0.055
15	55.0	152295	0.980	58	55.7	157	0.596
16	24.4	5718	0.0004	154	45.5	72	0.0099
17	24.4	5718	0.0007	155	47.0	81	0.0179
18	24.4	5718	0.0008	156	53.5	127	0.0625
19	21.3	1974	0.001	157	62.0	208	0.1851
20	27.0	9936	0.001	159	115.0	1857	2.8651
21	24.4	5718	0.0013	160	49.5	97	0.0106
22	28.5	12860	0.0035	161	53.5	127	0.0195
23	23.0	3864	0.004	162	54.0	131	0.0454
24	34.0	27030	0.019	163	68.5	288	0.2105
25	40.0	49560	0.021	164	81.5	501	0.5390
26	43.0	63984	0.024	165	100.0	952	1.4906
27	46.0	80730	0.081	166	115.0	1857	2.9316
28	37.0	37296	0.090	167	49.5	97	0.0871
29	46.0	80730	0.220	168	57.5	162	0.0223
30	46.0	80730	0.250				

In this table $f(u)=(u^2-u_{cr}^2)u$, and the run designations are taken from the original papers.

Comparison With Experimental Results

A number of datasets are available with which to test Eq.9.43. The results shown in Table 9.2 are taken from the compilations of Guy *et al* (1966) and Williams (1967). The experiment numbers in the table are the same as those listed in the original papers and the data has been converted from American units into the cgs system.

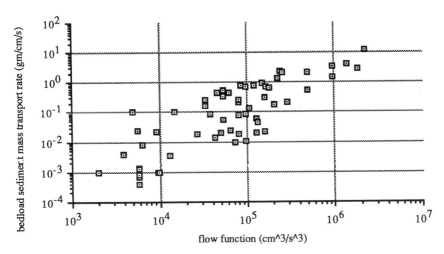

Figure 9.9 Experimental test of the bedload transport equations. Details in the text. (from Hardisty, 1990p).

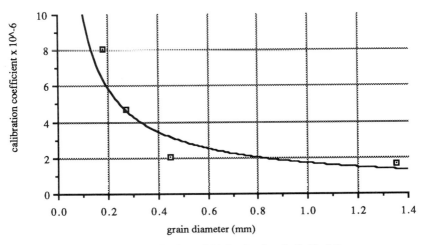

Figure 9.10 Functional relationship $k_b = f(D)$ for the data in Table 9.2.

Run numbers 1 to 15, 16 to 30 and 42 to 58 are from Guy *et al.* (1966) and are for quartz sand with D_{50}=0.18mm, 0.45mm and 0.27 mm with u_{cr} values of 16, 19 and 17 cm s^{-1} respectively. Runs 154 to 168 are from Williams (1967) and are for 1.35 mm quartz sand with an assumed u_{cr} of 22 cm s^{-1}. The results are plotted in Figure 9.9, which shows agreement with theory. The best fit line through the results in Figure 9.10 suggests a functional relationship between k_b and D for Eq.9.43:

$$k_b = 1.7109 \ D_{mm}^{-0.7703} \times 10^{-6} \qquad \qquad \text{Eq.9.44}$$

where kb has units of $g \ cm^{-4} \ s^{-2}$ and flow speeds and the threshold values have units of $cm \ s^{-1}$.

9.5(b) Slope Inclusive Bedload Rates

Despite the control which the surface slope exerts on geomorphological evolution through its effect on the sediment transport rate, very few theoretical or experimental results are presently available which quantify the effect. Thus, although the effect of slope is recognised, attempts to include a slope dependent transport function into geomorphological models are often less than satisfactory. For example Horikawa (1988) includes a sediment transport correction which, in the present format, is written:

$$i_b = i_0 - \varepsilon_b |i_0| \tan\beta \qquad \qquad \text{Eq.9.45}$$

where i_b is the bedload transport rate on a slope, i_0 is the horizontal bed value and ε_b is a correction factor. Horikawa (1988) suggests that a value of ε_b=10 is appropriate, but clearly the use of Eq.9.45 in the region $\tan\beta$>0.1 (β>6°) will reverse the direction of transport at any flow speed. A more rigorous correction was derived by Bagnold (1956) and involves the inclusion of the slope and dynamic friction angle terms which were subsumed in k_b to generate Eq.9.43 (Section 9.5(a)). Retaining these terms we write:

$$A_b = \frac{k_b\beta}{k_{bo}} = \frac{\tan\phi}{\cos\beta(\tan\phi-\tan\beta)} \qquad \qquad \text{Eq.9.46}$$

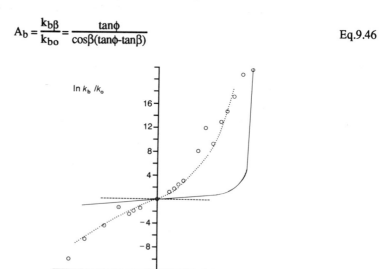

Figure 9.11 Experimental test of Eq.9.46 (After Hardisty & Whitehouse, 1988)

where $k_b\beta$ and k_{bo} are the calibration coefficients on a bedslope, β, and on a horizontal bed respectively. This function has been extensively utilised in aeolian work and in the marine environment by Allen (1982 a, b, and c) and Huthnance (1982).

Comparison With Experimental Results

Hardisty and Whitehouse (1988) performed experiments to test the applicability of Eq.9.46 and the results are given in Figure 9.11 showing that the bedload rate is greater than predicted by theory in the downslope case and less than predicted by theory in the upslope case. They attributed the discrepancy to a process which they called *impact induced gravity flow*, observing that the impacting grains energised the surface grain layer reducing its shear strength and resulting in additional downslope transport. Hardisty and Whitehouse (1988) were unable to analyse the process theoretically but did suggest that the following empirical equation fitted the preliminary results:

$$A_b = \left(\frac{\tan\phi}{\tan\phi \pm \tan\beta} \right)^7 \qquad \text{Eq.9.47}$$

where the positive sign is appropriate for upslope flow and the negative sign is appropriate for downslope flow. It might be suggested that a similar result can be obtained by allowing the angle of dynamic friction to decrease as in Figure 9.8, and that the apparent increased grain Reynolds Number is manifested by the impact induced gravity flow, but this does not fully account for the magnitude of the discrepancy.

9.6 The Threshold of Suspended Load Transport

Francis (1973) studied the paths of particles in flume experiments and found that, when the vertical components of the turbulent velocity were approximately equal to or exceeded the settling velocity of the grain, then the particles remain above the bed for longer periods and execute a wavy path in the body of the fluid, that is they move under the action of fluid transmitted forces and represent suspended load transport.

9.6(a) Horizontal Suspended Load Thresholds

Bagnold (1956) had postulated that suspension occurs when the upward directed components of the turbulent velocity fluctuations w'_{up} exceeds the fall velocity of the grains, w_s. Bagnold related the turbulent velocity fluctuations to the shear velocity, u_*, of the flow and estimated that the onset of suspension is given by:

$$w_s = 1.25 u_* \qquad \text{Eq.9.48}$$

Bagnold's hypothesis has recently been validated from a consideration of turbulence data by Leeder (1983). Since $u_* = \sqrt{(\tau/\rho)}$ (Eq.9.1) and $\tau = \rho C_{Dz} u_z^2$ (Eq.9.2) then Equation 9.48 can be written in terms of the flow velocity as:

$$w_s = 1.25 \, u_z \sqrt{C_{Dz}} \qquad \text{Eq.9.49}$$

Alternatively the criterion can be transformed to a Shield's curve. Substituting Bagnold's criterion (Eq.9.48) into the Shields entrainment function (Eq.9.21) gives, for quartz density grains:

$$\theta = \frac{0.4\ w_s^2}{gD}$$

Eq.9.50

Though McCave (1970) argues on grounds of the turbulence distribution near the bed that the constant should be 0.19. The threshold criterion is utilised by Hardisty (1990q) and shown in Figure 9.12.

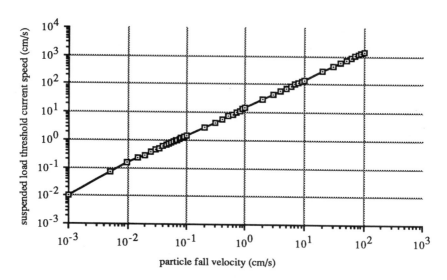

Figure 9.12 Bagnold's suspended load threshold criterion for U_{100} with $C_{D100}=0.003$.

9.6(b) Slope Inclusive Suspended Load Thresholds

It appears that, neglecting hydrodynamic controls on the turbulent velocity field close to the bed, there is no dependence of the threshold conditions for suspension on the gradient of the bed. This is simply because neither the upward directed components of the turbulent velocity fluctuations w'_{up} nor the fall velocity of the grains, w_s, which lead to the threshold criterion in Eq.9.48 are dependent upon the bedslope. This is in contrast to the case for purely bedload transport detailed earlier, and also does not take account of the effect of slope enhanced bedload transport leading to complex grain collision and hence vertical grain momentum which will effectively decrease the suspension threshold criteria. Such an analysis is however complicated, and cannot be attempted at the present time. It will be assumed, instead, that since the suspended sediment profile reference concentration is to be linked to the bedload rate (Section 9.7) and that since this rate is corrected for bedslope effects as detailed above, then no additional terms are required.

9.7 The Rate of Suspended Load Transport

At flow speeds in excess of the threshold conditions detailed above, sediment particles are maintained in suspension by exchange of momentum from the fluid to the particles. The suspended load transport rate is then predicted either by assuming a diffusion gradient for this momentum exchange and relating the concentration at various heights to the known concentration at a particular height. The result is a diffusive formula which can adequately describe the suspended load rate provided that a reasonable estimate of the reference concentration is available. The most satisfactory solution to the reference concentration problem is to utilise a measure of the bedload rate defined earlier.

9.7(a) Horizontal Suspended Load Transport Rates

The diffusion approach is described by Dyer (1986) and Hill *et al.* (1988) on which this section is based. It assumes steady state conditions for which the flux of settling particles (Cw_s) is equal to the vertical turbulent flux given by the product of the concentration gradient and a kinematic eddy diffusion coefficient:

$$C\, w_S = K_S \frac{\partial C}{\partial z} \qquad\qquad \text{Eq.9.51}$$

where C is the concentration of suspensate having a fall velocity w_S, z is the height above the bed and K_S is the diffusion coefficient. Since von-Karman (1934) argued that the diffusion coefficient for mass should take the same form as the diffusion coefficient for momentum, K_m, and assuming a linear shear stress distribution from surface to bottom, then K_S can be expressed as:

$$K_S = \beta_* \, \kappa \, u_* \, z \left(1 - \frac{z}{h} \right) \qquad\qquad \text{Eq.9.52}$$

Where u_* is the shear velocity (Eq.9.1), κ is again von-Karmen's constant (≈ 0.4), h is the water depth, z is the height above the bed and β_* relates the diffusion coefficients for the solid and the momentum, $K_S = \beta_* K_m$. Substituting for K_S and integrating gives the relative concentration profile:

$$\frac{C_z}{C_a} = \left(\frac{a(h-z)}{z(h-a)} \right)^{\frac{w_S}{\beta_* \kappa u_*}} \qquad\qquad \text{Eq.9.53}$$

where C_z and C_a are the concentrations of suspensate at a height z and at a reference height a respectively. This equation was first derived by Rouse (1938) and is therefore usually called the Rouse Equation and examples for different values of the exponent are shown in Figure 9.13. It is usually assumed that β_* has a value of about unity (although see Hill *et al*, 1988 for further details) and the problem then devolves to the determination of the reference concentration. Various authors (Einstein, 1950, Yalin, 1963) couple the reference concentration hypothetically to a concentration within the bedload or saltation load layers, and this approach is followed by Hill *et al.* (1988). Their experiments are described in the following section.

In order to translate the concentration profile into a suspended load transport rate we must evaluate the sum of the product of the concentration and the flow velocity throughout the water column:

$$i_s = \int_{z=0}^{z=h} u_z C_z \, dz \qquad\qquad \text{Eq.9.54}$$

Dyer (1986) notes that, if the concentration profile is expressed in terms of the Rouse Equation (Eq.9.53) and the velocity profile is expressed in terms of the von-Karman Equation (Eq.9.3) then Eq.9.54 cannot be integrated analytically. Instead he assumes a deep flow, $\beta*=1$ and relatively coarse sediment with a reference height at z_0 to show:

$$i_s = \frac{\kappa\gamma_1\rho_s u*^3 z_0}{(\kappa u* - w_s)^2} \frac{\tau - \tau_c}{\tau_c}$$

Eq.9.55

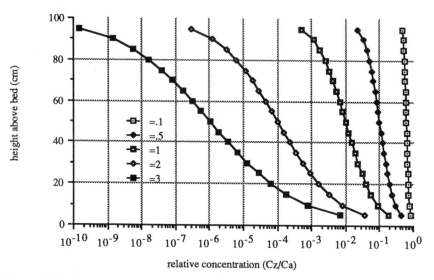

Figure 9.13 Typical suspension profiles for various values of the exponent (from Hardisty, 1990r).

Comparison With Experimental Data

The comparison of the Rouse equation with laboratory and field measurements divides into two inter-related problems. On the one hand the concentration profile has been shown to agree reasonably well with the form of Eq.9.53 shown in Figure 9.13 for a given reference concentration. Secondly, the relationship between the reference concentration and the flow parameters is required if the suspended load rate is to be predicted without prior measurements of the reference concentration. Such is the case in most of the numerical models discussed in later chapters. Both the full expression for the concentration profile, Eq.9.53, and the approximation for the suspended transport rate, Eq.9.55 require a reference concentration, and in both cases it is assumed to be linearly related to the excess bed shear stress. In the former case, Hill *et al.* (1988) utilise:

$$C_a = \gamma \frac{\tau - \tau_{cr}}{\tau_{cr}}$$

Eq.9.56

where the reference concentration is that at the top of the 'wake' layer some 2 cm above the bed in a laboratory flume. Those authors quote a value of $\gamma=1.3 \times 10^{-4}$ and also quote field values of 1.95×10^{-3} (Smith and McLean, 1977a); 2.40×10^{-3} (Smith and McLean, 1977b); 5×10^{-4} to 3×10^{-3} (Glenn, 1983 referenced in Hill *et al*, 1988); 1.6×10^{-5} (Wiberg and Smith, 1983); 1.5×10^{-2} (Kachel and Smith, 1986); 2×10^{-5} to 5×10^{-4} (Drake and Cacchione, 1988, referenced in Hill *et al.*, 1988). Alternatively Dyer (1980) obtained a value of $\gamma_1=0.78 \times 10^{-4}$ and Dyer (1986) quotes another value of $\gamma_1=0.84 \times 10^{-4}$, both obtained in a tidal flow over a rippled sand bed at the level of the bed roughness (i.e. about 1-2 cm). This parameter can be used in Eq.9.55 to determine the suspended sediment transport rate from the modified Rouse profile. The result of Eq.9.53 with 9.56 are programmed into Hardisty (1990r).

9.7(b) Slope Inclusive Suspended Load Rates

In the absence of any precedents for the effect of bedslope on the suspended load transport rate, it will be assumed that, since the reference concentration is given by Eq.9.56 and since this is effectively a bedload function, then the reference concentration will be influenced by the bedslope in the same manner and to the same degree as is the bedload function. That is:

$$C_a = A_b \, \gamma \frac{\tau - B_b^2 \tau_{cr}}{\tau} \qquad \text{Eq.9.57}$$

9.8 Conclusions

It is apparent that no single, satisfactory theory is presently available which predicts the bedload or suspended load mass transport rates in terms of the flow parameters, and current research in this area is detailed in Chapter Twelve. However this Chapter has presented the derivation of functional formulae based upon Bagnold's work which contain many of the fundamental requirements, particularly the explicit inclusion of both threshold and gravity terms. The resulting formula, in terms of a u_{100} mean flow speed at 1 m above the bed are:

$$i_t = i_b + i_s$$

$$= A_b \, k_b \, (u^2 - B_b^2 u_{bcr}^2)u \; + \; \int_{z=0}^{z=h} u_z C_z \, dz \qquad \text{Eq.9.58}$$

where for bedload:

$$A_b = \left(\frac{\tan\phi}{\tan\phi \pm \tan\beta} \right)^7 \qquad \text{Eq.9.59}$$

$$k_b = 1.7109 \, D_{mm}^{-0.7703} \times 10^{-6} \qquad \text{Eq.9.60}$$

$$B_b = \sqrt{\frac{\tan\phi - \tan\beta}{\tan\phi} \cos\beta} \qquad \text{Eq.9.61}$$

$$u_{100cr} = \sqrt{\frac{(128.2 \, D)^{-1.69}}{\rho C_{D100}}} \qquad \text{Eq.9.62}$$

for D<0.08 cm else:

$$u_{100cr} = \sqrt{\frac{(43.48 \, D)^{-0.9}}{\rho C_{D100}}} \qquad \text{Eq.9.63}$$

and for suspended load:

$$u_z = u_* \frac{1}{\kappa} \ln \left(\frac{z}{z_o} \right)$$

Eq.9.64

$$C_z = C_a \left(\frac{z(h-a)}{a(h-z)} \right)^{\frac{w_s}{\beta_* \kappa u_*}}$$

Eq.9.65

$$C_a = A_b \gamma \frac{\tau - B_b^2 \tau_{cr}}{\tau}$$

Eq.9.66

where A_b and B_b are as above, k=0.4, β_*=1, γ takes Hill *et al.*'s value of 1.3×10^{-4} and the particle fall velocity is given by Stokes Law (Section 12.2):

$$W_s = \frac{1}{18} \frac{\rho_s - \rho}{\mu} gD^2$$

Eq.9.67

This general structure will be utilised to examine orthogonal sediment transport in the following chapter. The formulae are programmed into Hardisty (1990s).

chapter ten

ORTHOGONAL SEDIMENT TRANSPORT

10.1 Introduction

One of the most surprising aspects of beach work to the new student is the extreme paucity of results on the onshore and offshore movement of sediment beneath waves. When one considers that literally hundreds of formulae have been developed for the unidirectional sediment transport which was described in the preceeding chapter, and tens of formulae exist for the quasi-unidirectional longshore transport which will be described in the following chapter, it is strange that this, possibly the most important chapter in the whole book, must rely on only a handful of theoretical papers, backed by even fewer laboratory experiments and hardly any published sets of field data. The explanation for this inconsistency lies in the fact that sediment transport is difficult to measure at the best of times, and when the flows are rapidly accelerating, decelerating and reversing, it is well nigh impossible. Nevertheless orthogonal transport formulae are presented here with details of much of the data which is already available together with some new data from recent experiments conducted by J.M.Woodruff.

10.2 Orthogonal Bedload Thresholds

Particle entrainment within the oscillatory boundary layer involves explicitly unsteady forces. As well as the lift, drag and the body forces which were derived in Chapter Nine, there are others related to the horizontal pressure gradients and to the acceleration of the fluid past the bed. Although of theoretical importance, analytical criteria for plane bed thresholds in unsteady flow have, not suprisingly, met with but little success. All theoretical formulae start from a consideration of the balance of forces on a grain of sediment that rotates about a pivot point. It was shown in Section 9.4(a) that the fluid forces tend to dislodge the grain from the bed, whereas gravity tends to return the particle to its niche position. Eq.9.21 showed that the Shields entrainment function (θ_{cr}) is a good indicator of threshold conditions and is dependent upon the grain Reynolds number, Re_g:

$$\theta_{cr} = \frac{\tau_{cr}}{(\rho_s - \rho)gD} = f(Re_g) = f\left(\frac{u_* D}{\mu}\right) \qquad \text{Eq.10.1}$$

Table 10.1

Some Bedload Threshold Formulae for Orthogonal Flows

Reference	Formula	Comments
1.Bagnold (1946)	$u_{maxcr}=2.38 \ \rho'^{2/3} D^{0.433} T^{1/3}$	Oscillating bed, SI units
2.Sato and Kishi (1954)	$u_{maxcr}=5.06 \sqrt{\rho'gD}$	
Manohar (1955) 3.laminar flow:	$u_{maxcr}=0.025 \ \rho'gDT^{1/2}\tan\phi u^{-1/2}$	
4.turbulent flow:	$u_{maxcr}=7.45 \ \rho'^{0.4}g^{0.4}u^{0.2}D^{0.2}$	
5.Kurihara $et\ al.$ (1956)	$u_{maxcr}=1.95\sqrt{\rho'gDtan\phi}$	Referenced in Sleath (1984)
6.Larras (1956)	$u_{maxcr}=95\rho'^{1/3}u^{1/2}T^{-1/2} + W$	W=fall velocity
7.Vincent (1957)	$u_{maxcr}=0.0012(W/D)$	SI units
8.Eagleson and Dean (1959)	$u_{maxcr}=0.016 \ \rho'gDT^{1/2}u^{-1/2}$	
9.Goddet (1960)	$u_{maxcr}=0.33 \ \rho'^{2/3}g^{2/3}D^{1/4}T^{3/8}u^{1/24}$	
10.Ishihara and Sawaragi (1962)	$u_{maxcr}=0.054 \ \rho'^{2/3}g^{3/4}D^{1/4}T^{1/2}$	for $D'\leq35$
11.Sato $et\ al.$ (1962)	$u_{maxcr}=0.39 \ \rho'^{2/3}g^{2/3}D^{1/3}T^{1/3}$	
12.Bonnefille and Pernecker (1966)	$u_{maxcr}=0.063\rho'^{5/6}g^{5/6}D^{1/2}\sqrt{T}u^{-1/6}$	$D'\leq18$
13.	$u_{maxcr}=0.0087\rho'^{16/15}g^{16/15}D^{6/5}\sqrt{T}u^{-19/30}$	$D'>18$

In this table, based partially on Sleath (1984), $\rho'= \left(\dfrac{\rho_s-\rho}{\rho}\right)$ and $D'=(\rho'g/u^2)^{1/3}D$

Table 10.1 (continued)

Some Bedload Threshold Formulae for Orthogonal Flows

Reference	Formula	Comments
14. Horikawa and Watanabe (1967)	$u_{maxcr}=0.39\ f_w^{-1/2}\sqrt{\rho'gDtan\phi}$	
15. Carstens *et al*. (1969)	$u_{maxcr}=1.37\ \sqrt{\rho'gD}$	Flow at z=0.6D
16. Silvester and Mogridge (1970)	$u_{maxcr}=0.034\ r'^{3/4}g^{3/4}D^{1/3}\tau^{1/2}u^{-1/18}$	
17. Chan *et al*. (1972)	$u_{maxcr}=0.148\ r'^{3/4}g^{3/4}D^{1/4}\tau^{1/2}$	
18. Komar and Miller (1973, 74)	$u_{maxcr}=0.24\ \rho'^{2/3}g^{2/3}D^{1/3}\tau^{1/3}$	for D'≤12.5
19.	$u_{maxcr}=1.05\rho'^{4/7}g^{4/7}D^{3/7}\tau^{1/7}$	for D'>12.5
20. Hallermeir (1980)	$u_{maxcr}=0.14\ \rho'^{3/4}g^{3/4}D^{1/4}\tau^{1/2}$	for D≤0.7mm
21.	$u_{maxcr}=2.83\sqrt{r'gD}$	for E= ≥1
22.	$u_{maxcr}=\dfrac{E^2u_{max2} + d^2u_{max1}}{(E^2+d^2)}$	for E<1 & u_{max} >0.35 where u_{max1} and u_{max2} are from Eq.20 & 21 & d=2-u_{max1}/0.35 (mks units)
23. Lenhoff (1982)	$u_{maxcr}=0.859\dfrac{D'(1+0.092logD')_{n}}{D(f_w)^{1/2}}$	

In this table, based partially on Sleath (1984), $\rho'=\left(\dfrac{\rho_s-\rho}{\rho}\right)$ and D'=$(\rho'g/u^2)^{1/3}$D

In steady flow, τ_{cr} is the critical value of the bed shear stress at which the grains of sediment first begin to move, but in oscillatory flow it ought to be taken as the total horizontal force per unit area of the bed rather than just the shear stress (Sleath, 1984, p256). Since the shear velocity is proportional to the root of the bed stress it follows that:

$$\frac{u_* D}{\mu} = \sqrt{\frac{\tau_{cr}}{\rho}} \frac{D}{\mu} = \sqrt{\frac{\tau_{cr}}{(\rho_s - \rho)gD}} \sqrt{\frac{(\rho_s - \rho)gD^3}{\rho \mu^2}}$$

Eq.10.2

so that Eq.10.1 could also be written:

$$\theta_{cr} = f_2 \left(\frac{\rho_s - \rho}{\rho} \frac{gD^3}{\mu^2} \right)$$

Eq10.3

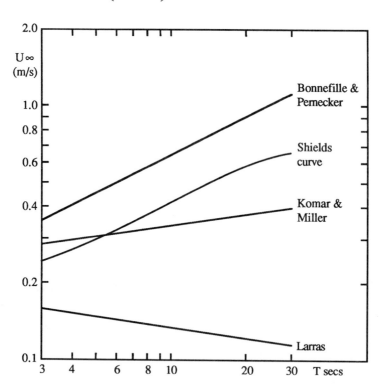

Figure 10.1 Threshold curves for formulae in Table 10.1 for fine sand (After Sleath, 1984)

In order to make use of Eq.10.3 it is necessary to evaluate the function f_2. The analyses are discussed in more detail in Chapter Twelve, but here we require functional formulae and must resort to empirical criteria. Because of the importance of the problem, there have been many attempted solutions, and some of these are listed in Table 10.1.

Comparison With Experimental Data

Some of the variety of solutions offered in Table 10.1 are plotted in Figures 10.1 and 10.2 for two grades of sand in fresh water at 20°C. Komar and Miller (1973) utilise the data of

Bagnold (1946) to show that, for grain diameters less than about 0.5mm (medium sands and finer), the threshold is best given by the equation:

$$u_{cr} = \sqrt{0.405 \frac{(\rho_s - \rho)gD}{\rho} \left(\frac{d_o}{D}\right)^{0.25}} \qquad \text{Eq.10.4}$$

which for quartz density sediment in water at 20°C, and in cgs units gives:

$$u_{cr} = 25.5 \ (Dd_o)^{0.25} \qquad \text{Eq.10.5}$$

where d_o is the orbital diameter of the nearbed orthogonal currents, given by linear theory (Chapter Six) as:

$$d_o = \frac{H}{\sinh(kh)} \qquad \text{Eq.10.6}$$

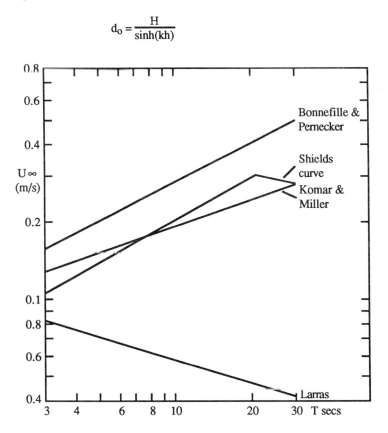

Figure 10.2 Threshold curves for coarse sand (After Sleath, 1984).

Replacing the nearbed orbital diameter in Eq.10.5 with Eq.10.6 gives a functional expression:

$$u_{cr} = 25.5 \left(\frac{DH}{\sinh(kh)}\right)^{0.25} \qquad \text{Eq.10.7}$$

141

For sediments having diameters larger than 0.5mm (coarse sands and coarser) Komar (1976) suggests that the empirical curve of Rance and Warren (1969) is more appropriate and gives:

$$u_{cr} = \sqrt{0.46\,\pi.\frac{(\rho_s - \rho)gD}{\rho}\left(\frac{d_o}{D}\right)^{0.25}} \qquad\qquad \text{Eq.10.8}$$

which for quartz density sediment in water at 20°C, and in cgs units gives:

$$u_{cr} = d_o^{0.125}D^{0.375} \qquad\qquad \text{Eq.10.9}$$

Again, replacing the near bed orbital diameter in Eq.10.8 with Eq.10.6 gives a functional expression:

$$u_{cr} = D^{0.375}\left(\frac{H}{\sinh(kh)}\right)^{0.125} \qquad\qquad \text{Eq.10.10}$$

These results are programmed into Hardisty (1990t) and are shown in Figures 10.1 and 10.2.

10.3 Orthogonal Bedload Rates

There have been numerous investigations of sediment movement under waves in the laboratory. The waves are almost invariably sinusoidal waves with a single frequency and consequently they may not be particularly good representations of the natural situation, but nevertheless an appropriate transport model is developed in the present section. The oscillatory nature of the flow, and therefore of the transport, demands a certain refinement of the transport terminology which was introduced earlier. Two types of orthogonal transport formulae are used:

Peak Flow Transport Formulae: relate the sediment transport rate to the amplitude of the oscillatory flow (that is to u_{in} or u_{ex} in Chapter Six which, since symmetrical first-order flow is generally considered, is referred to as u_{max}).

Integrated Flow Transport Formulae: relate the sediment mass transport rate to the instantaneous flow velocity u(t). This is clearly preferable since the effects of higher order, asymmetrical flows are more properly included in the analysis.

In either of these types of formulae, there are three possible transport rates which can be predicted:

Instantaneous Transport Rate: is the continually varying dry weight transport rate across unit width of the bed parallel to the wave crest at any instant during the passage of the wave. It is symbolised by i(t), and may be subscripted for bedload, suspended load or total load as in Chapter Nine: $i_b(t)$, $i_s(t)$, $i_t(t)$.

Mean Transport Rate: is the average dry weight transport rate across unit width of the seabed parallel to the wave crest taken over a wave cycle. It is therefore the mean of the integral of the instantaneous transport rates. It is symbolised by $\bar{i}$ and is therefore defined as:

Table 10.2

Some Orthogonal Bedload Transport Formulae

Reference	Formula	Comments (units of rate)
Bagnold(1963)	$i_b(t) = \dfrac{ku^3}{\tan\phi - \tan\beta}$	
Madsen and Grant (1976)	$i_b = 12.5 \ W \ D \ S^3 \ u_{max}^6$	Sinusoidal bed expnt. $(cm^3 cm^{-1} s^{-1})$
Sleath (1978)	$i_h = 94p \ D^2 \ T^{-1} \ S^{1.5} \ (u_{max}^2 - u_{cr}^2)^{1.5}$	Sinusoidal bed expnt. $(cm^3 cm^{-1} s^{-1})$
Watanabe et al (1980)	$I_b = ASWD \ (u_{max}^2 - u_{cr}^2)$	Laboratory wave channel $A \approx 1 - 5$ $(cm^3 cm^{-1} s^{-1})$
Hallermeir (1980)	$I_b = 0.032 \ 2S \ u_{max}^2 / f_w$	About 700 data points from 20 references $(cm^3 cm^{-1} s^{-1})$
Shibayama and Horikawa (1982)	$I_b = 19 \ AWDS^3 u_{max}^6$	A=1 for flat beds, but depends on ripple and oscillation wave length $(cm^3 cm^{-1} s^{-1})$
Watanabe (1982)	$I_b = 7WDS^{1.5}(u_{max}^2 - u_{cr}^2)u_{max}$	Laboratory wave channel. $(cm^3 cm^{-1} s^{-1})$
Ishida et al	$i_b(t) = 0.000249 \ \dfrac{\rho}{T} u(t)^3$	Laboratory wave channel. $(gm \ cm^{-1} s^{-1})$
Bailard (1982)	$i_b(t) = \rho c_f \dfrac{e_b}{\tan\phi} \left(|u(t)|^2 u(t) - \dfrac{\tan\beta}{\tan\phi} |u(t)|^3 \right)$	

In this table $S = \dfrac{f_w \rho}{2(\rho_s - \rho)gD}$ where f_w is Jonsson's (1966) friction factor, and the Shield's criterion in the original papers has accordingly been replaced with $\theta = Su^2$ as discussed in the text.

$$\bar{i} = \frac{\int_0^T i(t)\, dt}{T}$$

Eq.10.11

and again may be subscripted to denote bedload, suspended load or the total load: $\bar{i}_b$, $\bar{i}_s$, or $\bar{i}_t$.

Net Transport: is the total dry weight of sediment which is transported across unit width of the seabed parallel to the wave crest during the passage of the wave. It is symbolised by the upper case I and is defined by:

$$I = \int_0^T i(t)\, dt$$

Eq.10.12

Clearly the net transport is also equal to the product of the mean transport rate and the wave period:

$$I = \bar{i}\, T$$

Eq.10.13

Again the net transport may be susbcripted to represent bedload, suspended load or total load: I_b, I_s or I_t.

Although each of these types of formulae and transport rate have their uses, it is desirable to identify and to utilise an integrated flow formulae which predicts the nett mass transport per wave in order to estimate the geomorphological beach response. The following sections will introduce examples of each type, but identify and calibrate a net mass transport, integrated flow formula.

10.3(a) Peak Flow Formulae

Two of the most frequently quoted laboratory experiments are those of Kalkanis (1964) and Abou-Seida (1965). In both experiments a sand bed was oscillated within a water flume, the beds containing recessed trays to collect the bedload sediment moved in each half cycle. The results have been interpreted in various ways. Madsen and Grant (1976) use the results to demonstrate that the average bedload transport rate, $\bar{i}_b$, is proportional to θ^3 (i.e. to u^6 or τ^3). Hallermeier (1982a) uses these and other data sets to demonstrate that the mean transport rate is proportional to u_{max}^3/T. Vincent et al. (1981) relate the bedload concentration at any instant to $(\theta - \theta_{cr})$, i.e. to $(u^2 - u_{cr}^2)$, and then multiply this concentration by the instantaneous flow speed and integrate over the wave semi-cycle to evaluate the average transport rate (i.e. the result is proportional to $\int(u^2 - u_{cr}^2)u\, du$. for each semi cycle). Sleath (1982) suggests that the maximum bedload rate is 8/3 times the average value and also relates the rate to a threshold inclusive term $(\theta - \theta_{cr})^{3/2}$. These, and various other functions, are listed in Table 10.2 which was largely assembled from Horikawa (1988) and which also references formulae from the following papers which are in Japanese: Sunamura et al. (1978), Noda and Matsubara (1980), Kajima et al. (1982), and Yamashita et al. (1984). The Japanese language papers have been ommitted from the table, as has Sunamura (1980) which is referenced by the same source but deals only with offshore transport, Sunamura (1984) which deals only with the swash zone and Sunamura and Takeda (1984) which was specifically formulated for the prediction of bar migrations.

It is clear that most of the formulae in Table 10.2 have the same general structure but it is unlikely that any of them properly account for the full complexities of the orthogonal

144

transport problem. The issues are discussed further in Chapter 12, but in order to make progress here the bedload function of Bagnold (1956, Section 9.4) was examined with the laboratory data of Kalkanis (1964). The data were recalculated in cgs units and are evaluated in the form:

$$\overline{i_b} = k_b(u_{max}^2 - u_{cr}^2)u_{max}$$ Eq.10.14

where k_b is analogous to the calibration coefficient in Eq.9.43.

Table 10.3						
The Recalculated Bedload Data of Kalkanis (1964)						
Run	(a) u_{max} cm s^{-1}	(b) i_b	(c) u_{max} cm s^{-1}	(d) i_b	(e) u_{max} cm s^{-1}	(f) i_b
1	26.7	0.004	28.9	0.004	28.0	0.001
2	36.6	0.016	35.9	0.016	33.8	0.005
3	44.2	0.087	44.8	0.111	42.7	0.023
4	39.3	0.016	39.9	0.009	38.1	0.004
5	48.2	0.094	47.6	0.033	68.6	0.020
6	57.0	0.174	56.1	0.167	58.8	0.164
7	48.5	0.058	48.5	0.064	39.0	0.007
8	60.7	0.155	60.4	0.152	53.0	0.046
9	71.1	0.212	69.9	0.213	65.8	0.280

Columns (a) and (b) are for Kalkanis's Series A with a grain diameter of 0.16 cm and a calculated threshold of 16 cm s^{-1}; (c) and (d) are for Series B with 0.21 cm and 18 cm s^{-1} and (e) and (f) are for Series C with 0.28 cm and 21 cm s^{-1} respectively. Note also that umax, the maximum amplitude of the near bed velocity is referred to as uo, and ib as qB in the original paper, and that the units have been changed from fps to cgs here. The units of i_b are g cm^{-1} s^{-1}. Finally the original paper states that each run was "repeated a number of times and the measured quantities were averaged out" to provide the above datasets.

Comparison With Experimental Results

The formula in Eq.10.14 was chosen because, despite its relative simplicity, it contains the essential physical elements of a good model, because u_{max} the amplitude of the orbital velocity can be calculated from wave theory (Chapter Six), because u_{cr} the threshold flow speed can be evaluated from the work given earlier in this chapter and because it is the type of formula which was employed in the geomorphological detailed in later chapters.

The recalculated data of Kalkanis (1964) are listed in Table 10.3 for the twenty seven experimental runs listed in his Table IV. It should be noted that Kalkanis' "qb" is the time-averaged weight of sand displaced from two source samples into a central collecting tray and that "no distinction was made as to whether the material in the tray came over the right or left edge" (Kalkanis, 1964 p.C-2). We infer that this represents the sum of the transports over a single cross section, although if we have misinterpreted the ambiguity then the calibration below is an underestimate by a factor of two. The results for all three (similar) grain sizes are shown in Figure 10.3 and regression analysis gives:

$$\overline{i_b} = 0.665\ (u_{max}^2 - u_{cr}^2)u_{max}.\ 10^{-6}$$ Eq.10.15

The intercept, 0.001 g cm^{-1}s^{-1} is small. The structure of Eq.10.15 is very similar to Watanabe's (1982) formula (Table 10.2) and it is interesting to compare the two equations in more detail. Taking $\rho=1.00$ g cm^{-3} and $\rho_s=2.65$ g cm^{-3}, $g=981$ cm s^{-2}, $D=0.16$ cm, $u_{cr}=16$ cm s^{-1} and $f_w=0.005$ (Jonsson, 1966) then $S=0.97.10^{-5}$ (Footnote to Table 10.2) and $W=20$ cm s^{-1} (Chapter Nine). Therefore the $7WDS^{1.5}$ term in Watanabe's formula is equal to $1.85.10^{-6}$, and converting from the volume rate to the dry weight for $\rho_s=2.65$ and a concentration of 0.6 (Chapter Three) Watanabe's formula becomes:

$$\bar{i}_b = 1.53\ (u_{max}^2 - u_{cr}^2)u_{max}.\ 10^{-6} \qquad\qquad \text{Eq.10.16}$$

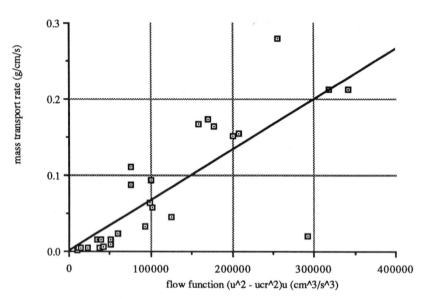

Figure 10.3 Data of Kalkanis plotted in the form of Eq.10.10
which is remarkably similar to Eq.10.15. These results suggest that, for the bedload transport of sand sized sediment, the Bagnold formula fits the Kalkanis data for symmetrical orthogonal flows utilising the peak orthogonal flow speed.

10.3(b) Integrated flow formulae

The peak flow approach which was developed in the previous section, and appears satisfactory for regular, sinusoidal flow, must be extended to account for asymmetrical orthogonal flows, in order to predict the instantaneous transport rate. The form of the solution is assumed to be analogous to Eq.10.14:

$$i_b(t) = k_{b(t)}\ (u(t)^2 - u_{cr}^2)u(t) \qquad\qquad \text{Eq.10.17}$$

and the net transport equivalent:

$$I_b = k_{b(t)} \int_0^T (u(t)^2 - u_{cr}^2)u(t)\ dt \qquad\qquad \text{Eq.10.18}$$

The integral cannot easily be evaluated, even for the relatively simple second order flow equations used here, and therefore a numerical solution is used in the following sections.

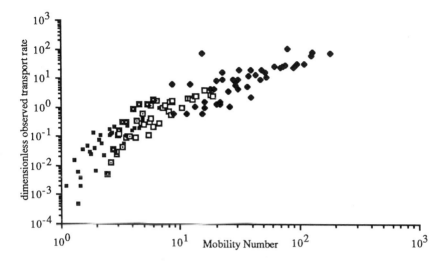

Figure 10.4 Results for Woodruff's horizontal, symmetrical oscillation experiments (dotted squares and closed squares) compared with those of Kalkanis (1964, open squares), Abou-Seida (1965, closed diamonds) and Sleath (1978, closed squares).

Comparison with experimental results

Experiments were conducted by Jason Woodruff during his Ph.D. work on an oscillating, perspex tray suspended in a laboratory flume channel from a trolley mounted above the flume and running on parallel rails. The trolley was connected by flexible wire around pegs on two rotating gear-wheels and then to a fixed point. The pegs were mounted in radial slots so that the amplitude of oscillation imparted by each gearwheel could be varied. The first gear wheel was directly driven through a variable speed gearbox from an electric motor. The second gearwheel was driven through a free gear from the first, but had exactly half the number of teeth of the first. The oscillations therefore represented a second order Stokes wave flow (Chapter Six) with the first gear wheel providing the fundamental and the smaller gear wheel providing the first harmonic. The displacement of the trolley was therefore calculated from the geometry of the experimental arrangement and was found to compare well with measurements of the flow of water along the bed with a three component ultrasonic current meter (Chapter Two) mounted on the trolley. Three small, adjacent slots were cut along the centre line of the trolley. More than three hundred and fifty experiments were carried out on different gradients, and with varying wave frequencies and fundamental and first harmonic amplitudes. In each experiment glass spheres with a mean diameter of 1mm were smoothed into the central slot, the trolley was oscillated for a given number of cycles, and then the weight of particles transported into each of the two outer slots was determined. The results of the symmetrical oscillation experiments were compared with other existing data sets in terms of the Mobility Number (Dyer, 1986):

$$M_b = \frac{\rho \, u_{max}^2}{(\rho_s - \rho)gD}$$

Eq.10.19

and are shown in Figure 10.4. Clearly the new, symmetrical results are comparable with existing data, and therefore further experiments were conducted with asymmetrical oscillations of the trolley.

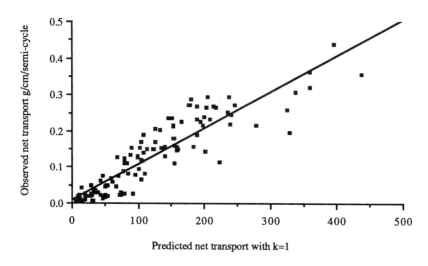

Figure 10.5 Woodruff's data for the short semi-cycle, downslope flow experiments.

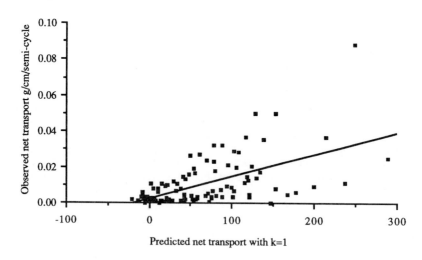

Figure 10.6. Woodruff's data for the short semi-cycle, upslope flow experiments.

The results are shown in Figures 10.5 to 10.8. Analysis of the results showed that:

$$i_b(t) = 1.356 \ (u(t)^2 - u_{cr}^2)u(t) \times 10^{-6} \qquad\qquad R=0.91 \qquad Eq.10.20$$

$$i_b(t) = 0.780 \ (u(t)^2 - u_{cr}^2)u(t) \times 10^{-6} \qquad\qquad R=0.84 \qquad Eq.10.21$$

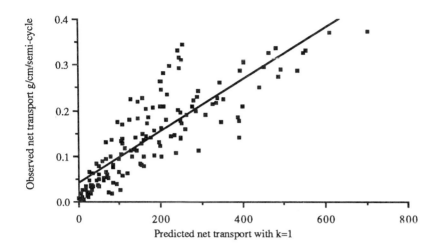

Figure 10.7 Woodruff's data for the long semi-cycle, downslope flow experiments.

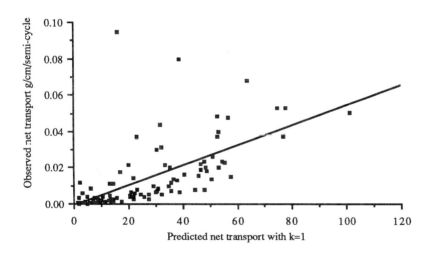

Figure 10.8 Woodruff's data for the long semi-cycle, upslope flow experiments.

$$i_b(t) = 0.753 \ (u(t)^2 - u_{cr}^2)u(t) \times 10^{-6} \qquad\qquad R=0.62 \qquad\qquad Eq.10.22$$

$$i_b(t) = 0.168 \ (u(t)^2 - u_{cr}^2)u(t) \times 10^{-6} \qquad\qquad R=0.59 \qquad\qquad Eq.10.23$$

$$i_b(t) = 0.764 \ (u(t)^2 - u_{cr}^2)u(t) \times 10^{-6} \qquad\qquad\qquad\qquad\qquad Eq.10.24$$

where Eq.10.20 to 10.23 are for the short semi-cycle, flow downslope case, the short semi-cycle flow upslope case, the long semi-cycle downslope flow case and the long semi-cycle upslope flow case respectively and Eq.10.24 represents the mean value for all experiments and all units are cgs.

It is apparent that these values for the calibration coefficient are similar to those obtained from the Kalkanis dataset for Eq.10.15 and to those obtained for the Watanabe formula in Eq.10.16. Furthermore, Vincent *et al.* (1981) obtained a similar result using bed stress rather than the velocity squared form of the equation. In conclusion, therefore, the integrated flow bedload transport function in the form of Eq.10.17 will be utilised in the SLOPES model developed in later chapters. The calibration coefficient will be taken to have a value of 1×10^{-6} g cm^{-4} s^{-2}. The formula is coded into Hardisty (1990u).

10.4 Orthogonal Suspended Load Thresholds

Although a significant proportion of the sediment transported about the beach system is moved in suspension, the processes are extremely complex and even Sleath (1984, p272) cautions that it would be prudent to treat currently available formulae as essentially empirical and to extrapolate them to non-laboratory conditions only with considerable caution. This is partly because, unlike the transport of bedload material, suspension transport beneath orthogonal flows can depend to an overriding extent on the vertical components of the oscillatory motion and upon the bedforms present. Inman (1957) first pointed out the importance of the suspended sand cloud over ripples and, more recently, observations of suspended sediment motion under waves have been reported by many wokers (e.g. Inman and Bagnold (1963) and Sunamura *et al.* (1978)). The sequence of events is shown in Figure 10.9. The cloud of suspended sand originates in the vortex which develops in the lee of the ripples during shorewards flow. The sediment cloud thus formed is transported upwards and backwards during the offshore flow, partially settling on the way.

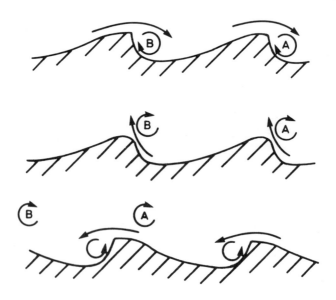

Figure 10.9 Diagrammatic sketch of suspension processes over ripples.

10.5 Orthogonal suspended load rates

The concentration of suspended sediment close to the seabed varies continuously through the wave cycle, as has indeed been shown experimentally with light absorption devices by Homma and Horikawa (1963), Homma et al. (1965), Horikawa and Watanabe (1967), MacDonald (1973) and Sleath (1982). However, of more practical interest is the mean concentration and concentration profile over the wave cycle because the orthogonal suspended load rate is the product of the velocity and concentration profiles (cf Section 9.7(a)). In direct analogy with the unidirectional case, the problem divides into determining the reference concentration, determining the concentration and velocity profiles and then determining the transport rate as the vector product of these two quantities.

10.5(a) Reference concentration beneath waves

A number of attempts have been made to determine the concentration of suspensate. Nielson (1979) suggests that the reference concentration, C_a, is given by:

$$C_a = k_s \, (\theta - 0.05) \frac{2}{\pi} \cos^{-1} \left(\frac{\theta}{\theta_{cr}} \right)$$

Eq.10.25

where k_s varies from 0.015 at the ripple trough to 0.028 at the ripple crest. The units of concentration are m^3 of sediment per m^3 of water. One of the few attempts to utilise this approach with field data, and to relate the bedload rate and the concentration profiles by linking the reference concentration to the bedload rate is reported by Vincent et al. (1982). They report results from an instrument systems which was deployed in 10m of water approximately 1km from the shore of Long Island. The system consisted of four Marsh-McBirney 511 two-component electromagnetic current meters (Chapter Two) mounted at heights of 15, 60, 100 and 120cm above the bed, an acoustic backscatter suspended sediment concentration meter which provided a vertical resolution of approximately 1cm and a wave pressure sensor. Vincent et al. (1982) follow an earlier paper by Vincent et al. (1981) to define an aereal concentration of bedload sediment, $C^*(t)$, which is the volume of sediment mobilised above unit area of the bed and therefore has units of $cm^3 \, cm^{-2}$. The aereal volume is then defined in terms of the volume bedload rate, $i_{bv}(t)$, and the instantaneous current, $u(t)$ (which is itself equal to the vector sum of the wave orbital and steady currents):

$$i_{bv}(t) = C^*(t) \, u(t)$$

Eq.10.26

Since the volume bedload transport rate is related to the dry mass bedload rate used here by:

$$i_b(t) = i_{bv}(t) \, \rho_s$$

Eq.10.27

then Eq.10.26 becomes:

$$C^*(t) = \frac{i_b(t)}{u(t) \, \rho_s}$$

Eq.10.28

Vincent et al. (1982) show that the reference concentration at a height of 1cm, here symbolised by C_a, is related to the mean aereal bedload concentration, C^*, by:

$$C_a = 330C^* + C_{ac}$$

Eq.10.29

where the units of C_a are mg/l and the constant, C_{ac}, appeared to represent a background value of, in this case, 3.8. Combining Eq.10.28 and 10.29 we have the required relationship between the bedload rate given earlier and the reference concentration:

$$C_a = 330 \frac{\overline{i_b}}{\overline{u}\, \rho_s} + 3.8$$

Eq.10.30

where $\overline{i_b}$ and $\overline{u}$ are mean values and have cgs units. Clearly this relationship forms a fundamental link in the sediment dynamic processes and will require further examination in the future, and comments concerning the absolute value of the coefficient of proportionality are included in the SLOPES modelling chapter later in this book. Eq.10.30 will, nevertheless, be used in the orthogonal system model which is developed in later chapters. For that work, the nearbed current speed and bedload rate associated with the reference concentration will be taken to be that pertaining to the maximum shorewards current generated by second-order wave theory and by the Stokes drift (Chapters Five and Six). Hardisty (1990v) presents a program to calculate the reference concentration from Eq.10.30, and typical results are shown in Figure 10.10.

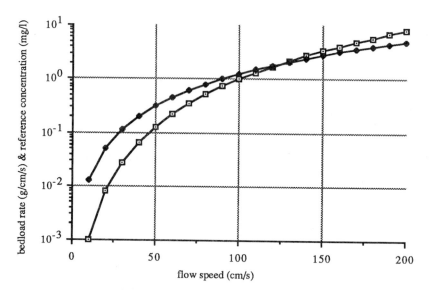

Figure 10.10   Bedload rates for medium sand (open squares, g cm-1 s-1) and the corresponding reference concentrations (closed diamonds mg l-1) (from Hardisty, 1990v).

10.5(b) Concentration profiles beneath waves

The determination of the concentration profile of suspended sediment beneath waves generally begins from the steady state mass and momentum diffusion equation (Sleath, 1984, p270) which was given in Chapter Nine as:

$$C\, w_s = K_s \frac{\partial C}{\partial z}$$

Eq.10.31

where C is the concentration of suspensate, w_s is the particle fall velocity, z is the height above the bed and K_s is an appropriate diffusion coefficient. If K_s is assumed to be constant, rather than proportional to the height, then Eq.10.27 integrates to give:

$$C_z = C_a e^{-\alpha_s z} \qquad\qquad \text{Eq.10.32}$$

where Sleath (1984) quotes expressions for α_s due to MacDonald (1973), Nielson (1979), and Sleath (1982). Vincent *et al.* (1982) show from field measurements that:

$$C_z = C_a \left(1 - A\ln\left(\frac{z}{z_a}\right) \right) \qquad\qquad \text{Eq.10.33}$$

where, for the bottom 100 cm of the water column $A=0.22$ where $z_a=1$cm and the reference concentration is given by Eq.10.30. Hardisty (1990v) presents a program to calculate the concentration profile from Eq.10.33, and typical results are shown in Figure 10.11, which seem to be in general agreement with data in the literature.

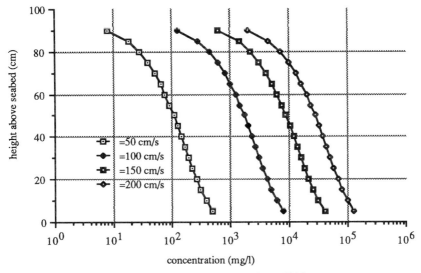

Figure 10.11 Suspended sediment profiles (from Hardisty, 1990v)

10.5(c) Suspended transport rates under Stokes drift

The calculation of the suspended sediment transport rate derives from the evaluation of the integral:

$$i_s = \int_o^h C_z u_z \, dz \qquad\qquad \text{Eq.10.34}$$

which is best evaluated numerically from C_z given by Eq.10.33 with C_a from Eq.10.30. If, however, this is driven by a sinusoidal flow field then the result will be zero, and therefore it follows that the transport rate can again only be predicted from, for example second order theory. The drift current, u_z, which is associated with Stokes waves was discussed in Section 6.4, and was there given as:

$$u_z = \frac{1}{4}\left(\frac{\pi H}{T}\right)\left(\frac{\pi H}{L}\right)\frac{1}{\sinh^2(kh)}$$

$$\{2\cosh\left(2kh\left(\frac{z}{h}-1\right)\right) + 3 + 2kh\left(3\frac{z^2}{h^2} + 4\frac{z}{h} + 1\right)\sinh 2kh$$

$$+ 3\left(\frac{\sinh^2 kh}{kh} + \frac{3}{2}\right)\left(\frac{z^2}{h^2} - 1\right)\} \qquad \text{Eq.10.35}$$

The suspended load transport rate is therefore obtained by integrating Eq.10.30 with C_z as above and u_z from Eq.10.35. Hardisty (1990v) evaluates the integral and typical results are shown in later chapters.

10.6 Slope inclusive effects

Although few datasets are presently available, we shall proceed here on the assumption that the movement of bedload is affected by bedslope beneath waves in the same manner, and to the same extent as is the movement of bedload in a unidirectional flow (cf Section 9.4(b) and 9.5(b)), and that the movement of suspended load is affected by a bedslope because the reference concentration has been taken equal to the concentration of the bedload as detailed above. The corresponding equations are given below.

10.7 Conclusions

The net sediment transport per centimetre width of seabed normal to the orthogonal line per wave will be modelled by:

$$I = \int_0^T \left(A_b k_b (u(t)^2 - B_b^2 u_{bcr}^2) u(t) + \int_o^h C_z u_z dz \right) dt \qquad \text{Eq.10.36}$$

where for bedload:

$$A_b = \frac{\tan\phi}{\cos\beta\left(\tan\phi + \frac{u_t \tan\beta}{|u_t|}\right)} \qquad \text{Eq.10.37}$$

$$k_b = 1.129 \times 10^{-6} \qquad \text{Eq.10.38}$$

$$B_b = \sqrt{\frac{\tan\phi + \frac{u_t \tan\beta}{|u_t|}}{\tan\phi}\cos\beta} \qquad \text{Eq.10.39}$$

$$u_{cr} = 25.5\left(\frac{DH}{\sinh(kh)}\right)^{0.25} \qquad \text{Eq.10.40}$$

for D<0.05 cm, else:

$$u_{cr} = D^{0.375} \left(\frac{H}{\sinh(kh)} \right)^{0.125}$$

Eq.10.41

and for suspended load:

$$C_z = C_a \left(1 - A \ln \left(\frac{z}{z_a} \right) \right)$$

Eq.10.42

where $z_a = 1 \text{cm}$, $A = 0.22$ and the reference concentration is given by:

$$Ca = 330 \frac{\overline{i_b u}}{rs} + 3.8$$

Eq.10.43

The flow profile, u_z, will be taken to be given by the Stokes drift current:

$$u_z = \frac{1}{4} \left(\frac{\pi H}{T} \right) \left(\frac{\pi H}{L} \right) \frac{1}{\sinh^2(kh)}$$

$$\{2 \cosh \left(2kh \left(\frac{z}{h} + 1 \right) \right) + 3 + kh \left(3 \frac{z^2}{h^2} + 4 \frac{z}{h} + 1 \right) \sinh 2kh$$

$$+ 3 \left(\frac{\sinh^2 kh}{kh} + \frac{3}{2} \right) \left(\frac{z^2}{h^2} - 1 \right)\}$$

Eq.10.44

This structure will be used to model orthogonal sediment transport in Section E. The full equations have been coded into Hardisty (1990w), and some examples of the solutions are shown in leter chapters.

chapter eleven

LONGSHORE SEDIMENT TRANSPORT

11.1 Introduction

In comparison with the shorenormal transport of sediment discussed in the preceeding chapter, shoreparallel sediment transport has received far more widespread attention, largely because it manifests itself wherever the movement is prevented by the construction of jetties, breakwaters and groynes. Such structures act as dams to the shoreparallel transport, causing a build up of the beach on the updrift side and a corresponding erosion in the down drift direction. Such consequences have often had serious and deleterious local effects, the former resulting in the infilling of harbour and river mouths, whilst severe cases of the latter have undermined and damaged many man made structures. There has, therefore, been considerable applied interest in quantifying shoreparallel transport and the results can be usefully transferred to geomorphic work.

Shoreparallel transport has been termed longshore transport, littoral drift or littoral transport. In the nineteenth century it was commonly believed that tidal currents and coastal ocean currents were primarily responsible for longshore transport. We now know that the wave induced longshore currents which were dealt with in Chapter Seven are the chief cause of the sediment movement; the other currents are only effective under exceptional circumstances.

Although only effective close to or within the breaker zone and thus less relevent to the present consideration of the orthogonal system, the background theory of longshore sediment transport is described in this chapter. A relationship between the longshore transport and the local waves is required. A series of purely empirical and deterministic solutions have been developed and are reviewed in the following sections. Some of these predict the total load across a shorenormal traverse, whilst others detail the distribution of the longshore transport from the shoreline to the break-point and beyond. The chapter is based largely on Komar (1976, 1983), although fuller reviews can also be found in Savage (1962), Das (1971), King (1972), Greer and Madsen (1978), Hallermeier (1982b) and in the N.S.T.S. work in Seymour (1989).

11.2 Empirical Total Load Formulae

The empirical formulae (Table 11.1) rely on a presumed correlation between the longshore transport rate and a measure of the longshore component of the incident wave energy. We saw (Chapter Five) that the wave energy flux per unit crest line is the wave power, P:

$$P = E C n \qquad\qquad Eq.11.1$$

Where the wave energy, E is (Chapter Five) given by:

Table 11.1

Some Longshore Transport Rate Formulae

Komar and Inman (1970): $j = 0.28(EC_g\cos\alpha)_b \dfrac{v}{u_{max}}$

Transport rate related to wave energy flux and tested with tracer experiments for grain sizes of 0.175 and 0.6 mm, $v \approx 0.75$ ms^{-1}, α of 8° and 16°and tanβ of 0.034 and 0.138. Units of j are immersed weight.

Wave Power Approach: $Q_L = k_L P_L{}^m$

	k_L	m
Watts (1953)	6.2	0.9
Caldwell (1956)	1.2	0.8
Savage (1959)	0.219	1.0
Ijima et al. (1960)	0.130	0.54
Ijima et al. (1964)	0.060	1.0
	0.120	1.0
Ichikawa et al. (1961	0.131	0.8
Komar and Inman (1970)	0.778	1.0
Das (1971)	0.325	1.0
Shore Protection Manual (1984)	0.401	1.0

Bagnold (1963) $j(x) = k_B \dfrac{d(EC_g)}{dx} \dfrac{v(x)}{u_{max}}$

Energetics approach following the development of the tranport function detailed in Chapter Nine. The total rate is obtained by integrating the local rate across the width of the surf zone.

In this table, $j(x)$ is the local longshore transport rate, j is the integrated longshore mass transport rate across a line normal to the shore, Q_L is the corresponding volume transport rateE is the wave energy, C_g is the wave group velocity, α is the angle of the breakers to the shoreline, v is the mean longshore current and u_{max} is the amplitude of the nearbed orbital velocity. The subscript b refers to parameters at the breakpoint.

$$E = \frac{1}{8} \rho \ g \ H^2 \qquad\qquad\qquad \text{Eq.11.2}$$

the wave celerity in shallow water is:

$$C = \sqrt{gh} \qquad\qquad\qquad \text{Eq.11.3}$$

and n is given by:

$$n = \frac{1}{2}\left(1 + \frac{2kh}{\sinh(2kh)}\right) \qquad\qquad\qquad \text{Eq.11.4}$$

For waves approaching at an angle α to the shore, this power is firstly converted to a unit shoreline length basis by multiplying by $\cos\alpha$ and then resolved into a shoreparallel direction by multiplying by $\sin\alpha$ so that the longshore wave power, P_L, becomes:

$$P_L = E \ C \ n \ \cos\alpha \ \sin\alpha \qquad\qquad\qquad \text{Eq. 11.5}$$

Assuming that the density of water is 1000 kg m^{-3} and that n approaches unity in shallow water then substituting for E, C and n from Eq.11.2 to 11.4 into 11.1 yields:

$$P_L = 3862 \ H^2 \ h^{0.5} \ \cos\alpha \ \sin\alpha \qquad\qquad\qquad \text{Eq. 11.6}$$

A number of workers have used field and laboratory data to produce formulae of the form:

$$Q_L = k_L \ P_L{}^m \qquad\qquad\qquad \text{Eq. 11.7}$$

Where Q_L is the volume of longshore transport in m^3 day^{-1} and P_L has units of tonnes day^{-1} m^{-1}.

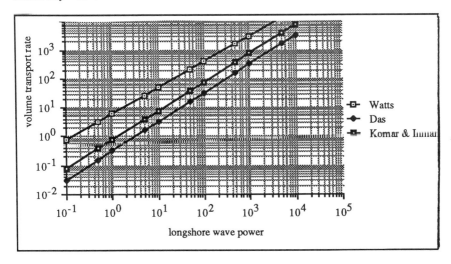

Figure 11.2 Empirical longshore transport equations based upon the wave power approach

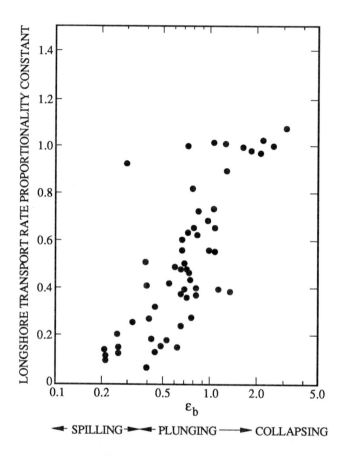

Figure 11.2 The relationship between the surf similarity parameter and the proportionality coefficient in the longshore transport formulae. (after Bodge, 1989).

Watts (1953) used the rate at which a bypassing plant had to pump sand past jetties at South Lake Worth Inlet, Florida to find $k_L = 6.2$ and m = 0.9. Caldwell (1956) used tracer results from Anaheim Bey, California to find $k_L = 1.2$ and m = 0.8. Savage (1959) summarised available field and laboratory data to find $k_L = 0.219$ and m = 1. Ijima *et al.* (1960) used field measurements from Japan to find $k_L = 0.130$ and m = 0.54, but revised this (Ijima *et al.* 1964) to $k_L = 0.060$ and m = 1; and later to $k_L = 0.120$ with m = 1. Ichikawa *et al.* (1961) also used Japanese data to find $k_L = 0.131$ and m = 0.8. Some of these papers are in Japanese and are included here from the references in Horikawa (1978). Komar and Inman (1970) summarised much of the data then available as shown in Figure 11.2 to find $k_L = 0.778$ with m = 1. Finally Das (1971) and the Shore Protection Manual (1984) attempt summaries and find that $k_L = 0.325$ and $k_L = 0.401$ respectively but with m = 1. Hardisty (1990x) computes the longshore transport rate in accordance with these various formulae and the results are shown in Figure 11.2.

These results show that there is some general agreement about the form of Equation 11.7 but that the coefficients of proportionality vary over two orders of magnitude between 0.06 and 6.2 with the exponents ranging from 0.54 to unity. The reason is that these empirical formulae are often site specific depending to a great extent on which wave height and water depth has been measured and upon the method by which the transport rate has been assessed. A better and more universal solution is based on the physical processes involved. Before considering the theoretical approach however, it should be noted that recent work (reviewed by Bodge, 1989) emphasises the relationship between k_L in Eq.11.7 and the type of breakers as shown in Figure 11.3. Specifically, it appears that the proportionality constant is more or less linearly related to the surf similarity parameter (Chapter Six) which is the ratio of the beach gradient to the root of the deep water wave steepness. It appears that a single value of the proportionality coefficient is inappropriate and that, in general, plunging or collapsing breakers lead to greater total longshore transport for a given level of longshore wave energy flux.

11.3 Deterministic Total Load Formulae

The above correlations are empirical and a more fundamental examination of the processes was carried out in a series of papers by Bagnold (1963), Inman and Bagnold (1963), Komar and Inman (1970) and is explained in detail by Komar (1976). This approach is far more satisfactory, since it relates the longshore transport to the longshore currents described in Chapter Seven. Firstly Inman and Bagnold (1963) pointed out that the longshore transport rate would be better expressed as an immersed weight j_L than a volume, Q_L where:

$$j_L = (\rho_s - \rho) \, g \, C \, Q_L \qquad\qquad \text{Eq.11.8}$$

Where ρ_s and C are the density and concentration of the sediment as detailed in Chapter Four. In line with the energetics model for unidirectional flows Bagnold argued that the longshore transport rate is given by:

$$j_L = k_L \, \tau_L \, v \qquad\qquad \text{Eq.11.9}$$

Where k_L is a dimensionless coefficient, τ_L is the longshore component of the stress exerted by the waves and v is the longshore current velocity which was given in Chapter Seven as:

$$v = \frac{5}{8} \frac{\pi \, \tan\beta}{c_f} u_{max} \sin\alpha \qquad\qquad \text{Eq.11.10}$$

and approximated by Equation 7.4 as:

$$v = 2.7 \, u_{max} \sin\alpha \qquad\qquad \text{Eq.11.11}$$

Komar (1976) argues that the longshore bed stress is the energy flux divided by the area over which it acts. Since this area is, for unit width, related to the mean near bed flow speed and this in turn to the maximum nearbed flow speed then:

$$\tau_L = \frac{K \, \text{Longshore Energy Flux}}{u_{max}}$$

$$= \frac{K \ E \ C \ n \ \cos\alpha}{u_{max}} \qquad\qquad \text{Eq.11.12}$$

Combining Eq.11.9 with the stress from Eq.11.12 and the longshore velocity from Eq. 11.11 we have:

$$j_L = K \ E \ C \ n \ \sin\alpha \ \cos\alpha \qquad\qquad \text{Eq.11.13}$$

Komar and Inman (1970 and Figure 11.3) found that the proportionality coefficient, K, is 0.77 so that the submerged weight longshore transport is given by:

$$j_L = 0.77 \ P_L \qquad\qquad \text{Eq.11.14}$$

or, in volumetric terms, for quartz sand density material of 60% packing in sea water:

$$Q_L = 79.2 \times 10^{-6} \ P_L \qquad\qquad \text{Eq. 11.15}$$

The similarity between the theoretical relationship in Eq.11.15 with the empircal results shown in Figure 11.4 is reassuring.

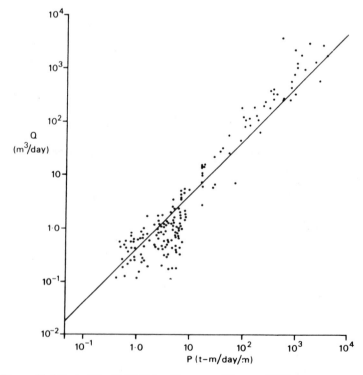

Figure 11.4 Test of Eq.11.15 (After Komar and Inman, 1970)

11.4 Cross Shore Transport Distributions

The preceeding analyses have defined the mean longshore transport rate across the full width of the surf zone. More detailed orthogonal modelling may well require the cross shore variation in the longshore transport rate to be determined (Figure 11.4). The problem is reviewed by Bodge (1989) upon which paper the following is based. Bodge finds that there is a wide range of available formulae which purport to predict the distribution of longshore transport across the surf zone. In general most of the models assume that sediment is locally mobilised (1) as a function of energy dissipation near the breakers, or (2) by the bed shear stress induced by the peak horizontal orbital velocities alone or (3) by the combined peak orbital velocity and longshore current. The mobilised sediment is then assumed to be advected downdrift by the local longshore current velocity. Many of the investigators have relied upon the expression for longshore current across the surf zone suggested by Longuet-Higgins (1970a and b) which was detailed in Chapter Seven. There is presently insufficient data available to enable a choice to be made between the various approaches and therefore, for the present work, the total load transport will be calculated from the semi-theoretical equation given earlier, and distributed according to the longshore current given by Longuet-Higgins' equation.

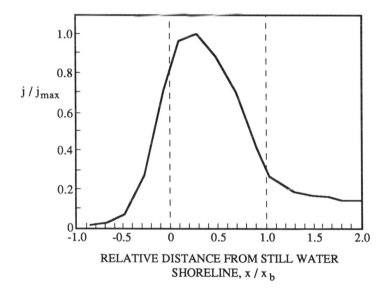

$$j / j_{max}$$

RELATIVE DISTANCE FROM STILL WATER
SHORELINE, x / x_b

Figure 11.4 Variation in the longshore transport rate across the breaker zone inferred from sediment volumes trapped updrift of a barrier on a laboratory beach (after Bodge, 1989)

This is effectively the same as Bodge's (1989) interpretation for which the longshore distribution is given by:

$$j(x) = \frac{25\pi}{128} k_B \, \rho \, g^{3/2} \, g^2 \, \frac{\tan\beta^2 \, \sin\alpha_b}{C} \, \sqrt{(h_b)} \, h \qquad \text{Eq.11.16}$$

where k_B is a proportionality coefficient, γ is again the ratio of the wave height to the water depth, $\tan\beta$ is the beach gradient, $\sin\alpha_b$ is the wave angle at the breakers, C is the local wave celerity and h_b and h are the breaker and local water depths. The resulting distributions are coded into Hardisty (1990y).

11.4 Conclusions

The mean longshore sediment transport rate can be modelled as:

$$j_L = 0.77 \ P_L \qquad\qquad \text{Eq.11.17}$$

where the longshore wave power is given by:

$$P_L = E \ C \ n \ \cos\alpha_b \ \sin\alpha_b \qquad\qquad \text{Eq.11.18}$$

Alternatively the distributed rate across the breaker zone can be modelled by:

$$j(x) = \frac{25\pi}{128} \ k_B \ \rho \ g^{3/2} \ g^2 \ \frac{\tan\beta^2 \ \sin\alpha_b}{C} \ \sqrt{(h_b)} \ h \qquad\qquad \text{Eq.11.19}$$

chapter twelve

ADVANCED SEDIMENT TRANSPORT

12.1 Introduction

The preceeding chapters have presented equations which allow the orthogonal and longshore sediment mass transports to be predicted in terms of the nearbed flow parameters, and these will be used in the orthogonal system models which are developed later in this book. They are, however, marinised versions of steady state fluvial and aeolian functions and as such will, at some time in the future, require modification to account for the unsteady nature of oscillatory flows beneath waves. The author has been working with J.P.Lowe on this problem through a research grant funded by the Natural Environment Research Council (GR3/6645) entitled "Higher frequency acoustic measurements of bedload processes in turbulent flow". This chapter presents some preliminary results from that work.

12.2 Particle Settling

The simplest case to consider is that of a sphere of diameter D settling under gravity in a stationary fluid. The acceleration phase is detailed later, but ultimately a terminal fall velocity, W_T, is reached when the gravitational force, F_g, is balanced by the fluid drag on the sphere due to pressure, F_p, and viscous, F_v, forces.

The gravitational force is equal to the particle weight minus its buoyancy:

$$F_g = \frac{\pi}{6} D^3 \rho_s \, g - \frac{\pi}{6} D^3 \, \rho_f \, g = \frac{\pi}{6} D^3 \, (\rho_s - \rho_f) \, g \qquad \text{Eq.12.1}$$

The pressure force can be approximated by a function of the stagnation point pressure. As the fluid approaches the sphere along the axial streamline, it slows and eventually stagnates at a point where it divides. The stagnation point pressure is given by Bernoullis theorem as $1/2\rho_f W^2$ where W is the sphere velocity relative to the fluid. The total pressure force is then proportional to the product of the pressure and the projected cross sectional area of the sphere, i.e. $\pi D^2/4$ where D is the sphere diameter. The constant of proportionality is a pressure drag coefficient, C_{DP} so that:

$$F_p = C_{DP} \frac{\pi}{4} D^2 \frac{1}{2} \rho_f \, W^2 \qquad \text{Eq.12.2}$$

The viscous force is proportional to the quadratic stress, $1/2 \, \rho_f \, W^2$, and the projected area of the particle. The second coefficient of proportionality is the viscous drag coefficient, C_{DV} so that:

$$F_v = C_{DV} \frac{\pi}{4} D^2 \frac{1}{2} \rho_f W^2 \qquad \text{Eq.12.3}$$

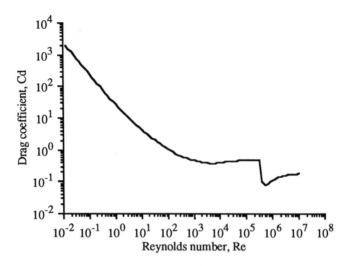

Figure 12.1 Variation in sphere drag with Re from the correlations in Clift et al. (1978, Table 5.2).

The relative magnitudes of the two contributions can only be studied empirically, so that C_{DV} and C_{DP} are combined into a single drag coefficient, C_D and $F_g = F_p + F_v$ is written:

$$(\rho_s - \rho_f) g \frac{\pi}{6} D^3 = C_D \frac{\pi}{4} D^2 \frac{1}{2} \rho_f W^2 \qquad \text{Eq.12.4}$$

Hence Eq.12.4 rearranges to predict the terminal velocity if the drag coefficient is known:

$$W_T = \sqrt{\left(\frac{3}{4} \frac{1}{C_D} \frac{\rho_s - \rho_f}{\rho_f} g D \right)} \qquad \text{Eq.12.5}$$

or Eq.12.4 rearranges to predict the drag coefficient if the terminal velocity is known:

$$C_D = \frac{4}{3} \frac{1}{W_T^2} \frac{\rho_s - \rho_f}{\rho_f} g D \qquad \text{Eq.12.6}$$

In practice, Eq.12.5 is employed to determine the terminal velocity from an estimate of the drag coefficient. The drag coefficient has been found empirically to depend upon the grain Reynolds Number, Re_g, (Chapter Twelve) which, for a falling particle, is defined by:

$$Re_g = \frac{\rho_f DW}{\mu} \qquad \text{Eq.12.7}$$

and numerous investigations have been conducted to determine the relationship between C_D and Re_g. The following review is based on Clift *et al.* (1978) and identifies six flow regimes as shown in Figure 12.1.

a) Creeping Flow Regime (Re_g<<1)
The viscous forces predominate and the drag coefficient is found to reduce to 24/Re. Therefore, within this regime:

$$C_D = \frac{24\mu}{\rho_f DW_T} \qquad \text{Eq.12.8}$$

whence, substitution into Eq.12.5 with Eq.12.7 gives a terminal flow velocity of:

$$W_T = \frac{1}{18}\frac{\rho_s - \rho_f}{\mu} g\, D^2 \qquad \text{Eq.12.9}$$

which is the well known Stokes Law for particle fall.

b) Oseen Regime (Re_g<0.01)
The Oseen approximation to the Stokes solution is:

$$C_D = \frac{24}{Re} + \frac{3}{16} \qquad \text{Eq.12.10}$$

c) Unseperated Flow Regime (0.01<Re_g<20)
Within this regime asymmetry in the flow is becoming increasingly apparent and the drag coefficient is given by:

$$C_D = \frac{24}{Re}\left(1 + \left(\frac{2.104}{16}Re^{(0.82 - 0.05\Lambda)}\right)\right) \qquad \text{Eq.12.11}$$

where:

$$\Lambda = \log_{10}(Re) \qquad \text{Eq.12.12}$$

d) Steady Wake Regime (20<Re_g<130)
At flows with a Reynolds Number above about 20 flow separation occurs, beginning at the rear of the sphere and developing a separation ring which moves forwards, and widens and lengthens the separation region. The pressure based form drag increases relative to the viscous skin friction. An empirical expression for the drag coefficient for Reynolds

Numbers between 20 and 260 is:

$$C_D = \frac{24}{Re}\left(1 + \left(\frac{3.096}{16} Re^{0.6305}\right)\right)$$

Eq.12.13

e) Wake Instability ($130 < Re_g < 400$)

At flows with a Reynolds Number above about 130 the diffusion and convection of votices within the wake can no longer keep pace with the generation of vorticity and discrete vortex shedding begins. An empirical expression for the drag coefficient for Reynolds Numbers between 260 and 1500 is:

$$C_D = 10^{(1.6435 - 1.1242\Lambda + 0.1558\Lambda^2)}$$

Eq.12.14

where Λ is given by Eq.12.12.

f) Inertial Region ($400 < Re_g < 350,000$)

The surface pressure distribution changes little and C_D depends primarily on form drag. It is insensitive to Re_g in this region and attains a value of $C_D = 0.445$ so that Eq.12.5 becomes:

$$W_T = 1.73\sqrt{\frac{\rho_s - \rho_f}{\rho_f} gD}$$

Eq.12.15

which is the well known Newton or Impact Law of particle fall.

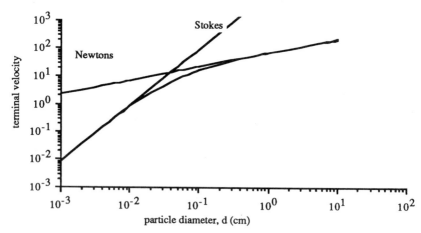

Figure 12.2 Comparison of results for terminal velocity with laboratory measurements.

12.3 Terminal Velocity

The preceeding section provides formulae for the drag coefficient of a non-accelerating particle. However the equations cannot be used to solve directly for the terminal velocity because, with the exception of regimes a and f, W_T occurs both on the left hand side of Eq.12.5 and on the right hand side within the Reynolds Numbers in the drag coefficients.

Instead resort must initially be made to iterative techniques to provide analytical solutions. Define a non-dimensional grain number, N_D, such that:

$$N_D = C_D Re^2 \qquad\qquad \text{Eq.12.16}$$

which from Eq.12.7 becomes:

$$N_D = \frac{4 \, \rho_f \, (\rho_s - \rho_f) \, gD^3}{3\mu^2} \qquad\qquad \text{Eq.12.17}$$

The standard drag curve (Figure 12.1) provides an empirical relationship between Re and C_D and analytical solutions for segments of the curve were given for each region above. These equations were used to generate a table of corresponding Re and C_D values, from which a table was constructed corresponding values of Re and N_D. Polynomial curve fitting techniques were applied to these data and generated the following four expressions:

For the region $N_D<73$:

$$Re_T = \frac{N_D}{24} - 1.7569 \cdot 10^{-4} N_D^2 + 6.9252 \cdot 10^{-7} N_D^3$$

$$- 2.3027 \cdot 10^{-10} N_D^4 \qquad\qquad \text{Eq.12.18}$$

For the region $73 < N_D < 580$

$$\log_{10} Re_T = -1.7095 + 1.33438 \Lambda - 0.11591 \Lambda^2 \qquad \text{Eq.12.19}$$

For the region $580 < N_D < 1.55 \times 10^7$

$$\log_{10} Re_T = -1.81391 + 1.34671\Lambda - 0.12427\Lambda^2 + 0.006344\Lambda^3 \qquad \text{Eq.12.20}$$

For the region $1.55 \times 10^7 < N_D < 5 \times 10^{10}$

$$\log_{10} Re_T = 5.33283 - 1.21728\Lambda + 0.19007\Lambda^2 - 0.007005\Lambda^3 \qquad \text{Eq.12.21}$$

where Λ is again as given by Eq.12.12. The procedure for determining the terminal velocity of any spherical particle is therefore:

i)   Determine the grain number, N_D from Eq.12.17

ii)   Determine the terminal Reynolds Number for the appropriate range of N_D from Eq.12.18 to 12.21

iii)   Determine the terminal velocity from Eq.12.7 which re-arranges to:

$$W_T = \frac{Re_T \mu}{\rho_f D} \qquad\qquad \text{Eq.12.22}$$

Comparison with experimental results

Experiments were performed in the terminal velocity of various spherical particles measured with high speed video camera. The results are compared with programs for the above equations given by Hardisty (1990z) by the asymptotic data in later figures in this chapter. There is clearly an acceptable agreement between the theory and the laboratory results. The terminal velocity curves for quartz particles in water are, therefore, given by the equations above as shown in Figure 12.2.

12.4 Acceleration Effects: Steady Drag

The movement of particles on the natural environment necessarily involves solid and fluid accelerations in anything except terminal fall in a stationary fluid or constant velocity transport in genuinely steady flow conditions. These additional effects are considered by firstly writing Newton's Second Law (F=ma) for a particle falling through stationary fluid from rest having an instantaneous velocity W, and by considering only the steady drag forces detailed above:

Particle Mass x Particle Acceleration = Net Gravity Force - Drag Force

$$\rho_s \frac{\pi}{6} D^3 \frac{dW}{dt} = (\rho_s - \rho_f) g \frac{\pi}{6} D^3 \; - \; C_D \frac{\pi}{4} D^2 \frac{1}{2} \rho_f W^2 \qquad \text{Eq.12.23}$$

Therefore the acceleration at any instant can be determined from:

$$\frac{dW}{dt} = \frac{\rho_s - \rho_f}{\rho_s} g - \frac{3}{4} C_D \frac{\rho_s W^2}{\rho_s D} \qquad \text{Eq.12.24}$$

12.5 Acceleration Effects : Unsteady Drag

The preceeding analysis underestimates the time and distance required for the particle to attain terminal velocity. Two additional terms are required to take full account of the development of the flow pattern around the sphere as it accelerates. These are the added mass term which represents the additional volume of fluid which is dragged along by the acceleration of the particle in the fluid and the Basset history integral term which represents the growth or decay of the boundary layer and wake on the sphere and is dependent upon the previous history of the motion.

The Added Mass Term

The added mass is a contribution which was noted by Dubuat (1786) and arises from the volume of water accelerated along with the particle. The force required to transport the additional fluid is:

Force = Added Mass Coefficient x Particle Mass x Acceleration

$$F_{AM} = C_A \cdot \frac{\pi}{6} D^3 \rho_f \cdot \frac{dW}{dt} \qquad \text{Eq.12.25}$$

The added mass coefficient was defined by Bessel (1826) as the ratio of the additional fluid mass to the particle mass and, in a manner analogous to the dependence of the drag coefficient C_D on Re, Iversen and Balent (1951) correlate C_A with an acceleration number, Ac:

$$Ac = \frac{|W| \, W}{\frac{dW}{dt} \, D}$$

Eq.12.26

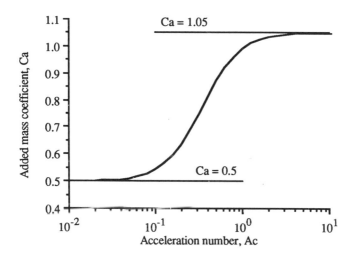

Figure 12.3 The variation of C_A with Ac from Eq.12.27.

The added mass of a perfect sphere at low Reynolds numbers is equal to half the mass of the fluid displaced by the sphere (C_A=0.5). In the same way as an empirical drag curve has been used to calculate C_D from Re, Odar and Hamilton (1964) proposed similar empirical relationships for C_A determined from measurements of the instantaneous drag on a sphere undergoing rectilinear simple harmonic motion in a viscous fluid:

$$C_A = 1.05 - \frac{0.066}{Ac^2 + 0.12}$$

Eq.12.27

It is noted that work by Lai and Mockros (1972) and Clift *et al.* (1978) provides a similar empirical analysis for disks and spheroids. The variation of C_A with Ac is shown in Figure 12.3.

The History Integral Term

The history integral describes the diffusion of vorticity away from the particle since the convective acceleration (spatial acceleration) is no longer insignificant. Vorticity generated by the particle as it moves is rapidly damped through time. The amplitude decreases exponentially as the time since generation increases (Batchelor, 1967, p256).

Previous accelerations are weighted by the elapsed time since the start of the acceleration, $\sqrt{t-t'}$ (Basset, 1888) and the force on the sphere is given by Landau and Lifshitz (1959) as:

Force = History coefficient x Protected Area x History Integral

171

$$F_H = C_H \cdot \frac{\pi}{4} D^2 \cdot \sqrt{\frac{\mu \rho_f}{\pi}} \int_0^t \frac{dW/dt}{\sqrt{t-t'}} \, dt' \qquad \text{Eq.12.28}$$

where t is the present time and t' is the time at the commencement of the acceleration. For low Reynolds numbers the value of C_H is a constant of 6.0 (Bassett, 1888) since the flow pattern around the sphere is essentially of the same symmetrical nature.

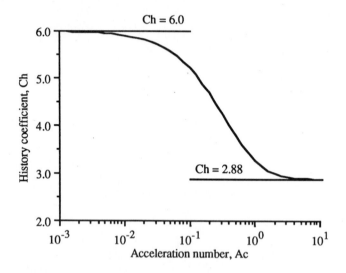

Figure 12.4. The variation of C_H with Ac from Eq.12.29

For higher values of Re Odar and Hamilton (1964) propose:

$$C_H = 2.88 + \frac{3.12}{(Ac + 1)^2} \qquad \text{Eq.12.29}$$

The variation of C_H with Ac is shown in Figure 12.4. The full equation for particle motion including these acceleration terms is therefore:

$$\frac{4}{3}\pi r^3 \rho_s \frac{du_r}{dt} = - C_D \frac{1}{2}\pi r^2 \rho \left| u_r \right| u_r - C_A \frac{4}{3}\pi r^3 \rho \frac{du_r}{dt}$$

$$- C_H r^2 \sqrt{\pi \rho \mu} \int_0^t \frac{du_r / dt'}{\sqrt{t - t'}} \, dt' + \frac{4}{3}\pi r^3 \rho \frac{du_f}{dt} \qquad \text{Eq12.30}$$

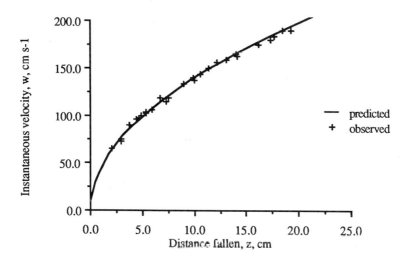

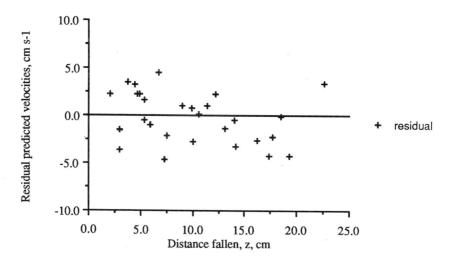

Figure 12.5 (a) Comparison of measured and predicted trajectories for a steel sphere falling through air; r=0.5cm, ρ_s=7.784 g cm-3 (b) Residual velocities of the results plotted in (a)

12.5 (a) Falling Ball Experiments

This section investigates the validity of Eq.12.30 for a sphere in a stationary fluid. The sphere is free-falling from rest, and therefore the experiments were designed to measure particle acceleration during the phase before terminal velocity was achieved.

The equation was tested with spheres of known size and density which were released in a still fluid so that the motion of the sphere alone could be studied.

173

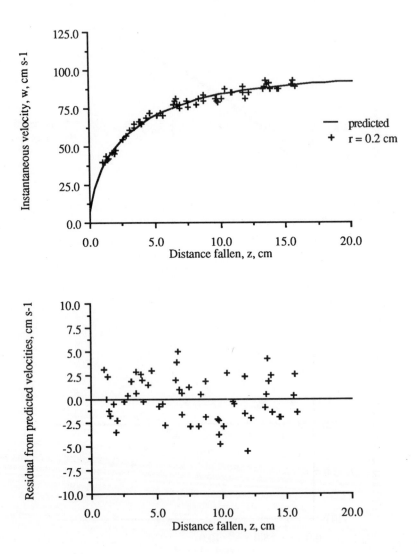

Figure 12.6 (a) Comparison of measured and predicted velocities and (b) residuals for steel spheres, r=0.2 cm, falling through water.

Spheres in the size range 0.2 to 0.5 cm in radius were chosen, reflecting the ultimate research aims in gravel size transport. Steel was chosen for the ease of availability of spheres in suitable sizes and for the compatibility with acoustic measurements of the impact noises made with the same experimental apparatus but not discussed here. The motion was recorded with a video camera. Release of the sphere was by a vertically mounted electromagnet and the sphere was then filmed with a fixed view video camera with the shutter speed of 1/1000 s, giving sharp still images of a moving object.

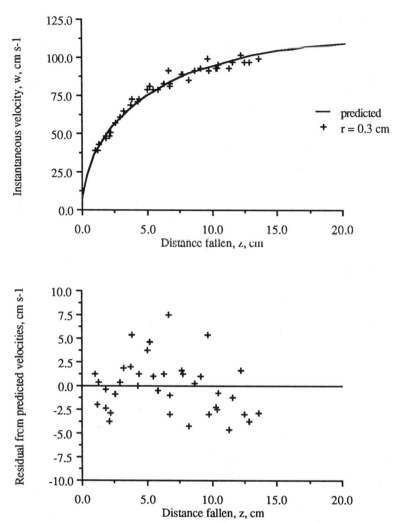

Figure 12.7 (a) Comparison of measured and predicted velocities and (b) residuals for steel spheres, r=0.3cm, falling through water.

Extra lighting was carefully arranged to emphasize the sphere and a graduated scale placed within the field of view to facilitate vertical position fixing. Using this system positions, velocities and accelerations of the sphere could be measured every 0.04 seconds. The accuracy of these measurements, made at a fixed frequency, were $\pm$ 2.5 cm s^{-1} (for the observed velocity range 0<W<250).

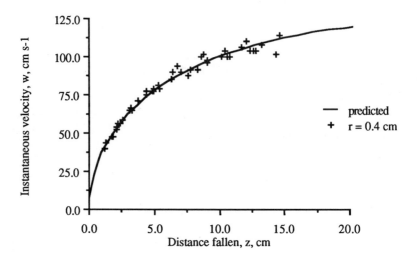

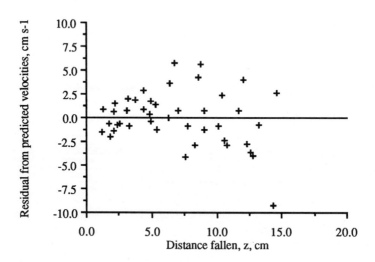

Figure 12.8 (a) Comparison of measured and predicted velocities and (b) residuals for steel spheres, r=0.4cm, falling through water.

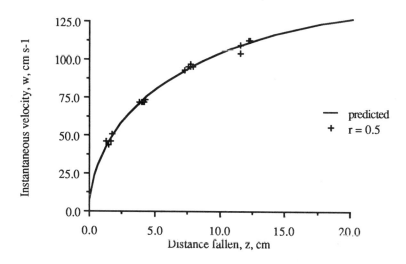

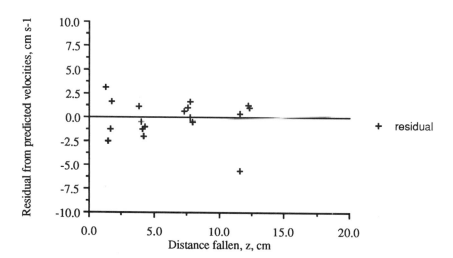

Figure 12.9 (a) Comparison of measured and predicted velocities and (b) residuals for steel spheres, r=0.5cm, falling through water.

Figures 12.5(a) and (b) show a set of results from trajectories of a steel sphere (r=0.5 cm) falling in air. A correlation analysis was conducted to compare the results with a numerical solution to Eq.12.30 in the form of a computer routine which calculated the particle accelerations at the same 0.04 s intervals. The calculation of the correlation coefficient between the empirical and theoretical results gave an R^2 value of 0.9923.

The results support the predictive capability of Eq.12.30 with all of the advective and convective terms. The complete solution is accurate within the requirements of sediment transport modelling, and, in particular, during the initial periods of high acceleration (Figure 12.10a). However there is some justification for leaving out the history term, which considerably increases the computation time of the trajectory since the history integral has to be calculated iteratively for each increment.

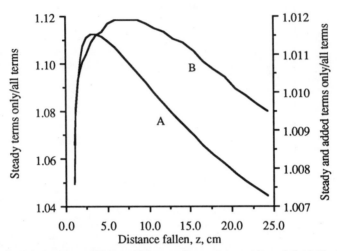

Figure 12.10 Velocity predictions of the steady state terms (A) and the full solution minus the history term (B) normalised against the full solution for the fall of the 0.5cm sphere in water.

The plot of residuals show that there is no discernible trend within the data; observed values being randomly distributed about the predicted values.

For the experiments conducted in water the Reynolds numbers will be lower and the acceleration numbers will be higher. Figures 12.6 to 12.9 show the results of the trajectories measured in water for a variety of sphere sizes. All indicate the predictive ability of equation 12.30, with no discernible trends in the plots of the residual velocities. Comparing with the predictions given by the steady state terms in equation 12.30 shows an over-estimation of the velocities by the simplified method with a maximum of about +12% for the case of the sphere with radius of 0.5 cm (Figure 12.10). The same diagram shows the effect of leaving out the history term and only keeping the added mass term, this gives a maximum over-estimation of the velocity of about 1.2%. The distinctive and similar shape of these two curves is due to the form of the force contributions changing with distance fallen (and so with time) as shown in Figure 12.11(a) for the Reynolds number and acceleration number and in Figure 12.11(b) for the force contributions for r=0.5cm.

It is apparent (Figure 12.10) that the maximum error introduced by neglecting the history integral in the velocity prediction is 1.2%, which is within the observational error of the experimental apparatus. Figure 12.11(b) shows that the 'added mass' term is the most significant throughout the initial stages of the acceleration (up to 0.1 seconds of the trajectory) where it accounts for between 75% and 45% of the total force whilst the history term, which follows the same trend as the added mass term, decreases from 22% to 15% of the total force over the same period. Leaving out both the added mass and the history terms gives errors in the calculation of the velocity of up to 12% during these initial stages of the trajectory.

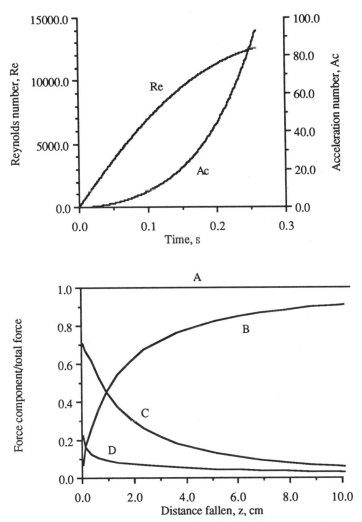

Figure 12.11 (a) Reynolds Number and Acceleration Number against distance from initial rest and (b) force contributions with distance for a steel sphere in water (r=0.5cm). Curve A is the total force, B is the steady state contribution, C is the added-mass term and D is the history integral term for a steel sphere in water (r=0.5cm).

12.5(b) Oscillating Ball Experiments

This section investigates the validity of Eq.12.30 for a sphere in a moving fluid. The type of flow considered has a simple and repeatable motion (approximately sinusoidal) such that the acceleration is smooth.

A major technical consideration in these experiments was the need to measure the flow velocity in the vicinity of the particle itself. A velocity sensor in a fixed position measuring flows in the Eulerian frame will not provide information about the flow directly affecting the moving particle. Instead a Lagrangian measure of the flow is required

179

so that the velocity of the parcel of fluid containing the solid particle is measured. Neutrally buoyant particles are often used in flow visualization experiments since the path of such a particle will be identical to the flow paths of the fluid (Hinze, 1975, p.154). A neutrally buoyant particle and a solid particle released at the same point and at the same time into the flow will provide information about the relevant fluid and solid particle motions so long as the solid particle remains within the volume of the fluid parcel which the neutrally buoyant particle represents.

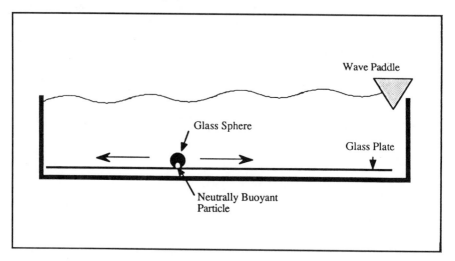

Figure 12.12. Oscillating ball experiment

Figure 12.12 shows the general arrangement of the experimental set up. To test the model a 0.5 cm radius silica glass sphere was placed on the glass bottom of a wave tank. Waves were generated by a mechanical paddle located at one end of the wave tank. A graduated scale was placed under the glass to facilitate horizontal position fixing.

The movement of the particle induced by the waves was recorded using a video camera. Measurements of the flow velocity were obtained by simultaneously recording the position of a neutrally buoyant foam particle placed next to the solid particle which then 'marked' a parcel of fluid. The camera shutter speed was set at 1/1000 of a second to give a sharp still image of a fast moving object. Using this system positions, velocities and accelerations of the solid particle and fluid parcel could be accurately measured every 0.04 s (25 Hz) to an accuracy of $\pm$ 0.625 cm s^{-1} (for the observed velocity range of 30 cm s^{-1}). Extra lighting was carefully placed to emphasize the spheres.

Figure 12.13 shows the velocities and accelerations for the fluid parcel and the solid particle respectively. Three waves are shown passing the sphere; one ending at 1.60s into the experiment, the second from 1.60s to 3.28s and the third from 3.28s to 4.96s. The time series of fluid velocities shows the waves generated by the paddle contain a significant second harmonic in the flows which gives an asymmetric wave with a longer duration but lower velocity negative flow in comparison with the positive flow beneath the wave crest. This can also be seen in the velocity time series of the solid particle but to a lesser extent due to the inertia of the particle damping out small fluctuations in the velocity. There is a suggestion of a time lag between the peak velocities of the fluid parcel and the solid particle of about 0.04s, although a greater time resolution is required to investigate this more fully. The time series of the accelerations also shows a damping of the small scale fluctuations and a similar time lag.

The results suggest that the flows measured are the same as those moving the solid particle for most of the time. The video record of the experiment lasted 20 seconds before the neutrally buoyant sphere moved out of the field of view, however after the passing of the third wave (approximately 5.2 seconds into the run) the distance between the two particles was such that the flow measured was no longer representative of the flow moving the sphere.

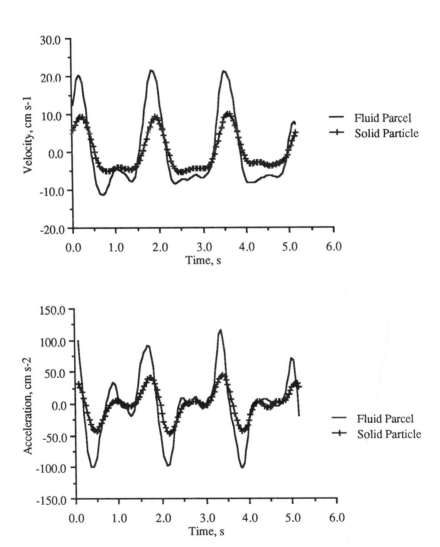

Figure 12.13 Fluid and particle velocities and accelerations.

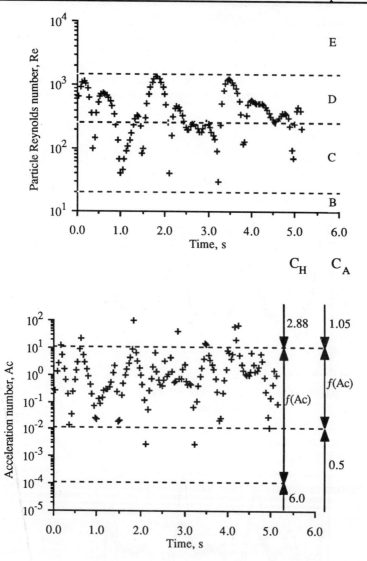

Figure 12.14   (a) Time series of the particle Reynolds number. Regions B, C, D and E indicate correlation ranges of the standard drag curve of CLIFT et al (1978, table 5.2), in particular:

$$Range\ C,\quad C_D = \frac{24}{Re}\left[1 + 0.1935\ Re^{0.6305}\right] \qquad 20 \le Re \le 260$$

$$Range\ D,\quad C_D = 10^{(1.6435\ -\ 1.1242\ log_{10}\ Re\ +\ 0.1558\ log_{10}\ Re^2)}\quad 260 \le Re \le 1500$$

(b) Time series of the Acceleration number. Also shown are the the added mass and history coefficient regimes of ODAR and HAMILTON (1964),

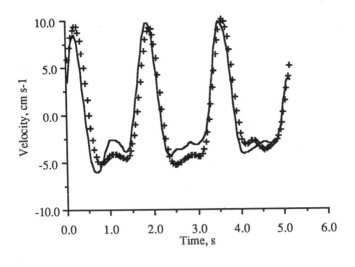

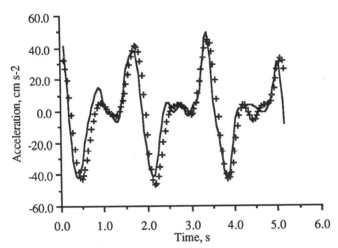

Figure 12.15 Comparison of the predictions of Eq.12.43 with the experimental results

For the flow the mean flow Reynolds number (u_f h/v, where h is the mean still water depth, h=20.0 cm) was in the range of 266-43000 and the particle Reynolds number (u_r D/v, where D is the particle diameter, D=1.0 cm) ranged from 29-1360. The acceleration number, Ac, varied between 10^{-3} and 10^2 as shown in Figure 12.18.

Before attempting to model the trajectory of the solid sphere the ranges of applicability of the three empirical coefficients, C_D, C_A and C_H need to be looked at in detail. Figure 12.14(a) shows the time series of the instantaneous particle Reynolds number through the experimental run and indicates that the values of Re fall into the regimes C and D used by Clift *et al.* (1978, table 5.2). Solving the equations for these regimes gives a difference of 0.01% at Re = 260 with no sharp discontinuities either side (as can be seen in Figure

12.1). The drag forces calculated using these equations do not therefore show a sharp discontinuity which may arise from badly matching drag curve regimes. Figure 12.14(b) shows that the experimental data fall within the experimental region of Odar and Hamilton (1964) for the coefficients which depend upon the flow acceleration, and only rarely reaches the upper asymptote given by values of Ac greater than 10. Again there is no evidence of sharp discontinuities in either of the coefficients.

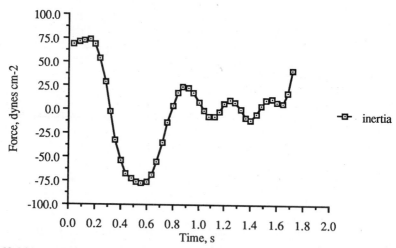

Figure 12.16 Inertial force contribution for the second wave

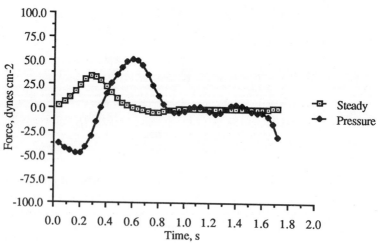

Figure 12.17 Steady and pressure force contributions for the second wave

Eq.12.30 was used to predict the positions, velocities and accelerations of the sphere, using the observed movement of the fluid parcel as inputs, and then comparing these with the observed movement of the solid sphere. Fluid velocities (u_f) and accelerations (du_f/dt) were calculated from the time series of fluid parcel positions. The sphere started at rest and

184

so initially $u_r = -u_f$ and $du_r/dt = du_f/dt$. Eq.12.30 was then solved for each time step by assuming the acceleration of the particle was the same as for the previous time step, solving the force balance, gaining a new estimate of the particle acceleration, and then using the new particle acceleration to solve again until the difference in the results was negligible. In this way the trajectory of the solid particle was reconstructed. The model is given by Hardisty (1990ac).

The close agreement between the observed and predicted results of the experiment can be seen in Figures 12.15(a) and (b) for the whole set of three waves (t = 0s to t = 4.96s). The other terms of Eq.12.30 are based on the instantaneous variables and so any measurement errors only affect that particular time period.

The component forces exerted on the sphere during the passage of the second wave are shown in Figures 12.16 to 12.18. As can be seen the contributions to the total force of each of the components is of importance. The wave can be split into two sections due to the asymmetry of the surface wave generated in the flume. Up to t = 0.9s the accelerations are high with high velocities so that all the forces are large. The effect of the second harmonic in the surface wave is to lessen the accelerations and velocities for the second half of the wave and so consequently the forces imposed on the solid particle. The history term in particular cannot be ignored in the description of this motion because it attains a similar magnitude to the other terms. This is in contrast to the results obtained in the earlier experiments for the free falling sphere.

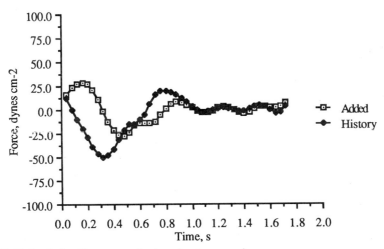

Figure 12.18 Analysis of force contributions for the second wave.

The prediction of the motion of a spherical particle in asymmetrically oscillating flow beneath surface waves by the momentum equation (Eq.12.30) has been shown to compare well with observations using photographic techniques and flow measurement with a neutrally buoyant particle. The analysis shows that, unlike the case of a free falling sphere in a stationary fluid detailed in the preceeding section, the results require the inclusion of the Bassett history integral term in the computations. However, as has already been shown, the Bassett history integral term is complex, and requires a slow numerical method. Although this is doubtless the way forward in the theoretical sense, this chapter will conclude by considering some approximations which may be applied in order to achieve a workable solution for applications in unsteady flow fields. It should be borne in mind that these solutions ought to be equally applicable to the case of both oscillatory flow beneath waves and turbulent flow in a unidirectional system.

12.6 Discussion

This Chapter has attempted to present one direction in which the theoretical analysis of the two phase, sediment transport problem could proceed. It has shown that the sediment transport formulae which depend upon the instantaneous values of the flow speed, or the bed stress, do not account for the acceleration terms which (at least for coarse sediment beneath waves) may be significant. It is unlikely that an amenable solution will, in the short or medium term, be developed to the full equation given above not least because it has been applied to spherical particles on an (almost) frictionless surface through the use of Lagrangian flow descriptors. None of these simplifications can be applied in the real world and we shall not, therefore, see such an unsteady description of the sediment transport processes incorporated into orthogonal models before the beginning of the next century. However the analysis presented is one of a number of converging routes which address the problem of predicting sediment mass transport beneath waves.

An alternative, and more pragmatic, development lies in the application of the so called "bedload spectrum" technique to sediment transport beneath waves. We have already seen (Chapters Five and Six) that useful predictive capabilities are being achieved in random waves with the transformation of the surface elevation spectrum to the near-bed velocity spectrum using relatively low order wave theory. It is possible, therefore, that a second tranform function could be developed to relate the flow power spectrum to the sediment transport power spectrum. Such a development would then be tested with some of the high frequency bedload and suspended load devices which are presently being developed, and would allow the analysis of the sediment transport processes to move out of the time domain into the frequency domain, wherein, of course, most of the important developments in wave mechanics and flow turbulence have been achieved. It is likely, finally, that the bedload spectrum achieved in this way will, in practice, show a high frequency cut off (at least for coarser grains) due to the acceleration effects described in this chapter.

SECTION D

MORPHODYNAMICS

chapter thirteen

BASIC MORPHODYNAMICS

13.1 Introduction

The preceding twelve chapters of this book have organised, reviewed and tested the theories which model the various processes operating on the orthogonal profile. The remaining chapters deal with the morphological results of those processes: the two and three dimensional form of the orthogonal system. Much of the preceding information is well precedented and has been taken from published papers and text books. Choices have been made, based upon published laboratory and field experiments and some additional data have ben collected and described. The remainder is, however, less clearly defined and many of the ideas will be but lightly referenced. Beach geomorphology, like the companion fields which deal with the form of hillslopes, rivers and other environments, has made slow and painful progress through the hierarchy of models which was outlined in Chapters One and Two.

In the present chapter, the manner in which the processes combine to form the beach are described. The relationship is based upon the mass continuity equation which is formally derived in the following section. Subsequently the effect of time on the equation and hence on the two dimensional orthogonal form is examined. It becomes apparent that simplified, analytical solutions provide useful insights into the operation of the system as a whole, whilst numerical solutions a necessary for a detailed exmination of the system. There are few introductory texts to the science of morphodynamics which compare with basic books on physics or chemistry or fluid mechanics. In this sense the present chapter seeks to define the principles upon which geomorphological analyses of different systems can be based. The reader is nevertheless referred to Thornes and Brunsden (1977) and to Thorn (1988) for the more general details.

13.2 The Continuity Equation: Spatial Solutions

The accumulation of detrital sediments is impossible without sediment transport. In transport which, however, is uniform (that is spatially invariate) and steady (that is temporally invariate) the transport rate changes with neither distance nor with time and the bed beneath cannot vary vertically.

This axiom is known as the mass continuity equation and states that the change in the height of the bed is proportional to the change in the transport rate. Consider Figure 13.1 which represents an idealised section at an instant in time. Consider X_1 to X_2 being of length δm m and that all cases represent transport per width δy m. Let i_1 be the sediment transport rate (in kg m^{-1}s^{-1}) entering the central volume and i_2 that leaving it.

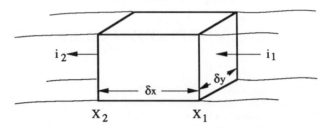

Figure 13.1 Definition diagram for the continuity equation in the spatial domain.

If, of course, $i_1 = i_2$ then there is neither accretion nor erosion and the bed level remains constant at a height z. If, however, $i_1 <> i_2$ then in unit time (1 second) more sediment leaves than arrives ($i_2 > i_1$) or arrives than leaves ($i_1 > i_2$) and the bed elevation changes. Consider the latter case where i_1 kg enters the volume in one second but only i_2 kg leaves then a mass of $i_1 - i_2$ is deposited in each second causing an increase in solid volume of $(i_1 - i_2)/\rho_b$, where ρ_b is the bulk density of the sediment (see section 3.5).

	$\partial i/\partial t = 0$	$\partial i/\partial t \neq 0$
$\partial i/\partial x \neq 0$	NON-UNIFORM AND STEADY	NON-UNIFORM AND UNSTEADY
$\partial i/\partial x = 0$	UNIFORM AND STEADY	UNIFORM AND UNSTEADY

Figure 13.2 Summary of mass continuity terminology.

Consider that this volume is equally distributed over the bed area, which in this example is δxδy causing a height increase of volume/area = $(i_1 - i_2)/(\rho_b \delta x \delta y)$. Consider unit length, and unit width of channel (δx=1, δy=1) so that, in general:

$$\text{Height Change per Second} = \frac{i_1 - i_2}{\rho_b} \qquad \text{Eq.13.1}$$

This is frequently written in differential terms as:

$$\frac{\partial z}{\partial t} = -\frac{1}{\rho_b}\frac{\partial i}{\partial x} \qquad \text{Eq.13.2}$$

where the rate of height change is $\partial z/\partial t$ and the rate of change between i_1 at X_1 and i_2 at X_2 is $\partial i/\partial x$. The minus sign ensures, sensibly, that accretion is represented by a positive $\partial z/\partial t$ and erosion by a negative $\partial z/\partial t$. The term $\partial z/\partial t$ is called the geomorphic response of the surface, and Eq.13.2 represents the geomorphic response due to spatial changes in the sediment transport rate. By analogy with the hydraulic literature we may refer to a transport rate which does in fact change with distance ($\partial i/\partial x <> 0$) as non-uniform, and to one which is constant ($\partial i/\partial x = 0$) as uniform. Erosion, that is a movement of sediment from the bed to the flow, is denoted by $\partial z/\partial t < 0$ whereas $\partial z/\partial t > 0$ represents deposition, a transfer from flow to bed. The terms are summarised in Figure 13.2.

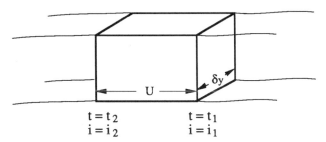

$$t = t_2 \qquad t = t_1$$
$$i = i_2 \qquad i = i_1$$

Figure 13.3 Definition diagram for the continuity equation in the temporal domain.

13.3 The Continuity Equation: Temporal Solutions

Consider separately now the case where the transport rate is constant at any instant at all points, but that the rate changes from moment to moment. The situation is similar to that above and is shown in Figure 13.3. Let i_{t1} be the transport rate at a point on the channel bed at a time t_1 and let i_{t2} be the rate at time t_2, δt seconds later. The passage of time means that the point will have moved a distance downflow U which is usually taken equal to the bedload velocity. The extra sediment entering (or leaving) the flow has therefore come from (or gone to) a bed area of width δy and of length U, that is an area $U\delta y$. The change in mass of mobile sediment is given by ($i_{t1}-i_{t2}$), and therefore the rate at which this mass is changing is given by ($i_{t1}-i_{t2}$)/δt. Since this mass will be deposited onto (or eroded from) an area of $U\delta y$, and therefore the mass change per unit time per unit area is given by ($i_{t1}-i_{t2}$)/($\delta t U\delta y$). As before this volume is assumed to be equally distributed over the area, and is converted to a change in bed elevation by dividing by the sediment bulk density.

$$\text{Height change per second} = \frac{1}{\rho_b}\frac{i_{t1}-i_{t2}}{\delta t U\delta y} \qquad\qquad \text{Eq.13.3}$$

In the limit and for unit width of channel ($\delta y=1$) we have:

$$\frac{\partial z}{\partial t} = -\frac{1}{\rho_b}\frac{1}{U}\frac{\partial i}{\partial t} \qquad\qquad \text{Eq.13.4}$$

where $\partial i/\partial t$ represents the rate of change of ($i_{t1}-i_{t2}$). Again, by analogy with the hydraulic literature, we may refer to a transport rate which does in fact change from one moment to the next as unsteady and one which is constant through time as steady. Erosion results when $\partial z/\partial t < 0$ and accretion when $\partial z/\partial t > 0$. These terms are also summarised in Figure 13.2.

13.4 The Continuity Equation: Combined Solution

In reality, the transport of sediment can be non-uniform and unsteady. The full continuity equation is then given by combining Eq.13.2 and Eq.13.4:

$$\frac{\partial z}{\partial t} = -\frac{1}{\rho_b}\left(\frac{\partial i}{\partial x} + \frac{1}{U}\frac{\partial i}{\partial t}\right)$$

Eq.13.5

Geomorpholgy involves substituting expressions for the transport rates into Eq.13.5 and solving for the changes in the surface form. Before proceeding to examine some of these substitutions in more detail, it is useful to consider some examples of the continuity equation. The equation was first formalised by Exner (1920) and is described by, for example, Allen (1970b) and Carson and Kirkby (1972). Swift and Ludwick (1976) present a marine application, quoting:

$$\frac{\partial \eta}{\partial t} = -\varepsilon\frac{\partial q}{\partial x}$$

Eq.13.6

where "η is the bed elevation relative to a datum plane, t is time, ε is a dimensional constant related to sediment porosity and q is the weight rate of bed sediment transport per unit width of streamline path and x is distance along the streamline". Apart from the symbols, Eq.13.6 differs from Eq.13.5 because of the choice of the streamline path (i.e. a line along which the transport is steady) and because of the use of ε for $1/\rho_b$. They apply the equation to map erosion and accretion on the Nantucket Shoal off the eastern United States. Additional examples are given in Chapters Fifteen and Sixteen below.

13.5 Geomorphological Equilibrium

There continues to be some discussion about what is meant by geomorphological equilibrium, and three different types of geomorphological equilibrium have been recognised. Firstly, in a situation where surficial sedimentary surfaces are unconsolidated but not mobile, then an equilibrium exists, because there is no change in the surface form through time. Allen (1982a) has referred to this state as statically stable equilibrium, but a better term is perhaps Jago and Hardisty's (1984) use of zero transport equilibrium. An example of the application of zero transport to beach models was the development of the null point hypothesis in the 1950's (Chapter Fifteen). It is also clear that, at least in plan, many beaches are in zero transport equilibrium, because nearshore wave refraction results in breakers developing parallel with the coastline, and therefore there is no longshore transport. A second state, wherein there is continued movement of sediment but no alteration in the geomorphological form, was referred to as dynamically stable equilibrium by Allen (1982b), but again we shall here use Jago and Hardisty's (1984) term of zero net transport. We shall see that orthogonal profiles can exist in states of zero net transport, which is represented by $\partial z/\partial t=0$ when $i<>0$ in the equations above. Finally, a state which is of interest to geologists and sedimentologists but does not directly concern us here is when sediment is steadily accumulating so that the thickness of the sedimentary record is directly and linearly related to time. This may be referred to as constant response equilibrium and is represented by $\partial z/\partial t$ as a non-zero, positive constant in the equations above. In the following chapters we shall be largely concerned with identifying and analysing states of zero net transport.

13.6 System Stabilities

Process response models do not allow for the existence of equilibrium states. A change in the process produces a changed response. In feedback modelling however, the output is re-presented to the processes and the system responds in one of three ways. The possibilities are illustrated by Figure 13.4 and are detailed as follows:

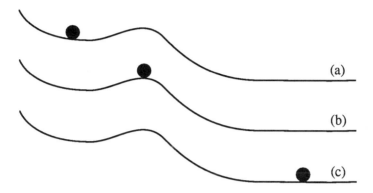

Figure 13.4 System stabilities, (a) stable equilibrium, (b) unstable equilibrium and (c) neutral equilibrium.

a. Stable Equilibrium

Here, when the ball is displaced from its equilibrium position, the system reponds by returning the ball towards equilibrium. This is called negative feedback or homeo-stasis and the system is said to be stable.

b. Unstable Equilibrium

Here, when the ball is displaced from its equilibrium position, the system responds by continuing to move away from the equilibrium. This is called positive feedback and the system is said to be unstable.

c. Neutral Equilibrium

Here, when the ball is displaced from its equilibrium position, the system does not respond and the ball remains in its new position. This is called neutral equilibrium.

13.7 The Response Function

The change in the surface elevation, $\partial z/\partial t$, is given by the continuity equation and is called the response function. The three classes of equilibrium which were outlined in the previous section have direct analogues in terms of the response function and the local elevation and these define the stability of the landform. They are shown in Figure 13.5 and are summarised as follows:

a. Stable Equilibrium

If $\partial z/\partial t$ is decreasing through the equilibrium value of $\partial z/\partial t = 0$ as the elevation increases, then the system is stable because any value in the region $z > z_e$ has a negative $\partial z/\partial t$, that is erosion occurs which moves the system towards z_e (i.e. A to B). Conversely any value in the region $z < z_e$ has a positive $\partial z/\partial t$, that is accretion occurs which again moves the system towards z_e (i.e. C to B). The equilibrium value z_e is therefore a point attractor and the system is stable.

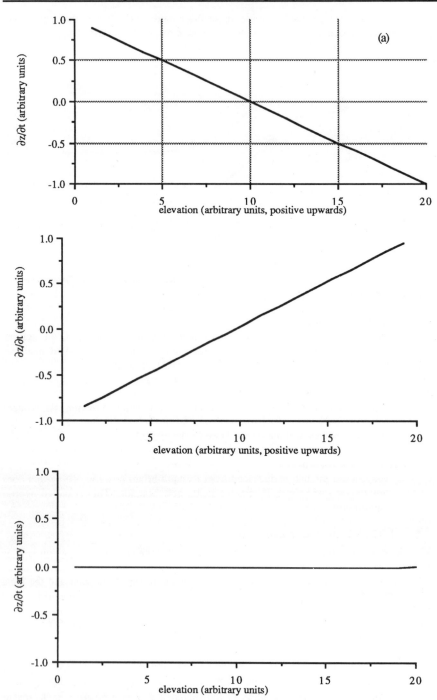

Figure 13.5 Response function stabilities: stable equilibrium (upper), unstable equilibrium (centre) and neutral equilibrium (lower).

Perturbations of z from z_e such as local accretion or erosion invoke a response which re-establishes the equilibrium. Additionally the steeper the gradient of $\partial z/\partial t$ versus z then the more rapidly will the geomorphology return to stability after being perturbed. This is depicted in Figure 13.5(a).

b. Unstable Equilibrium

If $\partial z/\partial t$ is increasing through the equilibrium value of $\partial z/\partial t=0$ as the elevation increases then the system is unstable because any value in the region $z>z_e$ has a positive $\partial z/\partial t$, that is accretion occurs which moves the system away from z_e (i.e. D from E). Conversely any value in the region $z<z_e$ has a negative $\partial z/\partial t$, that is erosion occurs which again moves the system away from z_e (i.e. F from E). The equilibrium value z_e is therefore a point repellor and the system is unstable. Perturbations of z_e, such as local accretion and erosion invoke a response which destabilises the equilibrium. Additionally the steeper the gradient of $\partial z/\partial t$ versus z, then the more rapidly will the geomorphology destabilise after being perturbed. This is depicted in Figure 13.5(b).

c. Neutral Equilibrium

Thirdly, if $\partial z/\partial t$ is always zero and neither erosion nor accretion result from a perturbation in the surface elevation then the geomorphology is in a state of neutral equilibrium. This is depicted in Figure 13.5(c).

13.8 Relaxation Time

The geomorphological relaxation time, T_R, is the time taken for the system to achieve equilibrium, and is defined as:

$$T_R = \frac{\text{change in elevation}}{\text{rate of elevation change}} = \frac{z - z_e}{\partial z/\partial t} \qquad \text{Eq.13.7}$$

where z is the elevation of a point and z_e is the equilibrium elevation. In general $\partial z/\partial t$ is a function of z and therefore:

$$T_R = \int_z^{z_e} \frac{1}{\partial z/\partial t}\, dz \qquad \text{Eq.13.8}$$

Quick and Har (1985) apply the concept to the rate of retreat of a shoreline. They observed the approach to equilibrium of laboratory beaches under the effect of constant wave attack and found that the ultimate "equilibrium profile" was approached asymptotically. They suggested that the shoreline retreat can be described by an exponential relationship:

$$x_t = (1 - e^{-bt})\, x_e \qquad \text{Eq.13.9}$$

where x_e is the equilibrium retreat and x_t is the retreat after a time t.

13.9 System Stability and Analytical Solutions

Thornes and Brunsden (1977) give three examples of the use of the mass continuity equation in geomorphological modelling, and these examples are described here because they illustrate the differences between *first* and *second* order solutions and between *linear* and *non-linear* systems. In the first, they consider a hillslope in which denudation (the rate of erosion) is assumed to be proportional to the local elevation:

$$\frac{\partial z}{\partial t} = -c_1 z \qquad \text{Eq.13.10}$$

where c_1 is the constant of proportionality. Scheidegger (1961) had suggested this elementary model, and it is not absurd if one assumes that the denudation is proportional to the precipitation and that precipitation increases linearly with altitude. Eq.13.10 is a differential equation of the first order because the equation contains a derivative and because that derivative is the first derivative of some function $z=f_o(x)$. The function describes the original form of the land surface in terms of the horizontal distance x. The solution of a differential equation is an expression for the dependent variable (in this case the elevation z) which does not involve any of its derivatives. Integration of Eq.13.7 provides the solution:

$$z = f_o(x) \, e^{-c_1 t} \qquad \text{Eq.13.11}$$

Suppose, for example, that $z=23x$ describes the original landform, then the solution is given by:

$$z = 23 \, x \, e^{-c_1 t} \qquad \text{Eq.13.12}$$

which can be evaluated for given values of x at any time t provided that measurements have been made to determine the value of the constant of proportionality, c_1. In a second example, Thornes and Brunsden (1977) use the work of Culling (1960) in which the denudation of the slope is assumed to be proportional to the local curvature:

$$\frac{\partial z}{\partial t} = c_2 \frac{\partial^2 z}{\partial x^2} \qquad \text{Eq.13.13}$$

where c_2 is the coefficient of proportionality. This is a second order differential equation because it involves the second differential of the dependent variable. In a third example, Thornes and Brunsden (1977) quote another model from Scheidegger (1970):

$$\frac{\partial z}{\partial t} = z \left(1 + \left(\frac{\partial z}{\partial x} \right)^2 \right)^{0.5} \qquad \text{Eq.13.14}$$

This example is different from the two others because neither of those involved powers or products of the dependent variable, z, or of its derivatives. They are therefore called linear models whereas Eq.13.14 is non-linear because the derivative $\partial z/\partial x$ is raised to a power, it is squared. Linear differential equations are the only ones for which complete analytical theory exists and for which general analytical solutions can be obtained. Non-linear differential equations, such as Eq.13.14, do not however have an analytical solution. Instead they must be converted into difference equations which are then solved by numerical techniques on the computer. The complexities of the orthogonal processes described in the preceding chapters negate the formulation of any solution to the orthogonal profile problem of the type shown by Eq.13.10 or 13.12. Instead, the solution must be formulated in numerical terms. One such solution is introduced in a later section and is presented by the SLOPES model in Chapter Seventeen.

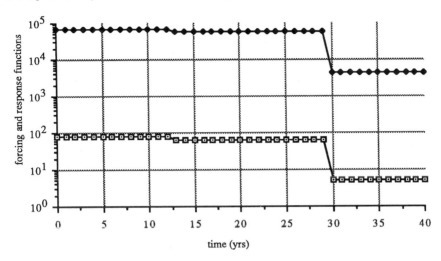

Figure 13.6 Surface elevation of a zero order system (closed diamonds) responding to changes in the forcing function (open squares) using Eq.13.15.

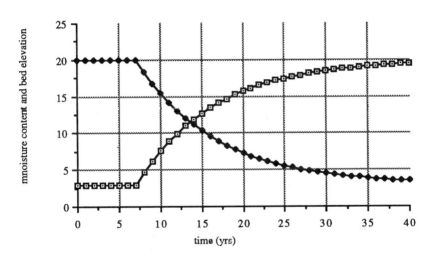

Figure 13.7 Soil moisture content (in arbitrary units, open circles) for a first order system described by equation 13.16 with a constant of 0.1 and $M_{sat}=20$ and an analogous first order elevation and erosion (closed diamonds) with the same, arbitrary constants with a negative $\partial z/\partial t$.

The concepts of systems and system stability which were described in the preceding sections are applied to analytical solutions for geomorphological systems by defining the appropriate differential equations. Three examples are again taken from Thornes and Brunsden (1977) because they are illustrative of the technique for zero, first and second order systems. In a zero order system the dependent variable (the elevation z) is independent of time and therefore we have:

$$z_t = c_3 \, x_t \qquad\qquad \text{Eq.13.15}$$

which states that the response or output (z_t) from a process (c_3) simply multiplies the forcing function or input (x_t) by a constant (the transfer function). The timing of change is invariable in this system and there is thus an instantaneous response to input variation. The relaxation time is therefore zero for a zero order system as illustrated by Figure 13.6 in which the one year relaxation times are an artifice of the modelling environment.

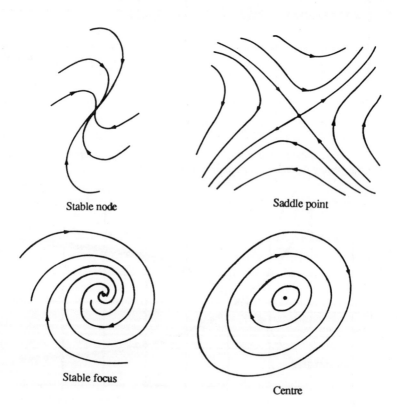

Stable node

Saddle point

Stable focus

Centre

Figure 13.8 Four types of equilibrium points for non-linear systems.

In a first order system the equation is differential in the first order so that the response to a change in the independent variable (the forcing function) produces a response which gradually damps down to a fixed level which is steady. An example in geomorphology is provided by the relationship between precipitation (the forcing function) and the soil moisture (output) through the infiltration rate (the transfer function):

$$\frac{\partial M}{\partial t} = P_t \, (M_{sat} - M_t) \qquad\qquad \text{Eq.13.16}$$

or in terms of the geomorphic response function a first order erosional system is:

$$\frac{\partial z}{\partial t} = c_4(z - z_e)$$

Eq.13.17

where M is the soil moisture, P_t is the precipitation and M_{sat} and M_t are the saturated and the instantaneous soil moisture contents respectively and z is the surface elevation, c4 is a constant and z_e is the equilibrium surface elevation. These types of system are illustrated by Figure 13.7.

Finally, in a second order system such as:

$$b_1 \frac{\partial^2 z}{\partial t^2} + b_2 \frac{\partial z}{\partial t} + b_3 \, z = c_4$$

Eq.13.17

the response will be damped to a steady state again, but particular values of the coefficients b_1, b_2 and b_3 lead to an equilibrium in the same way as for the first order systems or by damped oscillations.

13.10 System Stability and Numerical Solutions

In a numerical solution to a geomorphological problem the mass continuity equation must be set up in a computer model. The procedures involve the definition of the process elements. The processes are then operated over a pre-existing geomorphology and the resultant spatial and temporal changes in the sediment transport rate are used to account for sediment and hence to erode or to accrete the landform surface. The analysis of the results involves the erection of a minimal process system to undertake process sensitivity analyses on the model and of variable independent parameters to undertake parameter sensitivity analyses of the model and thus to determine system stabilities. These procedures are detailed in later chapters. The results, even for relatively simple systems, can be very enlightening and attempts are being made to identify various types of attractors. Some of the simpler solutions have been named and are illustrated in, for example Thornes (1983). A fairly comprehensive treatment of this field of non-linear systems analysis is given by Cook (1986) and four types of system are illustrated in Figure 13.8.

chapter fourteen

EMPIRICAL BEACH MODELS

14.1 Introduction

This, and the following two chapters, are a selective review of the literature which deals with mathematical beach models. The references have been organised into a chronological pattern, but it will be apparent that this pattern, though preferable, becomes difficult to maintain because beaches have been investigated by scientists in different disciplines each utilising a different approach to solve different, though related, problems. The result is that although the beginnings of a particular approach can be identified and clearly dated, it may well then be developed in a number of papers over the succeeding years. At the same time a different group may have initiated a different approach which itself generates a series through time. It is the task of the reviewer to order these overlapping developments so that the whole has a coherent structure, and so that the reader can perceive a series of stepping off points for the developments presented in the remainder of the book. The empirical models which are described in the present chapter divide into those which are static, in so much as they predict either the beach gradient or the orthogonal profile in terms of independent controls, or they are dynamic, in so much as they predict changes in the beach volume or the beach profile on a day to day basis. The theme which will emerge is that there have been a small number of isolated attempts to formulate functional form models, but that the only serious attempt to test the predictions against prototype time series (Seymour and Castel, 1989) yielded less than encouraging results.

14.2 Beach State Models

The descriptive beach models which were reviewed in Chapter Two had, since the early 1950s, been suggesting that the orthogonal system existed in one of a number of states, according to the incident wave regime and to the antecedant conditions. Initially these were crudely classified as Summer or swell versus Winter or storm states, the former being represented by a build-up and steepening of the profile, whilst the latter was represented by excess offshore transport and the formation of bars at the expense of the upper beach. This approach has been extended in a number of studies which, in the following forty years, sought to identify and to characterise typical beach forms or states. Sonu and van Beek (1971) and Sonu (1973) recognise six profile configurations divided into three groups: convex upwards, linear and convex downwards. Each group may or may not have an associated, superimposed bar. Accretive transitions were found to be associated with either bar growth or bar migration. Erosional transitions were

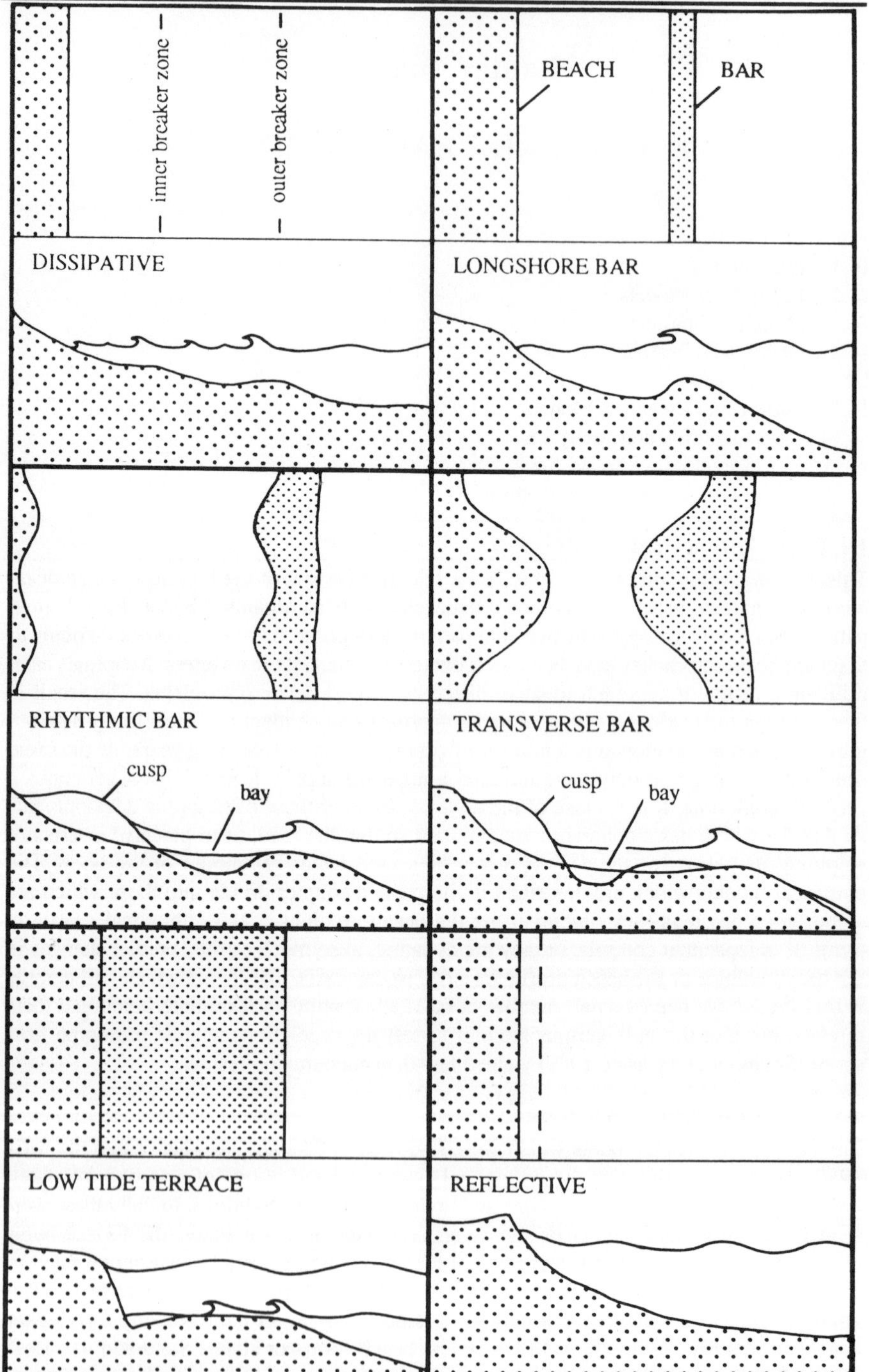

Figure 14.1 The six beach transition states recognised by the Australian School of beach morphodynamics (after Wright and Short, 1983)

found to be associated with either attentuation of the bar or bar destruction. These papers introduced the subaerial storage of sand into the cycles and emphasised that pre-existing morphology is an important factor in controlling subsequent profile changes.

The most important body of work which attempts to link such changes to the controlling wave climate (Pawka et al., 1976) has also been descriptive. For example, Wright (1976) identified six *morphology circulation* types ranging from State 1, a highly dissipative beach with a low angle and broad surf zone to State 6 represented by a narrow surf zone and a steep reflective beach. Chappel and Elliot (1979) extended this to seven types and attempted to introduce beach plan as well as profile form by the introduction of rip current circulations. Short (1979a and b) recognised subdivisions to distinguish fifteen beach types. This earlier Australian work demonstrated that microtidal beaches and surf zones may be dissipative, reflective or in any of several intermediate states depending on the local environmental conditions and on antecedent wave conditions (Wright et al., 1979a, b; Short, 1979a, b). Modes of beach cut, amplitudes and frequencies of standing waves and of possible edge waves (cf Chapter Six and Seven), and relative strengths and scales of rip and longshore currents are related to beach state (Wright, 1981a and b; Wright et al., 1982a). Depending upon the beach state, near-bottom currents show variations in the relative dominance of the incident waves, subharmonic oscillations and infra-gravity oscillations. They conclude that the beach morphology and rate of change of beach profiles are therefore also related to the environmental conditions and to the dominant beach configuration (Wright and Short, 1983, 1984). The different states recognised are summarised in Figure 14.1 and will be discussed in greater detail in later chapters. Wright et al. (1982b), extended the work through a complex series of process measurements on a macrotidal foreshore at Cable Beach, north western Australia and describe the greatest profile changes as being associated with short, within-tidal-cycle time scales. They also noted that, with certain exceptions (e.g. Gresswell, 1937; Watts and Dearduff, 1954; King and Barnes, 1964) little attention has been given to tidal range effects on profile evolution. Finally an interesting idea was proposed by Chakrabati (1977) to explain oscillations observed in beach states in West Bengal. He argued that a physical analogue is the well known mass-spring damped oscillations, and that the inertial mass was represented by the beach whilst the the spring action was provided by nearshore resonance of waves.

14.3 Shoreline Gradients

Various workers, over the last fifty years, have measured beach profiles in the laboratory and in the field, and have employed statistical or dimensional analyses to produce equations for either the beach gradient at the waters edge or for the full beach profile. The results provide formulae which relate the seabed gradient directly to the wave height and sediment size and inversely to wavelength. These trends are in accordance with the descriptive models detailed earlier. We consider the results of the two groups separately, detailing the shoreline gradient models in the present section and the full profile models in Section 14.4. Early experiments were conducted by Meyer (1933) in a laboratory wave tank with sand having a median diameter of 0.368 mm. The experiments identified a linear relationship between the tangent of the beach gradient at the water's edge, $\tan\beta$, and the ratio of the wave height to the wave length, H/L (cf Chapter Six). The results are quoted by King (1972) as being $\tan\beta$ equals 0.21 and 0.111 for H/L equals 0.008 and 0.080 respectively which suggests the equation:

$$\tan\beta = 0.32 - 13.75(H/L) \qquad\qquad \text{Eq.14.1}$$

Similar laboratory experiments with sand having a median diameter of 0.41 mm were conducted by Rector (1954) who recognised two gradients and gave the following equations:

$$\tan\beta = 0.07(H/L)^{-0.42}$$

Eq.14.2

for the region from the crest of the breakpoint bar to the foot of the swash slope and

$$\tan\beta = 0.30(H/L)^{-0.30}$$

Eq.14.3

for the foreshore slope above the still water level. More recently Doornkamp and King (1971) presented results which related the cotangent of the beach face slope to the wave parameters. Re-arranging their equation to predict the more conventional tangent of the angle gives:

$$\tan\beta = \frac{1}{6.1 + 3.36\log\left(\frac{H}{L}\right)}$$

Eq.14.4

These formulae have been coded into Hardisty (1990ae) and typical results are shown in Figure 14.2. It is clear that all of the models show an increasing beach gradient with a decreasing wave steepness, and also that all appear to suggest a "critical" value for the wave steepness, below which beaches steepen rapidly. A further empirical model for beach gradient on gravel beaches was reported by van Hijum (1974) and the results were presented as a series of data graphs.

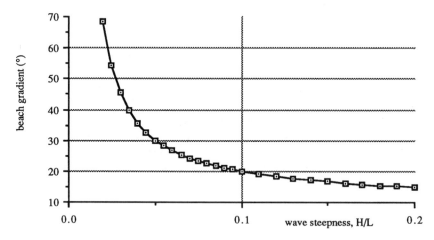

Figure 14.2 Beach gradient as a function of wave steepness using Rector's equation (from Hardisty, 1990ae).

The empirical models which attempt to incorporate the effect of grain size on beach gradient have used either the grain diameter, D, or the sediment fall velocity, W. Wiegel (1964) summarised data from "most of the beaches along the Pacific Coast of the United States" and produced the curves shown in Figure 14.3 which segregate beaches according to the degree of exposure. Alternatively Dalrymple and Thompson (1976) chose to employ W and developed the curve shown in Figure 14.4 which is based upon the experimental results of Rector (1954), Eagleson *et al.* (1963), Nayak (1970), Raman and Earattupzha (1972) and van Hijum (1974). Kemp and Plinston (1968) produced a similar curve to Figure 14.3 but utilised $H_b/TD^{0.5}$ instead of H/TW where H_b is again the wave height at the break point.

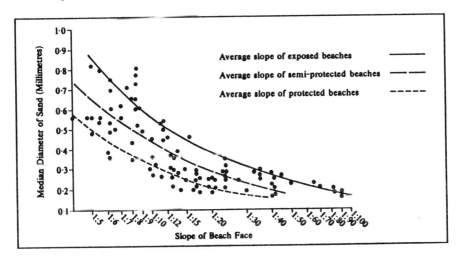

Figure 14.3 The relationship between grain size and gradient (after Wiegel, 1964)

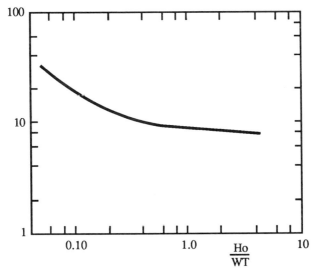

Figure 14.4 The variation of beach slope with the sediment parameter, H/WT (after Dalrymple and Thompson, 1976).

In a more sophisticated treatment of the problem using stepwise regression techniques, Harrison and Krumbein (1964) worked on the beach at Virginia Island in an attempt to predict the beach slope based on wave energy and the mean grain size. The measurements related to wave energy are included as process elements, and grain size and beach slope as response elements in the process-response model (Krumbein and Graybill, 1965). The dependent variable, nearshore bottom slope in the zone of shoaling waves, was measured at low tide. At the same time, four independent variables were measured, including wave height in deep water, wave period, wave angle and still water depth. Mean grain size was also included as a secondary independent variable because it was thought to be strongly correlated with the bottom slope and dependent on the wave energy. The prediction equation was:

$$\tan\beta = \beta_0 + \beta_1 M_z + \beta_2 T + \beta_3 H_\infty + \beta_4 \alpha + \beta_5 h + \text{constant} \qquad \text{Eq.14.5}$$

where the β coefficients were derived from linear regressions and $\tan\beta$ is the seabed gradient, M_z is the mean grain size, T is the wave period H_∞ is the deep water wave height, α is the wave angle and h is the still water depth. The results showed that the set of five independent variables accounted for some 78% of the variance in the data.

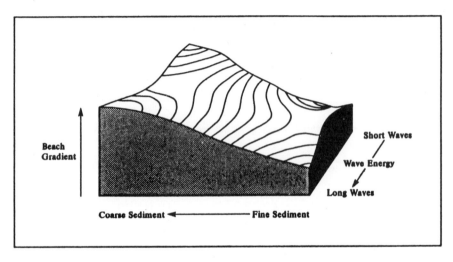

Figure 14.5 Shoreline gradients predicted by King's equation.

King (1972) reported similar results from twenty seven natural beaches in widely varying conditions. The variables tested were foreshore gradient, material size in phi units (cf Chapter Three), sand sorting, tidal range and a measure of exposure. The fetch distance adjusted for the orientation relative to the dominant waves was used as a measure of energy reaching the beach. The beach gradient correlated most highly with the sand size, while there was also a correlation, significant at 99%, between the measure of energy and the sand size. The relationships were shown by trend surface analysis using sand size and energy as independent variables in relation to beach gradient in the form of the logarithm of the cotangent of the foreshore slope. The surface accounted for 71.82% of the variability of the beach slope and had an equation:

$$\log(\cot\beta) = 407.71 + 4.2D - 0.71\log E \qquad \text{Eq.14.6}$$

where E is the measure of wave energy and D is the sediment grain size. This formula is also coded into Hardisty (1990ae) and the results are shown in Figure 14.5. An alternative approach has been taken by, for example, Hughes and Cowell (1987) who were interested in the height of the pronounced step which is often observed to form beneath the breakers on reflective beaches. They found, from repeated surveys at Pearl Beach in south eastern Australia, that:

$$Z = 0.32 H_b^{0.44} \, W_s^{0.21} \, \Delta W_s^{0.99} \, [1 - \exp(DW_s^2)] \qquad \text{Eq.14.7}$$

where Z was the step height, H_b the breaker height, W_s the sediment fall velocity at the step and ΔW_s the change in sediment fall velocity across the step and the correlation coefficient was 0.99.

In summarising these empirical shoreline gradient models, it is apparent that all support the finding that steep waves lead to shallow beaches and vice versa, but that additional allowance must be made for the sediment grain size, with coarser sediments appearing to be associated with steeper beaches.

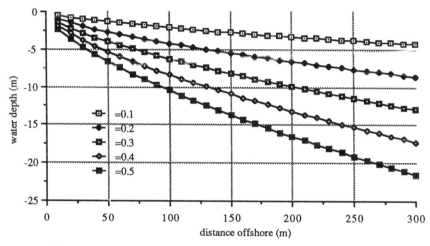

Figure 14.6 Orthogonal profiles predicted by power law formula for various values of the coefficient, a, and for b=0.65. Compare these with Dean's field survey results and with the NSTS satasets in the following Figures.

14.4 Orthogonal Profiles

Empirical relationships for the form of the full orthogonal profile, as opposed to the gradient of the beach at the water's edge, have been developed by Rector (1954) who found from laboratory experiments that the offshore zone profile was given by:

$$\frac{h}{x} = 0.223 \left(\frac{H_\infty}{L_\infty}\right)^{0.5} \left(\frac{D}{L_\infty}\right)^{-0.1} \left(\frac{x}{L_\infty}\right)^{-0.627(H_\infty/L_\infty)^{-0.005}} \qquad \text{Eq.14.8}$$

and that the foreshore zone was given by:

$$\frac{h}{x} = 0.07 \left(\frac{H_\infty}{L_\infty}\right)^{-0.42} \left(\frac{D}{L_\infty}\right)^{0.10} \qquad \text{Eq.14.9}$$

where h is the water depth at a distance x from the shoreline.

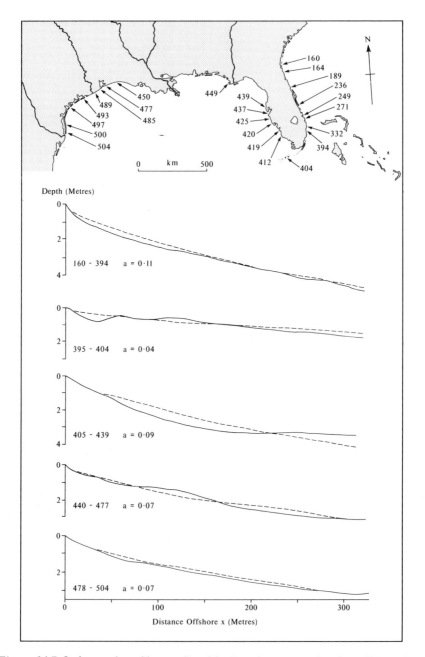

Figure 14.7 Orthogonal profiles predicted by Dean's equation for the U.S.East Coast obtained by averaging a large number of profiles from each area (after Dean, 1977).

The results began to quantify the generally concave upwards form of the beach profile which was discussed in earlier chapters and to recognise the relationship between sediment grain size and the gradient of the profile which was discussed in the preceeding section.

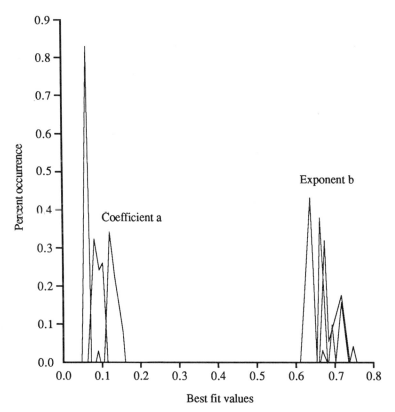

Figure 14.8 Histograms of the coefficient and exponent in Eq.14.9 from the NSTS data (after Seymour and Castel, 1989).

Field relationships which are analogous to the laboratory data used in Eq.14.8 and Eq.14.9 have been developed and Silvester (1974) references the analysis of Sitarz (1963) which shows that:

$$h = a \, x^b \qquad\qquad\qquad \text{Eq.14.10}$$

where

$$a = 0.95 \,/\, (g_s - 1)^{0.5} \, DH^{1.5} \qquad\qquad \text{Eq.14.11}$$

where b=2 and D and H are again the median grain diameter and wave height respectively, and g_s is the specific gravity of the sediment, $(\rho_s - \rho)/\rho$. This power law form for the orthogonal profile is coded into Hardisty (1990af) and typical results are plotted in Figure 14.6.

Dean (1977) examined a large number of profiles from the United States East and Gulf coasts (Figure 14.7) and found that, on average, the depth of the profile increased as $x^{2/3}$ out to three or four hundred metres from the shoreline. A reasonable fit to the data in the form of Eq.14.10 is obtained with a coefficient from 0.04 to 0.11, as shown by the broken lines in the diagrams. Seymour and Castel (1989) used profiles from the American Nearshore Sand Transport Study work on Pacific beaches at Santa Barbara and Scripps Beach in southern California and on Atlantic beaches at Rudee Inlet, Virginia to test Dean's empirical model between the breakers and the shoreline. The results are shown here (Figure 14.8) as smoothed histograms. The value of the coefficient, a, varied from about 0.065 for Scripps Beach to 0.13 for Virginia Beach and increased with increasing grainsize. The value of the exponent, b, ranged from 0.63 to 0.67 with a mean value of about 0.65. Hughes and Chiu (1978) showed that a further 464 beach profiles could be described by the type of power law given in Eq.14.10.

14.5 Erosion Models

Although apparently less demanding than the preceding attempts to predict profile gradient or form, much effort has been dedicated to predicting simply whether a given beach will erode (that is a measure of its sediment volume will decrease) or accrete (that is its sediment volume will increase. Additionally this change in volume is, at times, related to shoreline retreat. There have been an number of empirical models which have these objectives, and they are summarised in this section.

The earlier studies of beach gradient by Harrison and Krumbein (1964) and Harrison *et al.* (1965) were expanded by Harrison (1969) using data from a twenty six day time series in which seventeen independent variables were measured, some at four hourly intervals and some at low tide. Presenting the results as non-dimensionbal groupings, Harrison developed an equation for predicting ΔQ, defined as the volume of sand eroded from or deposited on the foreshore in one tidal cycle. The model incorporated five time lags which were separated by three hour intervals:

$$\Delta Q = 5.803 - 113.151 \sqrt{\frac{H_b}{gT^2}}_{0.0} - 52.111 \left(\frac{h_y}{x_b}\right)_{3.0}$$

$$- 7.269 \, (\alpha_b)_{6.0} + 2.724 \left(\frac{D}{z_b}\right)_{6.0} - 35.067(\tan\beta_b)_{12.0} \qquad \text{Eq.14.12}$$

where H_b/gT^2 is the breaker coefficient (cf Chapter Five), h_y is the hydraulic head, x_b is the breaker zone width, $\tan\beta_b$ is the beach gradient at the breakers, α_b is the wave angle at the breakers, D is the mean grain size and z_b is the trough to bottom distance immediately in front of the breakers. The subscripts for each dimensionless variable indicate the time lag that has the most influence on the predictor equation. Another statistical approach was reported by Hallermeier (1981) which sought to estimate the seaward limit of the mobile seabed in terms of typical and annual wave characteristics.

A recent review of erosion models was given by Seymour and King (1982) and Seymour and Castel (1989) who compared six models with the survey data obtained during the American Nearshore Sediment Transport Study project (cf also Seymour, 1986). The models are summarised in Table 14.1 and are all designed to predict either net onshore or offshore sediment transport as determined by changes in the beach volume due to orthogonal as opposed to longshore sediment transport. In the table, R is an arbitrary constant, H is the deep water significant wave height, W is the sediment fall velocity, T is the spectral peak period, g is the gravitational acceleration, $\tan\beta$ is the beach slope, ρ is the fluid density and D is the sediment grain diameter.

Table 14.1
Models for predicting the threshold of
cross-shore sediment transport (after Seymour and Castel, 1989)

Dean (1973)	$\dfrac{2RH}{WT}$	accretion<1<erosion
Sunamura and Horikawa (1974)	$\dfrac{1.845}{g^{0.33}} H \dfrac{\tan\beta^{0.27}}{(TD)^{0.67}}$	accretion<4-8<erosion
Short (1979a) height model	H	accretion<120cm<erosion
Short (1979a) power model	$\dfrac{\rho g^2 H^2 T}{16\pi}$	accretion<30KW/m<erosion
Hattori and Kawamata (1980)	$\dfrac{2H\tan\beta}{WT}$	accretion<0.5<erosion
Quick and Har (1985)	$\left(\dfrac{H}{WT}\right)_{final}$	accretion<$\left(\dfrac{H}{WT}\right)_{initial}$<erosion

The table has been taken from Seymour and Castel (1989) and the symbols have been changed to comply with the present usage and the Quick and Har parameter has been changed to compare the final value of their parameter with the previous value in order to retain the arithmetic sense of the threshold. Seymour and Castel (1989) compared the models with net volume change from day to day expressed in $m^3\ m^{-1}$ and they recommended a value of R=0.229 for the Dean model. The models were compared by calculating a skill factor for each parameter (the ratio of correct predictions of accretion or erosion against the total number of observations) and showed the following values: Dean' model: 0.57; Sunamura and Horikawas' model: 0.60 (with a threshold value of 13.2); Short's height model: 0.25 (later improved to 0.65 by reducing the threshold to 65cm); Short's power model: 0.63 (with a threshold of 43 w/cm); Hattori and Kawamata's model: 0.68 (with a threshold value of 0.064); Quick and Hars' model: 0.49 (with a threshold value of 1.01). These adjustments to the threshold values were calculated by Seymour and Castel as being required to balance the number of erosion and accretion events. Their conclusion is that the models which had been tested for their ability to predict the sense of cross shore transport were correct between half and two thirds of the time. This, as the main conclusion to a project designed to investigate nearshore sediment transport and which had cost some $4m, is hardly a glowing recommendation for the state of the science. Effectively it shows that, notwithstanding the impressive body of theory detailed in the preceeding chapters, the relationship between beach form and wave height is the most advanced predicor presently available! It should, however, be recalled that, in section 14.5, a threshold wave steepness was discussed which appeared to mark the boundary between steep and flat beaches. Since many of the models in Table 14.1 essentially represent H/L because T is related to L, then the same type of changes in beach steepness were apparently being identified by the models. This chapter will conclude with a consideration of three other models which attempt to extend the erosion predictions to the full profile form.

14.6 Orthogonal Response Models

There are three groups of empirical models which attempt to predict the response of the orthogonal profile to changes in the independent variables. In the first group, here called Eigenfunction analyses, the profiles are represented by a time x profile number x water depth multivariate matrix, from which modes of intercorrelation are calculated as the eigenvectors of the matrix. The second group is based upon defining a landward limit to the profile based upon wave run-up and then dividing the orthogonal into two curves separated by a point, the depth of which depends upon wave and sediment characteristics. This group is here called developing profile modelling because the shoreward curve is known as the developing or D profile. A single paper introduces the third type of orthogonal response model, and it is based upon the proposal that the profile pivots around a fixed point seaward of the breakers and that profile response is caused by changes in profile concavity and slope about this point.

14.6(a) Eigenfunction Analyses

A new development in the empirical analysis of orthogonal profiles appears to have been first presented by Winant et al. (1975), and to have then been applied by a number of others, including Vincent et al. (1976), Aubrey (1979), Dolan et al. (1977, 1979a and b), Hayden et al. (1979), Aranuvachapun and Johnson (1979), Aubrey et al. (1980), Bowman (1981), Eliot and Clarke (1982), Clarke and Eliot (1982, 1983), Clarke et al. (1984), Lins (1985), Aubrey and Ross (1985), Wright et al. (1987), and Clarke and Eliot (1988a and b), to a range of sites around the world. These papers attempted to show that profile variations through time can be accounted for using empirical eigenfunctions.

The details are complex and are given in full in some of the original papers, and in a review form by Fox (1985). The technique is referred to as empirical orthogonal function (EOF) analysis by Clarke and Eliot (1983) who described its application to the results of fortnightly beach surveys along eighteen profiles at Warilla Beach south of Sydney, Australia. They report that EOF analysis is comprehensively described by Morrison (1967) and by Halliwell and Moores (1979). Analysis of each time series decomposes the spatial structure into empirical modes, each of which represents a spatially coherent structure. Associated with each mode is a modal amplitude time series which describes how that mode varies in time. In the analysis of each of the profiles an autocovariance matrix is generated from the time series data. The matrix is decomposed into eigenvalues and associated eigen-functions. The ratio of the eigenvalue to the sum of the eigenvalues is the percentage of the variance attributed to that mode. Most of the variance is usually contained within the first few modes. Nodes (zero crossings) in each profile eigenfunction represent pivotal points through which sediment is transferred. The first four modes for the Warilla Beach results are shown in Figure 14.9. The first mode has no zero crossings (nodal points) and is referred to as the fundamental beach response. It identifies long-period cyclical changes for berm to bar morphological transfers of the type identified earlier in this chapter (e.g. Short and Wright, 1981). It appears that the beach profile can be divided into a mean beach function (which is independent of time), a bar-berm function (which shows seasonal variations) and a terrace function (which is also relatively stable through time), and that the sum of these accounts for variance in the beach profiles. Since the presence of nodes indicates pivotal points through which sediment is transferred, analysis of the longshore as opposed to orthogonal matrices indicates the longshore transport contribution to profile change.

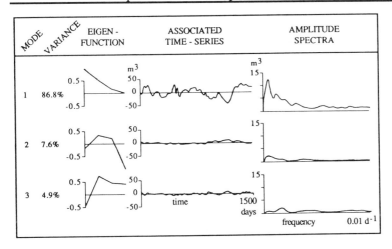

Figure 14.9 Eigenfunction modes, associated time series and power spectra for the Warilla Beach data (after Clarke and Eliot,1983)

Following Wright *et al.* (1987) the equation representing the profile height is:

$$z_{tx} = Z_x + \sum (\lambda_k)^{1/2} v_{tk} u_{tk}$$

Eq.14.13

where z_{tx} is the profile height at time t and point x, Z_x is the time averaged height at point x, λ_k is the kth eigenvalue of the datamatrix, v_{tk} is the kth temporal eigenvector (eigenfunction) and u_{xk} is the kth spatial eigenvector (eigenfunction).

An interesting development of the technique was given by Dolan *et al* (1978a and b, 1979a and b) and was concerned with the identification of spatial frequencies in nearshore bathymetric data. Bathymetric chart data were detrended by subtracting a second order surface from the original data so that the trend and fundamental harmonic (mode 0 and 1) of the surface was removed. The two dimensional power spectra of the detrended data was then computed. The orientation of the result showed, for example, that at Assateague Island on the Cape Hatteras Arc, there were spatial periodicities at 55.7° and 65.6° with wavelengths of 7.44 km and 2.12 km, and that the wavelengths increased in a shorewards direction. These periodicities were said to relate to the dynamics of the linear shoal field and to the presence of edge wave processes.

It appears that these developments offer a very powerful technique for the identification of appropriate spatial and temporal scales on the orthogonal profile, and they will doubtless be used to an increasing extent in the future. The relevance of eigen-vector analysis is re-considered in the final chapter of this book.

14.6 (b) Developing Profile Analysis

The second group of profile response models is an empirical-numerical scheme based on the work of Swart (1974, 1976). It has the following characteristics:

1. The upper limit of wave action (Figure 14.10) is calculated as a function of wave height and period and of the mean sediment size. The upper limit is time-varying, relative to the mean sea level, because of tides and storm surges. The backshore, above this upper limit, is assumed to

213

remain unchanged.

2. A lower limit for the developing profile (which Swart calls the D-profile) is defined by another empirical expression dependent upon the same parameters as the upper limit. In the intervening zone between the two limits both bedload and suspended load are expected to be significant.

3. The developing profile is expected to be driven towards, but not necessarily achieve, an equilibrium shape. The shape is defined purely on the basis of the median sediment size and, given constant water depth and wave conditions, the profile will approach the equilibrium exponentially (cf Chapter Thirteen).

4. Below the lower limit of the developing profile, bedload is assumed to dominate. Sediment is forced, in either direction, across the lower limit to satisfy continuity on the developing profile. The orthogonal profile shape below the lower limit is determined by iteratively calculating an equilibrium slope at each location following Eagleson *et al.* (1963).

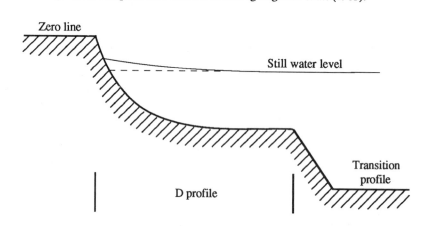

Figure 14.10 Schematic diagram of the three sections of the orthogonal profile used in the Swart model, the backshore, the developing profile, and the transition profile.

The Swart model has been evaluated under laboratory and field conditions by Swain and Houston (1983), Swain (1984) and Seymour and Castel (1989) and has been shown to accord well with the experimental results for beaches without bars (the condition for which it was formulated). A further analysis of the Swart model is given by Swain (1989), and it includes a clear enunciation of the governing equations. The following is based upon Swain's 1989 paper. The upper limit of wave action is taken to be a height h_o (Figure 14.10) above the still water level, where this maximum runup was determined empirically by Swart to be (in metric units):

$$h_o = 7650\ D \left(1 - \exp \left(\frac{-0.000143\ H_{m\infty}{}^{0.488}\ T^{0.93}}{D^{0.786}} \right) \right) \qquad \text{Eq.14.14}$$

where $H_{m\infty}$ is the maximum wave height in the spectrum, T is the wave period and D is the median sediment grain diameter. The depth, h_m, of the lowest point on the developing profile is also given empirically by:

$$h_m = 0.0063 \ L_\infty \left(\frac{4.347 \ H_\infty^{0.473}}{T^{0.894} \ D^{0.093}} \right)$$
Eq.14.15

where L_∞ is the deep water wave length (Chapter Five). The equations for the developing profile are then given for various conditions by Swart's booklet.

14.6(c) Pivotal Monotonic Analyses

The third group of profile response models is another empirical-numerical scheme based on the work of Quick and Har (1985) and tested with the American Nearshore Sediment Transport Study data by Seymour and Castel (1989). It has the following characteristics:

> 1. There is a certain depth, occurring just outside the breaker zone, at which there is no vertical change. The maximum sediment transport rate occurs at this depth which functions as a pivot point for profile re-alignment.
> 2. The slope at mean sea level is related to the wave sediment parameter following Dalrymple and Thompson (1976) as detailed in section 14.3 above.
> 3. The profile follows the power law suggested by Dean (1977) and detailed in section 14.3 above. The depth is, therefore, proportional to the offshore distance to the 2/3s power. Furthermore, the profile approaches this equilibrium state exponentially with an expected half-life on natural beaches of "a few hours".

The profile is then calculated by fitting an appropriate curve through the fixed pivotal point and the still water level. Seymour and Castel (1989), however, explored the assumptions on which this model is based and particularly questioned the prime assumption that profile response is caused by slope and shape changes. They report that their profile observations at Santa Barbara showed that some 97% of the variation about the mean position was caused by a horizontal translation of the profile without change in shape or slope, and that, at Virginia Beach, 92% of the variation was attributable to the same cause. They found, however, that at Scripps Beach about 85% of the variation was caused by a change in beach concavity and slope, but that the pivotal point about which the beach rotated was some 180cm above mean sea level, rather than about 100cm below as predicted by Quick and Har. Seymour and Castel conclude that, given these conceptual shortcomings and the computational effort required to include tides in the Quick and Har model, they could not justify a further evaluation of its predictions.

chapter fifteen

ANALYTICAL BEACH MODELS

15.1 Introduction

The number of published precedents decreases very rapidly when those beach models are sought which incorporate a theoretical approach, although to a certain extent this may simply mean that the theoretical models have not been tested and parameterised with empirical data. Such is the case with the null point hypothesis detailed in the preceding chapter which began, essentially, as an analytical model but was then empiricised with prototype data. Nevertheless, it is instructive to present and to compare three similar models which remain as purely analytical solutions to some combination of the process functions which have been developed in the preceding chapters, and which predict and explain the gradient of the orthogonal profile.

It will become apparent that none of the three models contain new insight into the problem. All, in one form or another, recognise that a horizontal seabed would represent a stable equilibrium in symmetrical or first order (Chapter Six) oscillatory currents, because any sediment transport induced by the onshore flow would be exactly balanced by return transport beneath the offshore flow. This axiomatic conclusion holds, of course, no matter which of the various transport functions, detailed in Chapter Nine, are considered to pertain, because all of those transport functions assume that the sediment transport rate is an instantaneous and unique function of the sediment parameters and the nearbed flow speed or bed stress. There can, therefore, be no lag or acceleration effects and a symmetrical flow leads, irrevocably, to a symmetrical sediment transport. All three analytical models are, however, concerned with the gradient of the seabed and it is obviously apparent to each of the authors that the inclusion of a gradient must lead, by an analogous inexorability, to a slope dependent sediment transport. Since we have seen (Chapter Nine) that an inclined surface results in an increase in the downslope transport and a decrease in the upslope transport for a particular flow condition, then the same symmetrical flow leads, necessarily, to excess downslope (usually offshore) transport until the slope is reduced to zero. All three models realise that this cannot be the case, for the orthogonal profile patently exhibits gradients which range up to the settling angle of the sediment, and they therefore attempt to stabilise the situation by the inclusion of an onshore dominated flow field. The first model attributes the required flow asymmetry to percolation losses, whilst the second and third appear content to rely upon the natural (and appropriately signed) flow asymmetry which results from the shoaling wave transformations detailed in Chapters Five and Six. Generally however, any process which, on a horizontal bed, leads to excess onshore transport provides the basis for this type of analytical solution, by balancing the gravitational effects with the flow asymmetry. Although the three models below propose and solve equations for such processes, the concepts behind the models should rightly be attributed to P.Cornaglia in 1898. Cornaglia's thesis is outlined in a speech by Munch-Peterson (1938, referenced by Komar,

1976) and it clearly contains both of the essential elements detailed above, for it seeks to balance the influence of asymmetrical waves trying to move sediment shorewards against gravity trying to pull sediment offshore in the downslope direction. The three models detailed below only differ in the sophistication of the sediment transport function and in the process to which they attribute the wave asymmetry.

15.2 Percolation Asymmetry: Inman and Bagnold (1963)

Inman and Bagnold (1963) were, perhaps, the first to attempt a theoretical analysis of the balance between gravity and flow asymmetry using a sediment transport function, although the null point hypothesis did, as we have seen in the preceeding chapter, attempt a similar route from the basis of simple threshold conditions.

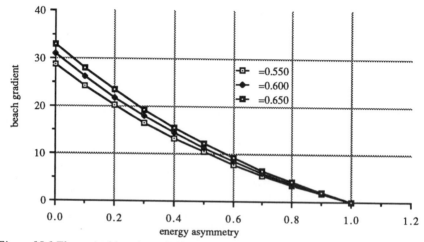

Figure 15.1 Theoretical beach gradient as a function of flow energy asymmetry predicted by Inman and Bagnold's model for three values of φ.

Inman and Bagnold consider firstly the energy expended by the swash and utilise the equality that the bedload sediment transport rate, $i_b=m_b u_b$ (Eq.9.39). Defining the distance moved by the swash sediment transport as x_1 for a swash sediment transport mass, m_1, then the Bagnold transport function, Eq.9.40, is written as:

$$\Delta E_1 = a \frac{\rho_s-\rho}{\rho_s} g\, m_1 x_1 \cos\beta\, (\tan\phi + \tan\beta) \qquad \text{Eq.15.1}$$

where ΔE_1 is the swash energy dissipation (the integral of the stream power over the swash duration) and the sign within parenthesis has been changed to account for upslope transport and a is a coefficient of proportionality analogous to e_b. Similarly the downslope transport beneath the backwash is given by:

$$\Delta E_2 = a \frac{\rho_s-\rho}{\rho_s} g\, m_2 x_2 \cos\beta\, (\tan\phi - \tan\beta) \qquad \text{Eq.15.2}$$

where ΔE_2 is the backwash energy loss, m_2 and x_2 are the backwash transports and distances and the sign has been corrected for downslope transport (Chapter Nine). Inman and Bagnold then argue that, if the profile is "mature", then onshore and offshore transports must equate:

$$m_1 x_1 = m_2 x_2 \qquad\qquad \text{Eq.15.3}$$

They also introduce the energy ratio, c:

$$c = \frac{\text{Backwash Energy Loss}}{\text{Swash Energy Loss}} = \frac{\Delta E_2}{\Delta E_1} \qquad\qquad \text{Eq.15.4}$$

Combining Eq.15.1, 15.2 and 15.3 yields:

$$\frac{\Delta E_2}{\tan\phi - \tan\beta} = \frac{\Delta E_1}{\tan\phi + \tan\beta} \qquad\qquad \text{Eq.15.5}$$

Thence solving for the equilibrium slope, $\tan\beta$, and substituting for c from Eq.15.4 yields:

$$\tan\beta = \tan\phi \frac{1-c}{1+c} \qquad\qquad \text{Eq.15.6}$$

Inman and Bagnold conclude by noting that the high percolation losses which are associated with coarse sediments would lead to a small c and therefore, from Eq.15.6, to a steep beach which is in accord with field observations (cf Chapter Fourteen). Conversely, small percolation losses, which are associated with a fine sediment, would lead to a large c and therefore, from Eq.15.7, to a flat beach which is also in accord with field observations. This approach is very sound but does not explain changes in the profile form in response to wave changes (in fact the profile gradient is independent of the waves), nor does it offer a whole profile model, only explaining the slope once the sediment and the associated energy deficit are present. Nevertheless the model has provided a useful insight into the problem and typical solutions are shown in Figure 15.1.

15.3 Hydrodynamic Asymmetry : Bowen (1980)

The most satisfactory solutions to systems models are achieved analytically, and the resulting equilibrium geomorphology is expressed in what is called phase space (Chapter Thirteen) as a function of some measure of the independent parameters. This representation should, when successfully completed, permit the workings of the model to be understood and disequilibriums to be addressed without a detailed enumeration of the time varying values of each parameter. This technique is discussed further in later chapters, and is introduced here through the beach profile model presented by Bowen (1980).

Bowen recognises not only the requirement for a slope dependent transport function but also that a flow asymmetry will then be required to stabilise a given non-horizontal seabed. He utilises Bagnold's (1963, 1966) sediment transport function and solves for equilibrium, with the condition that equal sediment masses must be transported shorewards and seawards at all depths on the orthogonal profile. He considers that the flow velocity, u, will consist of two components, the symmetrical orbital velocity, U_o, and a perturbation, U_1, and therefore expresses the flow asymmetry in the general form:

$$u = U_o + U_1 \qquad\qquad \text{Eq.15.7}$$

where the orbital velocity is taken to be a sinusoid with an amplitude u_o:

$$U_o = u_o \cos\omega t \qquad\qquad \text{Eq.15.8}$$

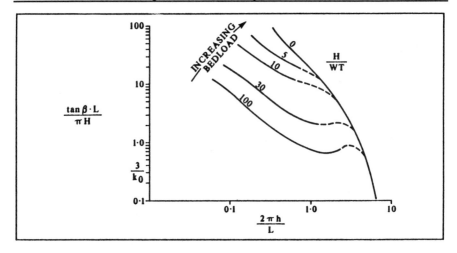

Figure 15.2 Bowen's phase-space diagram for equilibrium beach gradient as a function of the water depth, h; the wave height, H, the wavelength, L; the wave period, T and the sediment fall velocity w (after Bowen, 1980)

Bowen considers three possible reasons for the pertubation, U_1:

i $U_1=u_1$, a constant, a steady current;

ii $U_1=u_m\cos(m\omega t + \theta_m)$, the velocity field associated with a higher harmonic of the incoming wave, m=2, 3, 4 etc where u_m and θ_m are the amplitude and phase of the harmonic (Chapter Six).

iii $U_1=u_t\cos\omega_t t$, a perturbation due to a wave with frequency ω_t unrelated to ω.

and uses a general expansion for Eq.15.7 to show that, for case (i), $U_1=u_1$, and using a second order drift velocity for the perturbation (Chapter Six), the equilibrium seabed gradient under suspended load is given by:

$$\beta = \frac{5W\omega}{g \tanh(kh)}$$

Eq.15.9

where W is again the terminal fall velocity of the sediment. Bowen notes that Eq.15.9 contains the empirical sediment parameter $W\omega/g$ favoured by Dean (1973). Bowen integrates Eq.15.9 to give an equation for the orthogonal profile:

$$h \approx \frac{(7.5\ Wx)^{2/3}}{g^{1/3}}$$

Eq.15.10

which, he also notes, is remarkably similar to the empirical equation obtained by Dean's (1977) survey of orthogonal profiles around the United States (Chapter Fourteen). However, these models represent the simplest case, and Bowen goes on to include bedload and to concentrate on case (ii) wherein the flow function includes perturbations due to the generation of harmonics. He shows that the total expression for the equilibrium slope, for which the net onshore transport due to the perturbation balances the gravitationally enhanced offshore transport, is:

$$\beta = \frac{u_o}{C}\left(2+(\sinh(kh)^{-2})\right)\left(\frac{9}{5}e_s\frac{u_o}{W} + \frac{27\pi}{64}\frac{e_b}{\tan\phi}\right)\bigg/\left(\frac{4}{5}e_s\left(\frac{u_o}{W}\right)^2 + \frac{e_b}{\tan^2\phi}\right)$$ Eq.15.11

where, to recap, β is the equilibrium seabed gradient at a depth h, uo is the amplitude of the oscillatory current, C is the wave celerity, k is the local wave number, and e_b and e_s are the bedload and suspended load efficiency factors, W is the terminal fall velocity of the sedimentary particles, and tanϕ is the angle of dynamic friction of the sediment. Bowen plots results for the analytical model, Eq.15.11, as the normalised beach slope (β/ak_o) as a function of the nondimensional depth (k_oh) where a is the wave amplitude (=H/2) and k_o is the deep water wave number (=$2\pi/L_\infty$) with his phase space diagram shown in Figure 15.2. The independent controls are here the wavelength and height and the dependent, geomorphological axis is the nondimensional ratio of the seabed gradient and the wave height to the wavelength. A series of stable solutions (equilibria) are generated for a range of wave heights and periods together with the fall velocity of the sediment. The fall velocity depends upon the sediment grain size so the family of curves represents both increasing grain diameter and the tendency for material to be moved in the body of the flow. The sediment is expressed as the parameter aω/W in the Figure. As aω/W tends to zero (the fall velocity and hence the grain diameter increasing) then the equilibrium slope is a function of bedload only. For any particular value of aω/W there is a transition region from bedload dominance to fully developed suspension.

Bowen's model does not include any allowance for the shoaling transformations or details of the sediment dynamics but nevertheless this is an important solution to the problem. It clearly demonstrates that the fundamental mechanisms which lead to a dynamic profile equilibrium are the balance between onshore flow asymmetry, driving sediment shoreward, and the offshore enhancement of the sediment transport by downslope gravitational forces.

The third model which is presented here followed essentially the same developments, solving for balanced onshore and offshore sediment transport in terms of a flow asymmetry and a slope inclusive transport function. However it was used to examine the response of the beach to change and the behaviour of the system in disequilibrium.

15.4 System Stability : Hardisty (1986)
A similar approach to that detailed for the two preceeding models was followed in a series of papers by Hardisty (1986, 1987, 1990ag). In the first paper, only bedload transport was considered and an attempt was made to solve for the gradient of the beach at a nominal position close to the water's edge. Using the type of transport functions developed in Chapter Nine, the net onshore and offshore transports per wave, I_{in} and I_{ex}, were written as:

$$I_{in} = \frac{k_1 u_{in}^3 t_{in}}{\tan\phi + \tan\beta}$$ Eq.15.12

$$I_{ex} = \frac{k_1 u_{ex}^3 t_{ex}}{\tan\phi - \tan\beta}$$ Eq.15.13

A version of the second order wave theory was used to generate the flow velocities:

$$u_{in} = \frac{\gamma_b}{\sqrt{2gh_b(1+F)}}$$ Eq.15.14

$$u_{ex} = \frac{\gamma_b}{\sqrt{2gh_b(1-F)}}$$

Eq.15.15

where the asymmetry of the flow was controlled by $F=(0.01\gamma_b{}^3)/(H/L)^2$. In these equations k_1 is a calibration coefficient (found by Hardisty *et al.*, 1984 to be ≈ 12.8 kg m^{-4} s^{-2}), t_{in} and t_{ex} are the durations of the shorewards and seawards flows respectively, $\tan\phi$ is again the angle of dynamic friction of the sediment, $\tan\beta$ is the bedslope and γ_b is again the ratio of the wave height to the water depth, h_b, at the breakers. Solving Eqs.15.12 to 15.14 for the equilibrium bedslope by assuming zero net transport ($I_{in}=I_{ex}$), Hardisty (1986) obtained

$$\tan\beta = \frac{2F \tan\phi}{(1 + F^2)}$$

Eq.15.16

The results compared reasonably well with the empirical formulae of Doornkamp and King (1971) and of Dalrymple and Thompson (1976) which were given in Chapter Thirteen as shown in Figure 15.3.

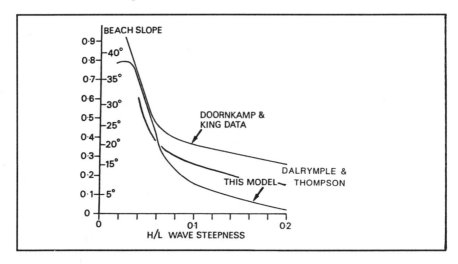

Figure 15.3 Comparison of the predictions of Eq.15.16 with the empirical models of Doornkamp and King (1971) and Dalrymple and Thompson (1976).

The analysis was extended by considering non-equilibrium conditions and plotting the response function, $\partial z/\partial t$ (Chapter Thirteen), which was calculated simply from the difference between the onshore and offshore transports per unit time:

$$\frac{\partial z}{\partial t} = \frac{I_{in} - I_{ex}}{T}$$

Eq.15.17

as shown in Figure 15.4. Hardisty suggested that the response function presentation begins to show why the beach system is stable. In the region to the right of the equilibrium curve, the beach slope is too steep and the wave steepness is too high for stability. The effect of this is that the waves produce a low velocity ratio, because the second harmonic is too small, which combines with the gravitational effects on the sediment transport rate so that offshore transport dominates. The transport ratio (I_{in}/I_{ex}) remains less than unity which in turn decreases the beach slope until the system returns to the equilibrium "valley" on the response surface. The opposite sequence of processes stabilises beach gradients which are too low for the incident wave steepness. It was also noted that the response surface was steeper below the equilibrium line than above it. This would suggest that the beach should respond more rapidly to a reduction than to an increase in wave steepnesses. Additionally, the model showed that low beach gradients are only weakly dependent on wave steepness which is in accord with the small changes observed on, for example, wide, flat low tide terraces (Jago and Hardisty, 1984). The model also shows an abrupt change in equilibrium gradient at a critical $H/L \approx 0.08$ which is similar to the values reported in some of the papers referenced in the previous chapter for the change from swell to storm profiles (e.g. King, 1972). Examination of the model suggests that this is because the flow asymmetry is exponentially rather than linearly related to wave steepness.

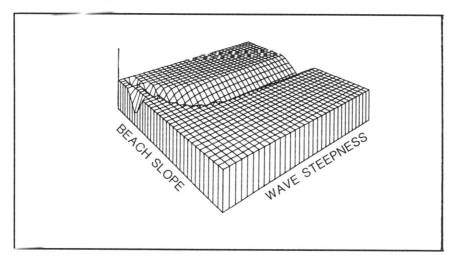

Figure 15.4 Response function plot of the disequilibrium states of the beach model for non zero net sediment transport.

In order to investigate the stability of the orthogonal profile in more detail, Hardisty (1990ag) obtained an alternative solution which did not require that the flow perturbation was due to the generation of higher harmonics, so that the peak onshore and offshore flow speeds were retained. This model utilised a slightly more sophisticated bedload function:

$$i_b = \frac{e_b \ C_d \ \rho \ u^3}{\tan\phi - \tan\beta\frac{u}{|u|}}$$

Eq.15.18

and assumed, by continuity, that $u_{in}.t_{in}=u_{ex}.t_{ex}$, (Chapter Six) to equate the net shorewards transport to the net seawards transport beneath a single wave. The resulting equations were solved for the equilibrium seabed gradient under bedload transport, β_{eb}, in terms of the velocity ratio $V_r=u_{in}/u_{ex}$ (Chapter Six):

$$\tan\beta_{eb} = \frac{\tan\phi(V_r^2 - 1)}{(V_r^2 + 1)}$$

Eq.15.19

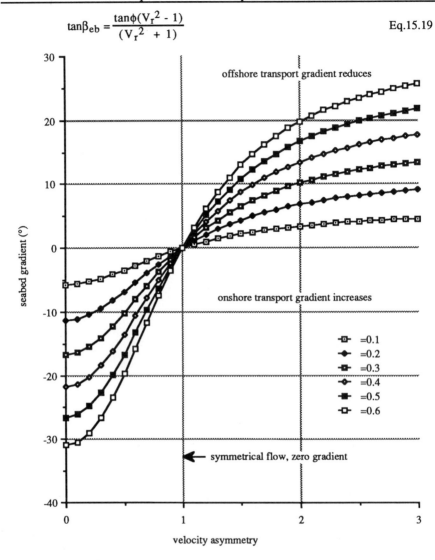

Figure 15.5 Equilibrium seabed gradient as a function of the flow asymmetry for different values of W/|U| in Eq.15.21. The results range from pure bedload (=0.6) to increasing suspended load dominance and show that the seabed gradient represents a stable attractor in equilibrium space.

In the later paper, Hardisty (1990ag) extended the analysis to include suspended load. The Bagnold type of function for suspended load transport which is analogous to Eq.15.18 is:

$$i_s = \frac{e_s \, C_d \, \rho \, u^3}{\dfrac{W}{|u|} - \tan\beta\dfrac{u}{|u|}}$$

Eq.15.20

It was assumed that the $W/|u|$ term, which should take the value $W/|u_{in}|$ for onshore transport and $W/|u_{ex}|$ for offshore transport may be replaced by a single $W/|U|$ term in both cases, where $U=(u_{in}+|u_{ex}|)/2$. This assumption is not physically unreasonable since the suspended sediment concentration profile outside the inner boundary layer has been shown to scale with a mean flow speed (Sleath, 1984), and was utilised in order to obtain amenable analytical solutions. The same analysis which was used for Eq.15.19 can then be applied to Eq.15.20 to yield the equilibrium seabed gradient under suspended load transport, $\tan\beta_{es}$:

$$\tan\beta_{es} = \frac{\frac{W}{|U|}(V_r^2 - 1)}{(V_r^2 + 1)} \qquad\qquad \text{Eq.15.21}$$

There is clearly considerable similarity between Eq.15.19 for bedload and Eq.15.21 for suspended load, and it was this similarity which prompted the further considerations detailed below. Figure 15.5 compares the bedload solution and the suspended load solution for a value $W/|U| = 0.1$. It is apparent that both $\tan\phi$ in the bedload solution and $W/|U|$ in the suspended load solution represent a measure of the erodibility of the sediment. In the former case, as the bedslope increases towards the angle of internal friction then avalanching occurs which is an infinite transport rate at zero flow speed. Conversely, in the latter case as the bedslope approaches $W/|U|$ then autosuspension occurs with the same result. In reality, the beach system attains an equilibrium under the action of both bedload and suspended load transport. The total onshore transport is given by summing Eq.15.18 and Eq.15.20:

$$I_{in} = \frac{e_b\, C_d\, \rho\, u_{in}^3 t_{in}}{\tan\phi + \tan\beta} + \frac{e_s\, C_d\, \rho\, u_{in}^3 t_{in}}{\frac{W}{|U|} + \tan\beta} \qquad\qquad \text{Eq.15.22}$$

where u_{in} is the peak onshore flow and t_{in} is the onshore flow duration. The analogous total offshore transport is given by:

$$I_{ex} = \frac{e_b\, C_d\, \rho\, u_{ex}^3 t_{ex}}{\tan\phi - \tan\beta} + \frac{e_s\, C_d\, \rho\, u_{ex}^3 t_{ex}}{\frac{W}{|U|} - \tan\beta} \qquad\qquad \text{Eq.15.23}$$

Writing again $I_{in}/I_{ex} = 1$ at equilibrium and solving for the equilibrium seabed gradient under total load transport, $\tan\beta_{et}$, Eq.15.22 and Eq.15.23 can be combined, and a little algebra yields:

$$\frac{e_b\, V_r^2}{\tan\phi + \tan\beta} + \frac{e_s\, V_r^2}{\frac{W}{|U|} + \tan\beta} = \frac{e_b}{\tan\phi - \tan\beta} + \frac{e_s}{\frac{W}{|U|} - \tan\beta} \qquad\qquad \text{Eq.15.24}$$

Neglecting second and higher order terms in $\tan\beta_{et}$ further manipulation yields:

$$\tan\beta_{et} = \frac{(V_r{}^2 - 1)(e_b \tan\phi \, \frac{W^2}{|U|^2} + e_s \, \frac{W}{|U|} \tan\phi^2)}{(V_r{}^2 + 1)(e_b \, \frac{W^2}{|U|^2} + e_s \, \tan\phi^2)} \qquad\text{Eq.15.25}$$

Senso stricto, Hardisty's (1990) analytical solution is given by Eq.15.25, but in order to introduce further realism, shoaling transformations, second order wave theory and a threshold criterion to show that the peak flow velocities and the velocity ratio increases shorewards were included. The model predicted the geomorphological and sedimentological trends of the prototype profile with the seabed gradient increasing exponentially shorewards and the grain size decreasing seawards. Further analysis showed that the model flattened the profile if either the deep water wave height increases, or the wave period decreases which is also in accord with field observations.

It is apparent that this analytical model correctly reproduced the general spatial and temporal changes which are observed on beach and nearshore orthogonal profiles and that all of these changes can be understood, albeit tenuously, by reference to the state-space representation of the relationship between the flow asymmetry and the seabed gradient. Decreased water depth or increased wave length, for example, increase the magnitude and the asymmetry of the flow which increases the equilibrium seabed gradient (Figure 15.5). The stability of the solutions shown in Figure 15.5 can be examined further because the equilibrium lines shown in the diagram represent the intersection of the I_{in}/I_{ex} surface with the $I_{in}/I_{ex}=1$ plane. Since $I_{in}/I_{ex}<1$ lies above the equilibrium lines on this plane, then offshore transport will dominate in that region leading to a reduction in the seabed gradient and a return to the equilibrium state. The suspended load and total load solutions are therefore, like the bedload solution, state-space attractors representing a stable equilibrium which, if perturbed, generate a system response which opposes the perturbation (cf Chapter Thirteen). Hardisty (1987) represented this stability for the bedload solution by plotting the modulus of the response function (Figure 15.3), suggesting that the system would "roll like a marble" into the stable equilibrium profile shape. The analogy is not strictly correct because the time history of the system depends upon the rate of change of the input and of the elevation, the sum of which will not necessarily correspond to the maximum gradient of the response surface.

These analytical representations of equilibrium solutions offer a broad understanding of the response of the system to change, but are restricted to rather simplified process functions. Hardisty (1990ag) concluded that "the equations generated by more comprehensive process functions cannot, however, be solved analytically and resort must then be made to iterative solutions, even though it is presently difficult to represent the solutions in a form which provides an acceptable understanding of the system's response". The remainder of this book is given over to numerical solutions of the equations, and to attempts to present the results in a comprehensive and comprehensible manner.

15.5 Discussion

There have, of course, been analytical models developed for the orthogonal system other than the ones described, and this review cannot, in any case, claim to be representative. For example Bailard and Inman (1980) and Bailard (1981) used essentially the same equations and techniques as Bowen (1980), but included a more sophisticated transport function. However the insight gained was no better than detailed above. Again Leont'ev (1985) detailed an analytical solution based upon the same Bagnold transport function but written:

$$i_b = \frac{e_b \tau u}{(K\tan\phi + \tan\beta)\cos\beta} \qquad\qquad \text{Eq.15.26}$$

$$i_s = \frac{e_s \tau u}{\left(\dfrac{W}{U_s} + \tan\beta\right)\cos\beta} \qquad\qquad \text{Eq.15.27}$$

Where K provides a sign for the transport being set equal to $|U_b|/U_b$ and U_b and U_s are the bedload and the suspended load particle speeds. However Leont'ev argues that the bedload transport is always on the onshore direction and is driven by asymmetries in the oscillatory current field, whilst offshore transport is always in the offshore direction and is driven by residual drift current. He thus solves for equilibrium, not by equating the onshore and the offshore transports, but by equating the time averaged values of the bedload and the suspended load transport rates at every point on the profile. The result is a concave upwards profile in the nearshore (presumably where bedload dominates) and convex in the offshore becoming ever steeper in deeper water. Despite this unrealistic result, Leont'ev continues to consider the effect of random and opposed to monochromatic waves and the formation of bars, finding some agreement with laboratory experiments.

chapter sixteen

NUMERICAL BEACH MODELS

16.1   Introduction
16.2   Fleming and Hunt (1976)
16.3   Davidsnon-Arnott (1980)
16.4   Dally and Dean (1984)
16.5   Watanabe (1988)
16.6   Martinez and Harbaugh (1989)
16.7   Discussion

16.1 Introduction

The use of computer simulation models in geology and geomorphology is well established (Dobson, 1967 and Harbaugh and Bonham Carter, 1970) and the 1970's saw early examples of their application to coastal problems. King and McCullagh (1971) used a weighted index calculation to show that a recurved spit could develop because of the interplay between two incident wave directions; Komar (1973, 1977) modelled the growth of a delta from sediment supplied at a river mouth and distributed by longshore currents, and Fox and Davis (1973) simulated nearshore storm cycles. There have also been a number of shoaling wave models and examples are given by Birkemeir and Dalrymple (1976), Southgate (1988) and Lakhan (1989). The engineering community has combined waves and sediment transport into a series of quite sophisticated "one-line" models. The name refers not to the complexity of the computer coding, but rather to the fact that these models are almost invariably concerned with longshore transport and with the resulting shoreline evolution. The output is, therefore, a single line which is a plan view of the water's edge. Examples of one line models are given by Price *et al.* (1973), Rea and Komar's (1975) study of hooked beaches, Horikawa *et al.'s* (1979) examination of the effect of dredging, Willis's (1977) study of longshore transport, Sasaki and Sakamoto (1978), Perlin and Dean's (1979) work on detatched breakwaters, Le Mehaute and Soldate (1977, 1980), Mimura *et al.* (1983), Kraus (1983), Hanson and Kraus (1986) and Hanson's (1989) description of the GENESIS program.

In this chapter, however, we wish to consider only those numerical models which specifically predict the form or evolution of the orthogonal profile. The author is aware of at least five numerical models which, at the present time, are apparently capable of generating these geomorphological predictions. It is likely that there are more models in existence, but the purposes of this chapter can nevertheless be served with those shown here. The models are both multi-national and multi-disciplinary in origin and objectives.

The first is British and is due to the engineers Fleming and Hunt (1976). It is concerned with a site specific, three-dimensional simulation which includes both bedload and suspended load transport. The second model is due to the Canadian Davidson-Arnott (1981) and specifically simulates the geomorphological evolution of the orthogonal profile. The third, due to the American engineers Dally and Dean (1984), is concerned with the transport of sediment as suspended load and predicts profile evolution over short time intervals. The fourth was developed in Japan and is reported by Watanabe (1988d). It is designed to facilitate engineering work in the nearshore environment. The fifth model is also American, and is reported by Martinez and Harbaugh (1989). It is concerned with the evolution of the three dimensional form of nearshore, clastic systems over the longer periods of geological time. Another model by Wang *et al.* (1975) is not detailed here because it proved difficult to trace and is said, by Dally and Dean (1984), to be similar to that of and Fleming and Hunt (1976). The work reported in Chappell and Elliott (1979) and Howd and Holman (1987) is discussed in section 16.7.

16.2 Fleming and Hunt (1976)

An early example of a numerical solution to the process equations is presented by Fleming and Hunt (1976b) using a development of an earlier, unidirectional sediment transport model given by Fleming and Hunt (1976a). The wave refraction elements of the model are relatively routine, but the sediment transport analysis is, in this case, particularly sophisticated and contains many of the elements of the transport functions used in later chapters in the present book. The sediment dynamics assume that the sediment is transported in a bedload region within which grains are supported by inter-granular collision and in a suspended load region wherein gravitational forces are overcome by fluid turbulence. These are, essentially, the same principles used by Bagnold and described in Chapters Nine and Ten. In the Fleming and Hunt (1976a, b) model the "dispersed shear stress", τ_d, taken to be responsible for bedload transport, is given by an excess stress function:

$$\tau_d = \tau_w - \tau_{cr} \qquad \qquad \text{Eq.16.1}$$

where τ_w is the mean wave stress which is taken to be proportional to the square of the peak, nearbed orbital velocity and τ_c is the critical shear stress. The velocity distribution within the bedload layer is taken to be given by a fluid form of the shear stress equation (Chapter Nine):

$$\tau_d = \mu_s \frac{du}{dz} \qquad \qquad \text{Eq.16.2}$$

where μ_s is said to be an "effective viscosity" of the fluid sediment mixture. The bedload transport rate is calculated by integrating the product of the velocity of the sediment with a bedload concentration function:

$$C = C_m \left(\frac{C_e}{C_m} \right)^{z/e} \qquad \qquad \text{Eq.16.3}$$

where C_m is the maximum grain concentration (assumed to be 0.65, cf Chapter Three), and C_e is the concentration at an interface height, $z=e$, and z is distance above the stationary grain layer. The suspended load concentration profile is taken to be given by another eddy diffusivity relation (similar to that given in Chapter Ten) where the eddy viscosity is not taken as constant, but rather is given by:

$$\frac{\varepsilon}{\varepsilon_e} = \left(\frac{z}{e} \right)^b \qquad \qquad \text{Eq.16.4}$$

where ε_e and ε are the eddy viscosities at the interface height e and at the height z, and b is a positive constant. The two concentration profiles are equated at the interface height to calculate the net sediment transport. In the full model, waves are refracted across a nearshore grid and the resulting bed stresses are used to drive the sediment transport functions. Within the breaker zone, defined by the ratio of the wave height to water depth as in Chapter Six, a longshore current based upon Longuet-Higgins (1970a and b) drives a longshore transport function. The results were used with field measurements of the seabed bathymetry and local wave regime to examine several different proposed breakwater configurations at a site in South Africa. This model exemplifies a very sophisticated presentation of the transport functions, and although not the objective of the work, orthogonal profile evolutions could have been examined from the numerical experiments.

16.3 Davidson-Arnott (1981)

Davidson-Arnott (1981) was perhaps the first author to attempt to include at least some of the processes in each of the three groups shown in Table 2.1 and, more importantly, to consider explicitly the generation of an equilibrium profile. The model represented 150 points spaced at 5m intervals, thus extending from the shoreline to 750m offshore. Wave height shoaling and the maximum orbital velocity at the bed were calculated using linear theory and allowance was made for energy loss due to seabed friction and to breaking. Sediment transport was driven by a second order mass transport current and an empirical function. It is interesting that, although using essentially monochromatic theory for each iteration (that is wave) in the model, the height and period were varied randomly by ±20% to "make the characteristics of the wave incident on the beach more realistic".

One of the problems which arises with any numerical models is the choice of initial geomorphological conditions. Davidson-Arnott is not the only author to choose an initially planar surface. A typical result showing the development of a barred profile due to transport towards the point of maximum wave height at the breakpoint is shown in Figure 16.1. These input parameters did not generate an equilibrium profile in this example, and the bar migrated seaward until it disappeared after 1000 iterations. Other conditions did, however, generate stable profiles and Davidson-Arnott uses the model to draw qualitative conclusions about the effect of tidal range, wave parameters and pre-existing geomorphology on the profile. Perhaps the model's most serious shortcoming is the lack of the bedslope control on the transport rate.

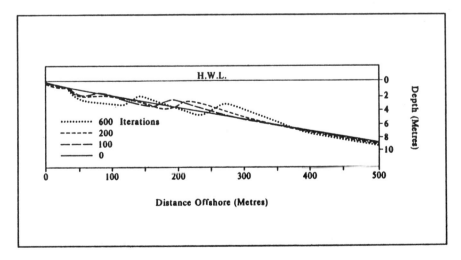

Figure 16.1 Example of Davidson-Arnott's model of nearshore bar formation for the first six hundred iterations showing the development and shoreward migration of a breakpoint bar on the orthogonal profile (after Davidson-Arnot, 1981)

231

16.4 Dally and Dean (1984)

This model computes the suspended load transport rate using a radiation stress approach and then uses mass continuity to accrete and to erode sediment from adjacent cells as the wave shoals from deep water to the shoreline. In detail, the time averaged net flux of suspended sediment (cf Chapter Ten) is given (using the sysmbols employed here) by Dally and Dean as:

$$I_s = \int_{z=-h}^{z=0} U_{sz}\, C_z \; dz \qquad\qquad 16.5$$

where U_{sz} is the mean horizontal velocity of the sediment and C_z is the sediment concentration profile. U_{sz} is taken to have two components, and the contribution of each varies according to the height above the bed. Within the lower layer, particles are assumed to be raised to an initial elevation by the passage of a crest, but to fall back to the bed before the arrival of the next crest. The motion of the sediment is therefore asymmetric even though first order, symmetrical wave theory is used, and oscillatory flow contributes to net sediment transport. In the upper layer, defined as being above a boundary at which particles return to the sea bed during the passage of each wave, the net oscillatory motion is zero and therefore does not contribute to sediment transport. In both layers the second order mean drift flow is taken to transport suspended sediment and hence to contribute to U_{sz}. Using first order theory, Dally and Dean find that the average horizontal particle velocity due to the oscillatory component was:

$$U_{sz} = \frac{H}{2\omega} \sqrt{\frac{g}{h}} \sin \frac{\omega(h+z_i)}{W} \qquad\qquad Eq.16.6$$

and due to the residual flow was:

$$U_{sz} = \frac{gh}{8\varepsilon} \frac{\partial H^2}{\partial x} \left(\frac{-3}{8} \left(\frac{z}{h}\right)^2 - \frac{1}{2}\left(\frac{z}{h}\right) - \frac{1}{8} \right)$$
$$+ \frac{u_o 2k}{2\omega} \left(\frac{3}{2} \left(\frac{z}{h}\right)^2 - \frac{1}{2} \right) - \frac{3Q}{2h} \left(\left(\frac{z}{h}\right)^2 - 1 \right) \qquad Eq.16.7$$

where Q was the net flow between the bottom and the bed calculated from stream function theory (Dean, 1974):

$$Q = \psi_w \frac{gHT}{h} \qquad\qquad Eq.16.8$$

where ψ_w was the surface stream function taken from tables for particular local wave steepnesses and relative depths. In Eq.16.6 the gradient of the radiation stress, $\partial H^2/\partial x$, is evaluated "using a model that quantifies the spatial changes in wave height due to shoaling, breaking, and reformation as the wave crosses the surf zone". It is regrettable that more details are not given for this model, because it effectively accounts for all of the hydrodynamic processes which have been described in Section B of the present book, and completely controls the results of the Dally and Dean model. Nevertheless, the suspended sediment concentration profile is taken to be Rousian (Chapter Nine, Rouse, 1938) and to be given by:

$$C_z = C_a \exp \left(\frac{-15W(z-z_a)}{h \sqrt{\frac{\tau}{\rho}}} \right)$$
Eq.16.9

in which $\sqrt{(\tau/\rho)}$ is evaluated separately for breaking and non-breaking waves. The model, therefore, computes the suspended sediment flux by determining the integral in Eq.16.5 from the mean particle velocities in the lower and upper layers (Eq.16.7 and 16.8 and from the concentration profile (Eq.16.9). The authors finally approach the central issue in the use of a concentration profile for the suspended sediment problem (which was discussed in Chapters Nine and Ten) which is the choice of an appropriate reference height, z_a, and concentration, C_a, for the Rousian profile. They find that: "determining the optimum reference elevation and sediment concentration for the surf zone fluid regime is a heretofore unsolved problem that will require extensive investigation. However, with the bottom of the water column somewhat arbitrarily chosen as the reference elevation, if the reference concentration is assumed to be unifrom across the surf zone, it can be absorbed by the time step of the computer model". In effect the concentration was inversely scaled with the time step to achieve qualitatively accurate results.

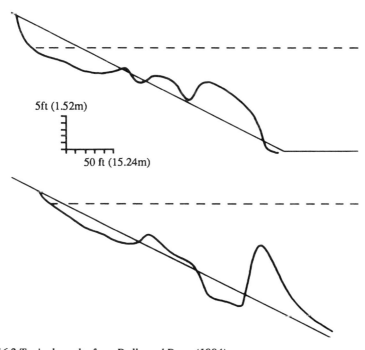

5ft (1.52m)

50 ft (15.24m)

Figure 16.2 Typical results from Dally and Dean (1984)

The model was run with a range of inputs on a constant initial profile of 1:15, and typical results are shown in Figure 16.2. The authors summarise their results in terms of the dependence of profile type on waves and sediment characteristics, the profile shape, and the rates of profile evolution:

Dependence of Profile Type on Wave and Sediment Characteristics
The results shown in Figure 16.2 evidence the prototype range from bar to berm profile,

and indicate that reducing the wave height or increasing the sediment fall velocity reduces the tendency for bar growth. Dally and Dean interpret this as being due to a reduction of sediment concentration in the upper layer of the water column. This is the layer in which there is return flow due to the second order drift current. There is, therefore, less offshore transport and bars are not constructed.

The Profile Shape
Dally and Dean note the comparison between the normal, monotonic and concave upwards profiles (Figure 16.2), and the barred profile which, they find, results directly from the wave breaking process. Multiple bars are generated when waves reform shorewards of an outer bar. They do not detail the bar growth processes, but it is apparently due to convergent transport following a mechanism similar to that discussed above (see also Dally and Dean's Figure 6). The generation of the monotonic, concave upwards profile from an initial plane surface, can only be due to the shorewards sediment transport rate increasing in a shorewards direction, due presumably to increasing drift current velocities.

The Rates of Profile Evolution
The authors do not detail the profile evolution rates quantitatively, but do observe that the rate of offshore transport, bar formation and upper beach erosion exceeds the subsequent rate of "recovery" towards a normal profile.

In conclusion, Dally and Dean (1984) have presented a very interesting numerical model, which has answered many of the questions about orthogonal system evolution posed earlier. Their emphasis on suspended sediment concentration, to the exclusion of bedload, may be unrealistic, as are the choice of reference concentrations and the use of monochromatic, linear theory in such shallow water. Nevertheless the profile evolution, and particularly the identification of the process of sediment transfer into the upper layer, within which there is a steady return flow, as being responsible for bar formation is an alternative to the gravitational effects invoked by the analytical solution discussed in the previous chapters, and is also in accord with most empirical results.

16.5 Watanabe (1988)
The theoretical basis for the three-dimensional bathymetric model which is here attributed to Watanabe (1988d) can be found in the first three sections of the edited volume by Horikawa (1988) and is summarised by Watanabe *et al.* (1986) and more fully by Watanabe's own chapters in Horikawa, Watanabe (1988a, b, c and d).

The model is based on a complex representation of the wave refraction, diffraction, breaking and reflection processes which is detailed in Watanabe and Maruyama (1986) and Watanabe (1988b). The solutions are based upon a finite difference method to the so called time-dependent mild slope equations, which essentially seek to balance the radiation stress with energy loss terms at all points in the nearshore, and the result provides an amplitude to the oscillatory current field and the velocity of the residual current field. Watanabe (1988b) shows that there is good agreement between this model and results for monochromatic waves in a laboratory wave tank. The computation of the sediment mass transport divides into consideration of the movement due to the residual currents predicted by the wave model and of the orthogonal transport due to oscillatory currents generated by those waves. The residual current sediment transport rates i_c (shore normal) and j_c (shore parallel are modelled in terms of the volume of sediment motion per unit area, Q_c, using:

$$i_c = Q_c U \qquad\qquad \text{Eq.16.10}$$

$$j_c = Q_c V \qquad\qquad \text{Eq.16.11}$$

$$Q_c = A_c \frac{\tau - \tau_{cr}}{\rho g} \qquad\qquad \text{Eq.16.12}$$

where τ and τ_{cr} are are the maximum and threshold values of the bed stress in the combined wave and current field and A_c is a non-dimensional coefficient "of the order of 0.1 to 1" (Watanabe, 1988c). Net sediment transport between cells in the three dimensional array is computed from these transport vectors. Similarly the purely wave induced transport is again computed from the excess stress equation (Chapter Nine and Ten):

$$i_w = Q_w u_o \cos\alpha \qquad\qquad \text{Eq.16.13}$$

$$j_w = Q_w u_o \sin\alpha \qquad\qquad \text{Eq.16.14}$$

$$Q_w = A_w \frac{\tau - \tau_{cr}}{\rho g} \qquad\qquad \text{Eq.16.15}$$

where u_o is the amplitude of the nearbed current given by linear theory from the amplitude of the surface elevation given by the wave model, and τ_{cr} is taken to correspond to a Shield's Parameter of 0.11 for fine sands or 0.06 for coarse sands.

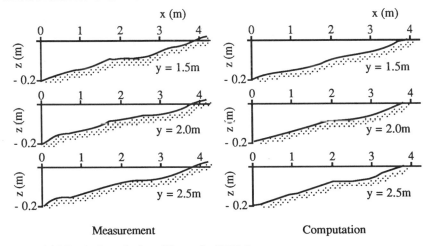

Measurement   Computation

Figure 16.3 Typical results from Watanabe (1988d)

The results of the model were compared with laboratory experiments in a wave tank (Watanabe, 1988d). The experiments consisted of running 4.5cm waves with a period of 0.87 s across an initial bathymetry composed of 0.2mm sand at an initial slope of 1/20. The geomorphological results of a comparison between the model and the laboratory profiles at t =2:37 with a 90s computation time interval are shown in Figure 16.3.

There is clearly some agreement between the response of the model and the prototype, although it is difficult to gauge the results from the descriptions given in the published reports. Again, it is apparent that the wave field model is extremely sophisticated, and yet has been matched to a relatively simplistic transport function. That the results compare well is probably due to the fact that the main bathymetric response appears to be the construction of a bar and then a concave profile by the model on a planar surface. The other work which has been reported here suggests that such is the almost inevitable consequence of almost any transport function which relates mass movements to the current speed, because it is largely governed by the energy loss during breaking, and therefore the ability of Watanabe's representation of the wave field predicates any shortcomings of the transport functions in the comparison with prototype profiles.

16.6 Martinez and Harbaugh (1989)

This solution consists of two separate computer programs which are linked together through a text editor. The first program, WAVE, is written in Fortran'77 and is a finite difference model which simulates nearshore circulation patterns due to waves and wind, and due to longshore and rip currents. The version of WAVE used by Martinez and Harbaugh was modified from a version reported by Ebersole and Dalrymple (1979) which itself included subroutines formulated by Noda *et al.* (1974). The output of WAVE was combined with another program called SEDSIM which is described by Tetzlaff and Harbaugh (1989), and itself consisted of two major subprograms called SEDCYC and SEDSHO. SEDCYC simulates erosion, transport and deposition whilst SEDSHO controls the graphical display of the results of the simulations. The operation of WAVE and SEDCYC are described in the following paragraphs from Martinez and Harbaugh (1989) because it has proven difficult to obtain copies of the other three references.

The WAVE program uses the radiation stress approach described earlier, and in its most general form, balances fluid accelerations with the applied force (or rate of change of momentum:

$$\frac{\partial U}{\partial t} = -g\frac{\partial \eta}{\partial x} - \frac{1}{\rho(h+\eta)}\left(\frac{\partial S_{xx}}{\partial x} + \frac{\partial S_{xy}}{\partial y} + \tau_{bx} - \tau_{sx}\right) \qquad \text{Eq.16.16}$$

$$\frac{\partial V}{\partial t} = -g\frac{\partial \eta}{\partial x} - \frac{1}{\rho(h+\eta)}\left(\frac{\partial S_{xy}}{\partial x} + \frac{\partial S_{yy}}{\partial y} + \tau_{by} - \tau_{sy}\right) \qquad \text{Eq.16.17}$$

where the components of the radiation stress are given (after Longuet-Higgins and Stewart (1964) cf Chapter Seven) by:

$$S_{xy} = \frac{E}{2}n\sin\theta \qquad \text{Eq.16.18}$$

$$S_{xx} = E\left(\frac{2kh}{\sinh(kh)} + \frac{1}{2}\right) \qquad \text{Eq.16.19}$$

$$S_{xx} = E\left(\frac{kh}{\sinh(kh)}\right) \qquad \text{Eq.16.20}$$

In these equations, as before, the x and y directions are parallel to and normal to the direction of wave advance, S_{xx} and S_{yy} are the radiation stresses or excess momentum fluxes along the x and y axes, k is the wave number, E is the wave energy taken as $E=1/8$ $\rho g H^2$, τ_s is the stress at the water surface due to the wind, τ_b is the shear stress at the seabed and the subscripted stresses refer to the two axes, h is the water depth measured positively upwards from the sea surface, α is the wave direction, n is the ratio of the wave group and phase celerities (Chapter Five), η is the mean water level, g is the gravitational acceleration and ρ is the fluid density. The depth mean average currents, U and V along the two co-ordinate axes are then calculated at given time intervals for a grid of points across

236

the bathymetric grid and are passed to the SEDCYC program.

The SEDCYC program is probably the most computationally demanding, yet conceptually simple, implementation of the sediment transport processes of any of the models described in this chapter. A two dimensional array representing five thousand fluid elements in the bathymetric grid is read from WAVE and contains the positions and velocities of each fluid element, together with its sediment load at each time increment. The program also calculates the local seabed gradient for each point.

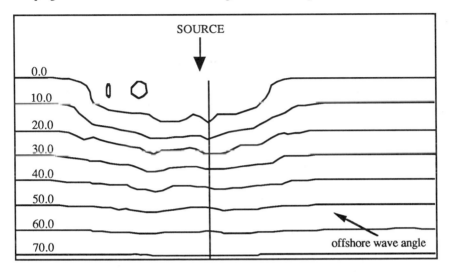

Figure 16.4 Typical results from Martinez and Harbaugh (1989)

Given this information the program calculates the transport capacity, λ_T, of each fluid element from:

$$\lambda_T = C_t \, \tau_o \, \frac{Q}{h}$$

Eq.16.21

where the bed stress, τ_o, is given by a quadratic stress law (Chapter Nine):

$$\tau_o = \frac{1}{8} \rho \, f \, Q^2$$

Eq.16.22

and C_t is a "transport coefficient (m^3 kg^{-1})" which "must be calibrated", Q is the vectorial mass transport velocity (U+V) from Eqs. 16.16 and 16.17 above, f is the Darcy-Weisbach friction factor ("set equal to between 0.01 and 0.08 in WAVE"), ρ is the fluid density and h is the water depth. It is apparent that, since Q is a vector current velocity, the transport function is effectively a cubic relationship with respect to flow speed and includes neither threshold nor slope inclusive effects and does not differentiate between bedload and suspended load transports. If the transport capacity exceeds a fluid element's sediment concentration, sediment is eroded from the nearest grid cell. If a fluid element's transport capacity is less than its sediment concentration, sediment is deposited at the nearest grid cell. The resulting change in bathymetry is returned to the WAVE program and the whole procedure is repeated.

The results of the combined WAVE and SEDSIM programs are used to demonstrate the difference between deltaic accretion over a period of 2000 years in a wave free and a wave controlled environment as shown in Figure 16.4. It can be seen that the control simulation without waves forms a small lobate delta which extends some 8km offshore, whereas the wave dominated delta is aggrades less and shows marked differences between its updrift eastern and downdrift western boundaries. Taking orthogonal profiles along the direction of wave advance it appears that the presence of the waves leads to a steeper nearshore gradient, but that both the wave dominated and the control profiles are concave upwards as has frequently been found by the empirical, analytical and numerical techniques described in the present and in earlier chapters.

16.7 Discussion

A computer based, numerical model for the orthogonal system has also been presented by Chappell and Eliot (1979) who developed solutions for nearshore bed disequilibrium and response based on a work distribution hypothesis. They simulate an equilibrium offshore profile by equating bed frictional losses to wave height changes and by regularising these in an onshore direction. Thence four equations give water depth with distance onshore and hence the profile. Although the model gives good results offshore from the breakers, Chappell and Elliot note that it is only applicable under Airy wave theory and thus only to the region beyond the breaker zone. Inshore the model is made to continue by assigning values of $H = P\ H_{\infty}$ to the 'effective' wave height in the surf zone where P is an iterative constant. The authors use the model to generate beach profiles having various designated values of P to show that a very narrow surf with surging breakers results in a steeper overall beach face of typical swell profile, whereas a lower value of P, associated with continually spilling breakers produces a flatter, storm type beach profile.

This summary of some of the various numerical models which have been developed, together with the empirical and analytical models which were detailed earlier, provides the introduction to experiments with a new numerical model for the orthogonal system in the next chapter.

SECTION E

NUMERICAL MODELLING

chapter seventeen

THE SLOPES MODEL

17.1 Introduction

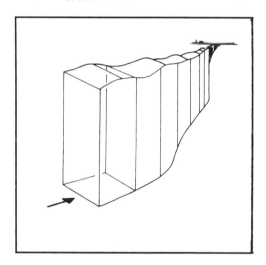

Figure 17.1 Schematic overview of the SLOPES model described in the present chapter.

Although the orthogonal system is one of the most dynamic geomorphological environments, responding constantly to hourly, daily and seasonal changes in the incident waves, and although numerous papers have been written which describe the profile, there have been few attempts to predict the profile from an understanding of the physics of the relevant processes. The process equations have been summarised in this book, and some analytical and numerical solutions have been described in the preceeding chapters. In order to investigate the orthogonal profile, the processes have been assembled into a micro-computer model running in Microsoft Excel. This chapter details the design and implementation of the model, and the results of sensitivity analyses are described in Chapters Seventeen and Eighteen.

17.2 The SLOPES Model

Numerical solutions to the process equations have been constructed using the Microsoft Excel package. The model, which is summarised in Appendix IV is called SLOPES, an acronym for the Shoreline and Orthogonal Process Emulation System. Excel is a spreadsheet environment and therefore the spatial scale (depth) is shown by Rows 3 to 39 with increments of 0.1 m, although the increments can be easily changed in the model. The columns, from left to right, in the model represent the process functions and are coded in the sequence described in Chapter Two (Table 2.1) and in the subsequent chapters. The model is designed so that each of the processes can be switched into or outoff the final simulations. The reasons for this are detailed in section 18.1. The model runs in Excel's "command macro" mode which allows the resulting sediment transport to feedback and alter the initial bathymetry before re-iterating all of the processes. Screen displays for the whole model are shown in Figure 17.2. Excel is a spreadsheet environment, and the operation of each column in the spreadsheet is described in this chapter.

COLUMN A. Program and revision number displays.

COLUMN B Blank spacer

COLUMN C-E Constants The wave period (5s), wave height (1m), wave direction (15°) are input. The pseudo constants are resident and include the grain size (0.1cm), the attenuation coefficient (0.75), the wave friction factor (0.01), $\tan\phi$ (0.6), ρ_s (2.65 gm cm^{-3}), ρ (1.0250 gm cm^{-3}) and the sediment conentration, C (0.63). Cell D5 calculates L_∞ from Eq.AII.1 (Appendix II), cell D7 calculates C_∞ from Eq.AII.2 and cell D16 calculates the sediment bulk density, ρ_b, from Eq.AII.3.

COLUMN F Blank spacer

COLUMN G and H The Control Panel. Permits the shoaling wave transformations, the flow characteristics and the sediment dynamic processes to be switched on and off. The default settings are H=H$_\infty$, u(t)=Airy Wave Theory no refraction and I = ,0,T i$_b$ dt with u$_{cr}$=0.

COLUMN I The depth of the seaward limit of each cell from h=0.2m to h=10.00 m at increments of 0.2m, plus the tidal depth given by cell $D$17.

COLUMN J The distance off (x) of the seaward limit of each cell computed from Eq.17.3 (Eq.AII.4).

COLUMN K The gradient of the cell calculated from h/x which is equivalent to Eq.17.4 (Eq.AII.5).

COLUMN L The initial sediment grain diameter taken from $D$12. This could, of course, be altered either during modelling or as a separate independent control.

COLUMN M Wave friction factor. Taken as a constant from $D$12. This could, of course, be altered either during modelling or as a separate independent control, or as a defined function of sediment grain size or bedform type.

COLUMN N The local wavelength is calculated iteratively from the separate "WAVELENGTH" macro (Figure 17.3) using the wave dispersion equation, Eq.5.8 (AII.6).

COLUMN O Calculates kh = $2\pi h/L$.

	A	B
1		
2	SLOPES	
3		
4	Program Number WM_005	
5	Version Number 30:7:90	
6		
7		
8		
9		
10		
11		
12	Software and Packaging is © Unico Geosystems Ltd, 1990	
13		
14	Unico Geosystems Ltd	
15	School of Earth Resources	
16	University of Hull, HU6 7RX, U.K.	
17		
18	Telephone 0482 465374	
19	Fax: (0482) 466340	
20		
21		
22		
23		
24		
25		
26		
27		
28		
29		
30		
31		
32		
33		
34		
35		
36		
37		
38		
39		

	C	D	E	F	G	H	I	J	K
1					CONTROL PANEL		h	distance	gradient
2	CONSTANTS		units				m	m	
3					DEFAULT : constant height	off	0.500	17	0.0294
4	Wave Period	5.0	seconds		Shoaling height	on	0.600	23	0.0167
5	Deep Water wave length	39.0	m		Frictional loss	on	0.700	29	0.0167
6	Deep water wave direction	15.0	degrees		Refraction loss	on	0.800	35	0.0167
7	Deep water wave celerity	7.8000	m/s		Breaking loss	on	0.900	42	0.0143
8	tidal phase	0.0000	°				1.000	49	0.0143
9	Deep water wave height	1.5000	m				1.100	56	0.0143
10	Wave friction factor	0.0100					1.200	64	0.0125
11	Attenuation coefficient	0.7500			DEFAULT : first-order flow	off	1.300	72	0.0125
12	sediment grain diameter	0.1000	cm		Second-order flow	on	1.400	81	0.0111
13	tan phi	0.6000			Second-order drift	on	1.500	89	0.0125
14	sediment density	2.6500	gm/cm^3				1.600	98	0.0111
15	sediment concentration	0.6300					1.700	108	0.0100
16	sediment bulk density	1.6695	gm/cm^3		DEFAULT : Integral bedload transport	off	1.800	118	0.0100
17	tidal height,	0.0000	m		Suspended load transport	on	1.900	127	0.0111
18	density of seawater	1.0250	gm/cm^3		Threshold inclusive transport	on	2.000	138	0.0091
19	M2	2.2320	m		Slope inclusive transport	on	2.100	148	0.0100
20	S2	0.7880	m				2.200	159	0.0091
21	tidal phase	-4.0000	°				2.300	170	0.0091
22	tidal height	0.0000	m		N.B. Defaults reset automatically		2.400	181	0.0091
23							2.500	192	0.0091
24	bedload k	0.000001	gm cm-4 s-2				2.600	204	0.0083
25	fluid viscosity	0.010	cgs				2.700	216	0.0083
26							2.800	228	0.0083
27	Time steps	24.000	hrs				2.900	240	0.0083
28	RJSW m	1.000					3.000	253	0.0077
29	gamma b	0.500					3.100	266	0.0077
30							3.200	279	0.0077
31							3.300	292	0.0077
32							3.400	305	0.0077
33							3.500	319	0.0071
34							3.600	332	0.0077
35							3.700	346	0.0071
36							3.800	361	0.0067
37							3.900	375	0.0072
38							4.000	389	0.0071
39							5.000	544	0.0065

	L	M	N	O	P	Q	R	S	T	U	V	W	X
1	grain diam.	friction		W'length	kh	h/L∞		cosh(kh)	sinh(kh)	tanh(kh)	cosh(2kh)	sinh(2kh)	tanh(2kh)
2	cm	factor		m									
3	0.10	0.0100		11.87	0.2646	0.0128		1.0352	0.2678	0.2587	1.1435	0.5546	0.4850
4	0.10	0.0100		12.88	0.2928	0.0154		1.0432	0.2972	0.2848	1.1766	0.6200	0.5269
5	0.10	0.0100		13.70	0.3211	0.0179		1.0521	0.3268	0.3107	1.2136	0.6877	0.5666
6	0.10	0.0100		14.38	0.3495	0.0205		1.0618	0.3569	0.3361	1.2547	0.7579	0.6040
7	0.10	0.0100		14.96	0.3780	0.0231		1.0724	0.3873	0.3612	1.3000	0.8306	0.6390
8	0.10	0.0100		15.46	0.4065	0.0256		1.0839	0.4181	0.3857	1.3496	0.9063	0.6715
9	0.10	0.0100		16.81	0.4111	0.0282		1.0858	0.4231	0.3896	1.3580	0.9188	0.6766
10	0.10	0.0100		17.14	0.4398	0.0308		1.0984	0.4544	0.4137	1.4130	0.9983	0.7065
11	0.10	0.0100		17.43	0.4685	0.0333		1.1119	0.4862	0.4373	1.4728	1.0813	0.7342
12	0.10	0.0100		18.57	0.4737	0.0359		1.1145	0.4920	0.4414	1.4841	1.0966	0.7389
13	0.10	0.0100		18.75	0.5025	0.0385		1.1291	0.5243	0.4644	1.5498	1.1840	0.7640
14	0.10	0.0100		19.78	0.5081	0.0410		1.1321	0.5306	0.4687	1.5632	1.2015	0.7686
15	0.10	0.0100		19.89	0.5370	0.0436		1.1479	0.5636	0.4910	1.6352	1.2938	0.7912
16	0.10	0.0100		20.83	0.5429	0.0462		1.1512	0.5704	0.4955	1.6507	1.3134	0.7956
17	0.10	0.0100		20.88	0.5718	0.0487		1.1682	0.6039	0.5170	1.7295	1.4111	0.8159
18	0.10	0.0100		21.74	0.5781	0.0513		1.1720	0.6113	0.5216	1.7473	1.4329	0.8200
19	0.10	0.0100		21.74	0.6070	0.0538		1.1902	0.6454	0.5423	1.8332	1.5364	0.8381
20	0.10	0.0100		22.53	0.6136	0.0564		1.1945	0.6532	0.5469	1.8535	1.5606	0.8420
21	0.10	0.0100		22.49	0.6425	0.0590		1.2139	0.6881	0.5668	1.9469	1.6704	0.8580
22	0.10	0.0100		23.23	0.6493	0.0615		1.2186	0.6963	0.5714	1.9698	1.6971	0.8615
23	0.10	0.0100		23.16	0.6781	0.0641		1.2392	0.7318	0.5906	2.0712	1.8138	0.8757
24	0.10	0.0100		23.84	0.6852	0.0667		1.2444	0.7406	0.5951	2.0969	1.8431	0.8790
25	0.10	0.0100		24.50	0.6924	0.0692		1.2498	0.7496	0.5998	2.1239	1.8738	0.8822
26	0.10	0.0100		24.39	0.7212	0.0718		1.2719	0.7860	0.6180	2.2356	1.9995	0.8944
27	0.10	0.0100		25.00	0.7287	0.0744		1.2778	0.7955	0.6226	2.2657	2.0331	0.8973
28	0.10	0.0100		25.60	0.7364	0.0769		1.2840	0.8054	0.6272	2.2972	2.0681	0.9003
29	0.10	0.0100		25.46	0.7651	0.0795		1.3077	0.8426	0.6444	2.4200	2.2038	0.9106
30	0.10	0.0100		26.01	0.7730	0.0821		1.3143	0.8529	0.6489	2.4549	2.2420	0.9133
31	0.10	0.0100		26.55	0.7810	0.0846		1.3212	0.8635	0.6536	2.4914	2.2819	0.9159
32	0.10	0.0100		26.38	0.8097	0.0872		1.3465	0.9018	0.6697	2.6264	2.4285	0.9247
33	0.10	0.0100		26.89	0.8179	0.0897		1.3540	0.9128	0.6742	2.6665	2.4719	0.9270
34	0.10	0.0100		27.38	0.8262	0.0923		1.3617	0.9242	0.6787	2.7084	2.5170	0.9293
35	0.10	0.0100		27.20	0.8548	0.0949		1.3886	0.9635	0.6938	2.8566	2.6759	0.9367
36	0.10	0.0100		27.66	0.8633	0.0974		1.3969	0.9753	0.6982	2.9026	2.7249	0.9388
37	0.10	0.0100		28.10	0.8720	0.1000		1.4055	0.9876	0.7027	2.9506	2.7760	0.9408
38	0.10	0.0100		28.53	0.8808	0.1026		1.4142	1.0000	0.7071	3.0002	2.8286	0.9428
39	0.10	0.0100		30.47	1.0309	0.1282		1.5810	1.2245	0.7745	3.9989	3.8719	0.9682

Y	Z	AA	A B	AC	A D	AE	AF	AG	AH	AI	AJ	AK
1	C local	n		H deep	H shoal	local dH fr	total dH fr	H-dH fr	incidence	refraction	H∞/Hs -	state
2	m/s			m	m	m	m	m	degrees	loss	dH f - dHr	
3	2.37	0.9771		0.00	1.94	0.05	0.28	1.67	4.52	0.03	1.64	broken
4	2.58	0.9722		0.00	1.87	0.03	0.23	1.64	4.90	0.03	1.61	broken
5	2.74	0.9669		0.00	1.82	0.02	0.20	1.62	5.22	0.03	1.59	broken
6	2.88	0.9612		0.00	1.78	0.02	0.18	1.60	5.48	0.03	1.58	broken
7	2.99	0.9550		0.00	1.75	0.02	0.16	1.60	5.70	0.03	1.57	broken
8	3.09	0.9485		0.00	1.73	0.01	0.14	1.59	5.89	0.03	1.57	broken
9	3.36	0.9475		0.00	1.66	0.01	0.12	1.53	6.41	0.02	1.51	broken
10	3.43	0.9406		0.00	1.65	0.01	0.11	1.54	6.53	0.02	1.51	broken
11	3.49	0.9333		0.00	1.64	0.01	0.10	1.54	6.64	0.02	1.52	broken
12	3.71	0.9320		0.00	1.59	0.01	0.09	1.50	7.08	0.02	1.48	
13	3.75	0.9244		0.00	1.59	0.01	0.08	1.51	7.15	0.02	1.49	
14	3.96	0.9229		0.00	1.55	0.01	0.08	1.47	7.54	0.02	1.45	
15	3.98	0.9150		0.00	1.55	0.01	0.07	1.48	7.59	0.02	1.46	
16	4.17	0.9134		0.00	1.52	0.00	0.06	1.45	7.95	0.02	1.44	
17	4.18	0.9052		0.00	1.52	0.01	0.06	1.46	7.96	0.02	1.45	
18	4.35	0.9035		0.00	1.49	0.00	0.05	1.44	8.29	0.02	1.42	
19	4.35	0.8951		0.00	1.50	0.00	0.05	1.45	8.29	0.02	1.43	
20	4.51	0.8932		0.00	1.48	0.00	0.05	1.43	8.60	0.02	1.41	
21	4.50	0.8846		0.00	1.48	0.00	0.04	1.44	8.59	0.02	1.43	
22	4.65	0.8826		0.00	1.46	0.00	0.04	1.42	8.87	0.02	1.41	
23	4.63	0.8739		0.00	1.47	0.00	0.04	1.44	8.84	0.02	1.42	
24	4.77	0.8717		0.00	1.45	0.00	0.03	1.42	9.10	0.02	1.40	
25	4.90	0.8695		0.00	1.44	0.00	0.03	1.40	9.36	0.02	1.39	
26	4.88	0.8607		0.00	1.45	0.00	0.03	1.42	9.32	0.02	1.40	
27	5.00	0.8584		0.00	1.43	0.00	0.03	1.40	9.55	0.01	1.39	
28	5.12	0.8561		0.00	1.42	0.00	0.02	1.39	9.78	0.01	1.38	
29	5.09	0.8472		0.00	1.43	0.00	0.02	1.41	9.73	0.01	1.39	
30	5.20	0.8448		0.00	1.41	0.00	0.02	1.39	9.94	0.01	1.38	
31	5.31	0.8423		0.00	1.40	0.00	0.02	1.38	10.15	0.01	1.37	
32	5.28	0.8334		0.00	1.41	0.00	0.02	1.40	10.08	0.01	1.38	
33	5.38	0.8309		0.00	1.40	0.00	0.01	1.39	10.28	0.01	1.37	
34	5.48	0.8283		0.00	1.39	0.00	0.01	1.38	10.47	0.01	1.37	
35	5.44	0.8194		0.00	1.40	0.00	0.01	1.39	10.40	0.01	1.38	
36	5.53	0.8168		0.00	1.39	0.00	0.01	1.38	10.58	0.01	1.37	
37	5.62	0.8141		0.00	1.38	0.00	0.01	1.38	10.75	0.01	1.36	
38	5.71	0.8114		0.00	1.38	0.01	0.01	1.37	10.91	0.01	1.36	
39	6.09	0.7663		0.00	1.37	0.00	0.00	1.37	11.67	0.01	1.36	

	AL	AM	AN	AO	AP	AQ	AR	AS	AT	AU	AV	AW	AX	AY	AZ
1	H local	surf scaling	Ursell		u0	u1	u2	u3	u4	u5	u6	u7	u8	u9	u10
2	m	parameter			Airy	Airy	Airy	Airy	Airy	Airy	Airy	Airy	Airy	Airy	Airy
3	0.30	55.3	335.09		0.0	0.0	0.0	0.0	0.0	0.0	0.0	0.0	0.0	0.0	0.0
4	0.33	192.4	254.87		0.0	0.0	0.0	0.0	0.0	0.0	0.0	0.0	0.0	0.0	0.0
5	0.39	224.5	211.90		0.0	0.0	0.0	0.0	0.0	0.0	0.0	0.0	0.0	0.0	0.0
6	0.44	256.5	178.85		0.0	0.0	0.0	0.0	0.0	0.0	0.0	0.0	0.0	0.0	0.0
7	0.49	387.4	150.80		0.0	0.0	0.0	0.0	0.0	0.0	0.0	0.0	0.0	0.0	0.0
8	0.55	430.4	130.37		0.0	0.0	0.0	0.0	0.0	0.0	0.0	0.0	0.0	0.0	0.0
9	0.60	473.5	127.47		0.0	0.0	0.0	0.0	0.0	0.0	0.0	0.0	0.0	0.0	0.0
10	0.65	667.6	110.21		0.0	0.0	0.0	0.0	0.0	0.0	0.0	0.0	0.0	0.0	0.0
11	0.70	723.2	97.11		0.0	0.0	0.0	0.0	0.0	0.0	0.0	0.0	0.0	0.0	0.0
12	1.48	1928.5	185.85		0.0	0.0	0.0	0.0	0.0	0.0	0.0	0.0	0.0	0.0	0.0
13	1.49	1530.2	154.79		0.0	0.0	0.0	0.0	0.0	0.0	0.0	0.0	0.0	0.0	0.0
14	1.45	1894.4	138.85		0.0	0.0	0.0	0.0	0.0	0.0	0.0	0.0	0.0	0.0	0.0
15	1.46	2354.1	117.77		0.0	0.0	0.0	0.0	0.0	0.0	0.0	0.0	0.0	0.0	0.0
16	1.44	2310.7	106.80		0.0	0.0	0.0	0.0	0.0	0.0	0.0	0.0	0.0	0.0	0.0
17	1.45	1884.6	91.84		0.0	0.0	0.0	0.0	0.0	0.0	0.0	0.0	0.0	0.0	0.0
18	1.42	2770.6	84.01		0.0	0.0	0.0	0.0	0.0	0.0	0.0	0.0	0.0	0.0	0.0
19	1.43	2307.6	73.14		0.0	0.0	0.0	0.0	0.0	0.0	0.0	0.0	0.0	0.0	0.0
20	1.41	2753.0	67.38		0.0	0.0	0.0	0.0	0.0	0.0	0.0	0.0	0.0	0.0	0.0
21	1.43	2776.2	59.27		0.0	0.0	0.0	0.0	0.0	0.0	0.0	0.0	0.0	0.0	0.0
22	1.41	2741.6	54.93		0.0	0.0	0.0	0.0	0.0	0.0	0.0	0.0	0.0	0.0	0.0
23	1.42	2763.2	48.75		0.0	0.0	0.0	0.0	0.0	0.0	0.0	0.0	0.0	0.0	0.0
24	1.40	3254.6	45.41		0.0	0.0	0.0	0.0	0.0	0.0	0.0	0.0	0.0	0.0	0.0
25	1.39	3221.3	42.38		0.0	0.0	0.0	0.0	0.0	0.0	0.0	0.0	0.0	0.0	0.0
26	1.40	3251.0	38.01		0.0	0.0	0.0	0.0	0.0	0.0	0.0	0.0	0.0	0.0	0.0
27	1.39	3221.0	35.62		0.0	0.0	0.0	0.0	0.0	0.0	0.0	0.0	0.0	0.0	0.0
28	1.38	3748.0	33.43		0.0	0.0	0.0	0.0	0.0	0.0	0.0	0.0	0.0	0.0	0.0
29	1.39	3783.9	30.26		0.0	0.0	0.0	0.0	0.0	0.0	0.0	0.0	0.0	0.0	0.0
30	1.38	3754.7	28.50		0.0	0.0	0.0	0.0	0.0	0.0	0.0	0.0	0.0	0.0	0.0
31	1.37	3727.7	26.87		0.0	0.0	0.0	0.0	0.0	0.0	0.0	0.0	0.0	0.0	0.0
32	1.38	3763.6	24.50		0.0	0.0	0.0	0.0	0.0	0.0	0.0	0.0	0.0	0.0	0.0
33	1.37	4336.6	23.18		0.0	0.0	0.0	0.0	0.0	0.0	0.0	0.0	0.0	0.0	0.0
34	1.37	3716.3	21.94		0.0	0.0	0.0	0.0	0.0	0.0	0.0	0.0	0.0	0.0	0.0
35	1.38	4351.8	20.14		0.0	0.0	0.0	0.0	0.0	0.0	0.0	0.0	0.0	0.0	0.0
36	1.37	4968.4	19.12		0.0	0.0	0.0	0.0	0.0	0.0	0.0	0.0	0.0	0.0	0.0
37	1.36	4270.4	18.16		0.0	0.0	0.0	0.0	0.0	0.0	0.0	0.0	0.0	0.0	0.0
38	1.36	4320.5	17.28		0.0	0.0	0.0	0.0	0.0	0.0	0.0	0.0	0.0	0.0	0.0
39	1.36	5264.8	10.11		0.0	0.0	0.0	0.0	0.0	0.0	0.0	0.0	0.0	0.0	0.0

	BA	BB	BC	BD	BE	BF	BG	BH	BI	BJ	BK	BL	BM	BN	BO	BP	BQ
1	u11		u0	u1	u2	u3	u4	u5	u6	u7	u8	u9	u10	u11		Stokes	
2	Airy		Stokes	Stokes	Stokes	Stokes	Stokes	Stokes	Stokes	Stokes	Stokes	Stokes	Stokes	Stokes		drift	
3	0.0		159.2	98.7	-27.3	-106.9	-79.6	8.2	54.6	8.2	-79.6	-106.9	-27.3	98.7		14.4	
4	0.0		133.9	86.2	-14.3	-81.3	-67.0	-5.0	28.6	-5.0	-67.0	-81.3	-14.3	86.2		13.5	
5	0.0		126.9	83.9	-7.6	-71.0	-63.4	-12.8	15.2	-12.8	-63.4	-71.0	-7.6	83.9		14.2	
6	0.0		120.6	81.7	-1.9	-62.2	-60.3	-19.5	3.7	-19.5	-60.3	-62.2	-1.9	81.7		14.8	
7	0.0		112.8	78.3	3.4	-53.0	-56.4	-25.2	-6.7	-25.2	-56.4	-53.0	3.4	78.3		14.9	
8	0.0		108.2	76.6	7.4	-46.7	-54.1	-29.9	-14.8	-29.9	-54.1	-46.7	7.4	76.6		15.3	
9	0.0		116.4	82.7	8.7	-49.5	-58.2	-33.1	-17.3	-33.1	-58.2	-49.5	8.7	82.7		16.6	
10	0.0		109.7	79.4	12.3	-42.5	-54.9	-36.9	-24.7	-36.9	-54.9	-42.5	12.3	79.4		16.5	
11	0.0		105.5	77.6	15.3	-37.4	-52.7	-40.2	-30.6	-40.2	-52.7	-37.4	15.3	77.6		16.6	
12	0.0		290.5	197.1	-3.6	-148.9	-145.3	-48.3	7.2	-48.3	-145.3	-148.9	-3.6	197.1		67.6	
13	0.0		248.7	173.2	9.1	-115.2	-124.4	-58.0	-18.3	-58.0	-124.4	-115.2	9.1	173.2		59.4	
14	0.0		228.6	161.5	14.7	-99.6	-114.3	-61.9	-29.4	-61.9	-114.3	-99.6	14.7	161.5		52.6	
15	0.0		201.2	145.3	21.7	-78.9	-100.6	-66.5	-43.4	-66.5	-100.6	-78.9	21.7	145.3		47.0	
16	0.0		187.7	137.3	24.7	-69.2	-93.9	-68.1	-49.4	-68.1	-93.9	-69.2	24.7	137.3		42.2	
17	0.0		168.4	125.5	28.6	-55.7	-84.2	-69.8	-57.1	-69.8	-84.2	-55.7	28.6	125.5		38.1	
18	0.0		159.0	119.6	30.2	-49.3	-79.5	-70.3	-60.3	-70.3	-79.5	-49.3	30.2	119.6		34.6	
19	0.0		145.0	110.8	32.2	-40.3	-72.5	-70.5	-64.3	-70.5	-72.5	-40.3	32.2	110.8		31.5	
20	0.0		138.0	106.3	33.0	-36.0	-69.0	-70.3	-65.9	-70.3	-69.0	-36.0	33.0	106.3		28.8	
21	0.0		127.4	99.5	33.9	-29.8	-63.7	-69.6	-67.8	-69.6	-63.7	-29.8	33.9	99.5		26.5	
22	0.0		122.1	95.9	34.2	-26.9	-61.1	-69.1	-68.4	-69.1	-61.1	-26.9	34.2	95.9		24.4	
23	0.0		113.9	90.4	34.5	-22.5	-56.9	-67.9	-69.0	-67.9	-56.9	-22.5	34.5	90.4		22.5	
24	0.0		109.7	87.5	34.5	-20.3	-54.8	-67.2	-69.0	-67.2	-54.8	-20.3	34.5	87.5		20.9	
25	0.0		105.8	84.9	34.4	-18.5	-52.9	-66.4	-68.9	-66.4	-52.9	-18.5	34.4	84.9		19.5	
26	0.0		99.7	80.6	34.2	-15.6	-49.9	-65.0	-68.4	-65.0	-49.9	-15.6	34.2	80.6		18.1	
27	0.0		96.6	78.4	34.0	-14.3	-48.3	-64.1	-68.0	-64.1	-48.3	-14.3	34.0	78.4		16.9	
28	0.0		93.7	76.3	33.8	-13.0	-46.8	-63.3	-67.6	-63.3	-46.8	-13.0	33.8	76.3		15.9	
29	0.0		88.9	72.9	33.3	-11.2	-44.5	-61.8	-66.6	-61.8	-44.5	-11.2	33.3	72.9		14.9	
30	0.0		86.5	71.2	33.0	-10.2	-43.2	-60.9	-66.0	-60.9	-43.2	-10.2	33.0	71.2		14.0	
31	0.0		84.2	69.5	32.7	-9.4	-42.1	-60.1	-65.4	-60.1	-42.1	-9.4	32.7	69.5		13.2	
32	0.0		80.4	66.7	32.1	-8.1	-40.2	-58.6	-64.2	-58.6	-40.2	-8.1	32.1	66.7		12.4	
33	0.0		78.4	65.2	31.7	-7.5	-39.2	-57.7	-63.5	-57.7	-39.2	-7.5	31.7	65.2		11.7	
34	0.0		76.6	63.8	31.4	-6.9	-38.3	-56.9	-62.7	-56.9	-38.3	-6.9	31.4	63.8		11.1	
35	0.0		73.5	61.4	30.7	-6.0	-36.7	-55.4	-61.5	-55.4	-36.7	-6.0	30.7	61.4		10.5	
36	0.0		71.8	60.2	30.4	-5.6	-35.9	-54.6	-60.7	-54.6	-35.9	-5.6	30.4	60.2		9.9	
37	0.0		70.3	59.0	30.0	-5.2	-35.1	-53.8	-60.0	-53.8	-35.1	-5.2	30.0	59.0		9.4	
38	0.0		68.8	57.8	29.6	-4.8	-34.4	-53.0	-59.2	-53.0	-34.4	-4.8	29.6	57.8		9.0	
39	0.0		54.4	46.4	25.2	-2.0	-27.2	-44.4	-50.4	-44.4	-27.2	-2.0	25.2	46.4		5.6	

	BR	BS	BT	BU	BV	BW	BX	BY	BZ	CA	CB	CC	CD	CE	CF	CG
1	u0	u1	u2	u3	u4	u5	u6	u7	u8	u9	u10	u11		Ab0	Ab1	Ab2
2		R	E	S	U	L	T	A	N	T						
3	173.5	113.1	-12.9	-92.5	-65.2	22.6	69.0	22.6	-65.2	-92.5	-12.9	113.1		0.954	0.954	1.052
4	147.4	99.7	-0.9	-67.8	-53.5	8.5	42.1	8.5	-53.5	-67.8	-0.9	99.7		0.973	0.973	1.029
5	141.1	98.1	6.6	-56.8	-49.2	1.4	29.4	1.4	-49.2	-56.8	6.6	98.1		0.973	0.973	0.973
6	135.5	96.6	13.0	-47.3	-45.5	-4.7	18.6	-4.7	-45.5	-47.3	13.0	96.6		0.973	0.973	0.973
7	127.7	93.2	18.3	-38.1	-41.5	-10.3	8.2	-10.3	-41.5	-38.1	18.3	93.2		0.977	0.977	0.977
8	123.5	91.9	22.7	-31.4	-38.8	-14.6	0.5	-14.6	-38.8	-31.4	22.7	91.9		0.977	0.977	0.977
9	133.0	99.3	25.3	-32.9	-41.6	-16.5	-0.7	-16.5	-41.6	-32.9	25.3	99.3		0.977	0.977	0.977
10	126.2	95.9	28.8	-26.1	-38.4	-20.5	-8.2	-20.5	-38.4	-26.1	28.8	95.9		0.980	0.980	0.980
11	122.1	94.2	31.9	-20.8	-36.1	-23.6	-14.0	-23.6	-36.1	-20.8	31.9	94.2		0.980	0.980	0.980
12	358.1	264.7	64.0	-81.3	-77.7	19.3	74.7	19.3	-77.7	-81.3	64.0	264.7		0.982	0.982	0.982
13	308.1	232.6	68.5	-55.8	-65.0	1.4	41.1	1.4	-65.0	-55.8	68.5	232.6		0.980	0.980	0.980
14	281.2	214.1	67.3	-47.0	-61.7	-9.3	23.2	-9.3	-61.7	-47.0	67.3	214.1		0.982	0.982	0.982
15	248.2	192.3	68.7	-31.9	-53.6	-19.5	3.6	-19.5	-53.6	-31.9	68.7	192.3		0.984	0.984	0.984
16	229.9	179.5	66.9	-27.0	-51.7	-25.9	-7.2	-25.9	-51.7	-27.0	66.9	179.5		0.984	0.984	0.984
17	206.5	163.6	66.6	-17.6	-46.1	-31.8	-19.0	-31.8	-46.1	-17.6	66.6	163.6		0.982	0.982	0.982
18	193.6	154.2	64.7	-14.8	-44.9	-35.7	-25.7	-35.7	-44.9	-14.8	64.7	154.2		0.985	0.985	0.985
19	176.5	142.3	63.7	-8.8	-41.0	-39.0	-32.8	-39.0	-41.0	-8.8	63.7	142.3		0.984	0.984	0.984
20	166.8	135.2	61.8	-7.2	-40.2	-41.4	-37.1	-41.4	-40.2	-7.2	61.8	135.2		0.985	0.985	0.985
21	153.9	125.9	60.4	-3.3	-37.2	-43.1	-41.3	-43.1	-37.2	-3.3	60.4	125.9		0.985	0.985	0.985
22	146.5	120.3	58.6	-2.4	-36.6	-44.6	-44.0	-44.6	-36.6	-2.4	58.6	120.3		0.985	0.985	0.985
23	136.4	112.9	57.0	0.1	-34.4	-45.4	-46.4	-45.4	-34.4	0.1	57.0	112.9		0.985	0.985	0.985
24	130.6	108.5	55.4	0.6	-33.9	-46.3	-48.1	-46.3	-33.9	0.6	55.4	108.5		0.986	0.986	0.986
25	125.3	104.4	53.9	1.0	-33.4	-46.9	-49.4	-46.9	-33.4	1.0	53.9	104.4		0.986	0.986	0.986
26	117.8	98.8	52.3	2.5	-31.7	-46.9	-50.3	-46.9	-31.7	2.5	52.3	98.8		0.986	0.986	0.986
27	113.5	95.4	51.0	2.7	-31.4	-47.2	-51.1	-47.2	-31.4	2.7	51.0	95.4		0.986	0.986	0.986
28	109.5	92.2	49.7	2.8	-31.0	-47.4	-51.7	-47.4	-31.0	2.8	49.7	92.2		0.987	0.987	0.987
29	103.8	87.8	48.2	3.7	-29.6	-46.9	-51.8	-46.9	-29.6	3.7	48.2	87.8		0.987	0.987	0.987
30	100.5	85.1	47.0	3.7	-29.3	-46.9	-52.0	-46.9	-29.3	3.7	47.0	85.1		0.987	0.987	0.987
31	97.4	82.6	45.8	3.7	-28.9	-46.9	-52.2	-46.9	-28.9	3.7	45.8	82.6		0.987	0.987	0.987
32	92.8	79.1	44.5	4.3	-27.8	-46.2	-51.8	-46.2	-27.8	4.3	44.5	79.1		0.987	0.987	0.987
33	90.1	76.9	43.4	4.2	-27.5	-46.0	-51.8	-46.0	-27.5	4.2	43.4	76.9		0.988	0.988	0.988
34	87.6	74.9	42.4	4.2	-27.2	-45.8	-51.7	-45.8	-27.2	4.2	42.4	74.9		0.987	0.987	0.987
35	83.9	71.9	41.2	4.4	-26.3	-45.0	-51.0	-45.0	-26.3	4.4	41.2	71.9		0.988	0.988	0.988
36	81.8	70.1	40.3	4.4	-26.0	-44.7	-50.8	-44.7	-26.0	4.4	40.3	70.1		0.989	0.989	0.989
37	79.7	68.4	39.4	4.3	-25.7	-44.4	-50.5	-44.4	-25.7	4.3	39.4	68.4		0.988	0.988	0.988
38	77.8	66.8	38.6	4.2	-25.4	-44.1	-50.2	-44.1	-25.4	4.2	38.6	66.8		0.988	0.988	0.988
39	60.0	52.0	30.8	3.6	-21.6	-38.7	-44.8	-38.7	-21.6	3.6	30.8	52.0		0.989	0.989	0.989

	CH	CI	CJ	CK	CL	CM	CN	CO	CP	CQ	CR	CS	CT	CU	CV
1	Ab3	Ab4	Ab5	Ab6	Ab7	Ab8	Ab9	Ab10	Ab11	kb	Bb	Bb	Bb	Bb	Bb
2											u0	u1	u2	u3	u4
3	1.052	1.052	0.954	0.954	0.954	1.052	1.052	1.052	0.954	0.000001	1.024	1.024	0.975	0.975	0.975
4	1.029	1.029	0.973	0.973	0.973	1.029	1.029	1.029	0.973	0.000001	1.014	1.014	0.986	0.986	0.986
5	1.029	1.029	0.973	0.973	0.973	1.029	1.029	0.973	0.973	0.000001	1.014	1.014	1.014	0.986	0.986
6	1.029	1.029	1.029	0.973	1.029	1.029	1.029	0.973	0.973	0.000001	1.014	1.014	1.014	0.986	0.986
7	1.024	1.024	1.024	0.977	1.024	1.024	1.024	0.977	0.977	0.000001	1.012	1.012	1.012	0.988	0.988
8	1.024	1.024	1.024	0.977	1.024	1.024	1.024	0.977	0.977	0.000001	1.012	1.012	1.012	0.988	0.988
9	1.024	1.024	1.024	1.024	1.024	1.024	1.024	0.977	0.977	0.000001	1.012	1.012	1.012	0.988	0.988
10	1.021	1.021	1.021	1.021	1.021	1.021	1.021	0.980	0.980	0.000001	1.010	1.010	1.010	0.989	0.989
11	1.021	1.021	1.021	1.021	1.021	1.021	1.021	0.980	0.980	0.000001	1.010	1.010	1.010	0.989	0.989
12	1.019	1.019	0.982	0.982	0.982	1.019	1.019	0.982	0.982	0.000001	1.009	1.009	1.009	0.991	0.991
13	1.021	1.021	0.980	0.980	0.980	1.021	1.021	0.980	0.980	0.000001	1.010	1.010	1.010	0.989	0.989
14	1.019	1.019	1.019	0.982	1.019	1.019	1.019	0.982	0.982	0.000001	1.009	1.009	1.009	0.991	0.991
15	1.017	1.017	1.017	0.984	1.017	1.017	1.017	0.984	0.984	0.000001	1.008	1.008	1.008	0.992	0.992
16	1.017	1.017	1.017	1.017	1.017	1.017	1.017	0.984	0.984	0.000001	1.008	1.008	1.008	0.992	0.992
17	1.019	1.019	1.019	1.019	1.019	1.019	1.019	0.982	0.982	0.000001	1.009	1.009	1.009	0.991	0.991
18	1.015	1.015	1.015	1.015	1.015	1.015	1.015	0.985	0.985	0.000001	1.008	1.008	1.008	0.992	0.992
19	1.017	1.017	1.017	1.017	1.017	1.017	1.017	0.984	0.984	0.000001	1.008	1.008	1.008	0.992	0.992
20	1.015	1.015	1.015	1.015	1.015	1.015	1.015	0.985	0.985	0.000001	1.008	1.008	1.008	0.992	0.992
21	1.015	1.015	1.015	1.015	1.015	1.015	1.015	0.985	0.985	0.000001	1.008	1.008	1.008	0.992	0.992
22	1.015	1.015	1.015	1.015	1.015	1.015	1.015	0.985	0.985	0.000001	1.008	1.008	1.008	0.992	0.992
23	0.985	1.015	1.015	1.015	1.015	1.015	0.985	0.985	0.985	0.000001	1.008	1.008	1.008	1.008	0.992
24	0.986	1.014	1.014	1.014	1.014	1.014	0.986	0.986	0.986	0.000001	1.007	1.007	1.007	1.007	0.993
25	0.986	1.014	1.014	1.014	1.014	1.014	0.986	0.986	0.986	0.000001	1.007	1.007	1.007	1.007	0.993
26	0.986	1.014	1.014	1.014	1.014	1.014	0.986	0.986	0.986	0.000001	1.007	1.007	1.007	1.007	0.993
27	0.986	1.014	1.014	1.014	1.014	1.014	0.986	0.986	0.986	0.000001	1.007	1.007	1.007	1.007	0.993
28	0.987	1.013	1.013	1.013	1.013	1.013	0.987	0.987	0.987	0.000001	1.006	1.006	1.006	1.006	0.994
29	0.987	1.013	1.013	1.013	1.013	1.013	0.987	0.987	0.987	0.000001	1.006	1.006	1.006	1.006	0.994
30	0.987	1.013	1.013	1.013	1.013	1.013	0.987	0.987	0.987	0.000001	1.006	1.006	1.006	1.006	0.994
31	0.987	1.013	1.013	1.013	1.013	1.013	0.987	0.987	0.987	0.000001	1.006	1.006	1.006	1.006	0.994
32	0.987	1.013	1.013	1.013	1.013	1.013	0.987	0.987	0.987	0.000001	1.006	1.006	1.006	1.006	0.994
33	0.988	1.012	1.012	1.012	1.012	1.012	0.988	0.988	0.988	0.000001	1.006	1.006	1.006	1.006	0.994
34	0.987	1.013	1.013	1.013	1.013	1.013	0.987	0.987	0.987	0.000001	1.006	1.006	1.006	1.006	0.994
35	0.988	1.012	1.012	1.012	1.012	1.012	0.988	0.988	0.988	0.000001	1.006	1.006	1.006	1.006	0.994
36	0.989	1.011	1.011	1.011	1.011	1.011	0.989	0.989	0.989	0.000001	1.006	1.006	1.006	1.006	0.994
37	0.988	1.012	1.012	1.012	1.012	1.012	0.988	0.988	0.988	0.000001	1.006	1.006	1.006	1.006	0.994
38	0.988	1.012	1.012	1.012	1.012	1.012	0.988	0.988	0.988	0.000001	1.006	1.006	1.006	1.006	0.994
39	0.989	1.011	1.011	1.011	1.011	1.011	0.989	0.989	0.989	0.000001	1.005	1.005	1.005	1.005	0.995

	CW	CX	CY	CZ	DA	DB	DC	DD	DE	DF	DG	DH	DI	DJ	DK	DL
1	Bb	Bb	Bb	Bb	Bb	Bb	Bb	ucrb		i0	Ca	Ca'		D	R	I
2	u5	u6	u7	u8	u9	u10	u11	cm s-1		gm/cm/s	mg/l	g/cm3	z=	0.10	0.30	0.50
3	1.024	1.024	1.024	0.975	0.975	0.975	1.024	4.3		4.9805	4	0.0004		0.00	-0.55	0.00
4	1.014	1.014	1.014	0.986	0.986	0.986	1.014	4.3		3.1115	3	0.0003		0.07	-0.41	-0.65
5	1.014	1.014	1.014	0.986	0.986	1.014	1.014	4.3		2.7323	2	0.0002		0.13	-0.35	-0.63
6	0.986	1.014	0.986	0.986	0.986	1.014	1.014	4.3		2.4167	2	0.0002		0.18	-0.28	-0.58
7	0.988	1.012	0.988	0.988	0.988	1.012	1.012	4.3		2.0331	2	0.0002		0.22	-0.21	-0.52
8	0.988	1.012	0.988	0.988	0.988	1.012	1.012	4.4		1.8374	2	0.0002		0.26	-0.16	-0.46
9	0.988	0.988	0.988	0.988	0.988	1.012	1.012	4.4		2.2959	2	0.0002		0.30	-0.12	-0.44
10	0.989	0.989	0.989	0.989	0.989	1.010	1.010	4.4		1.9652	2	0.0002		0.33	-0.06	-0.38
11	0.989	0.989	0.989	0.989	0.989	1.010	1.010	4.4		1.7801	2	0.0002		0.36	-0.02	-0.32
12	1.009	1.009	1.009	0.991	0.991	1.009	1.009	4.8		45.0746	16	0.0016		1.53	0.08	-1.12
13	1.010	1.010	1.010	0.989	0.989	1.010	1.010	4.8		28.6440	12	0.0012		1.43	0.20	-0.82
14	0.991	1.009	0.991	0.991	0.991	1.009	1.009	4.8		21.8328	10	0.0010		1.31	0.28	-0.60
15	0.992	1.008	0.992	0.992	0.992	1.008	1.008	4.8		15.0275	8	0.0008		1.23	0.34	-0.43
16	0.992	0.992	0.992	0.992	0.992	1.008	1.008	4.7		11.9541	6	0.0006		1.13	0.37	-0.30
17	0.991	0.991	0.991	0.991	0.991	1.009	1.009	4.7		8.6447	5	0.0005		1.06	0.39	-0.11
18	0.992	0.992	0.992	0.992	0.992	1.008	1.008	4.7		7.1416	5	0.0005		0.99	0.40	-0.11
19	0.992	0.992	0.992	0.992	0.992	1.008	1.008	4.7		5.4031	4	0.0004		0.93	0.41	-0.05
20	0.992	0.992	0.992	0.992	0.992	1.008	1.008	4.6		4.5722	3	0.0003		0.87	0.41	0.00
21	0.992	0.992	0.992	0.992	0.992	1.008	1.008	4.6		3.5892	3	0.0003		0.82	0.41	0.04
22	0.992	0.992	0.992	0.992	0.992	1.008	1.008	4.6		3.0959	3	0.0003		0.77	0.40	0.07
23	0.992	0.992	0.992	0.992	1.008	1.008	1.008	4.6		2.4979	2	0.0002		0.73	0.40	0.09
24	0.993	0.993	0.993	0.993	1.007	1.007	1.007	4.6		2.1951	2	0.0002		0.69	0.39	0.11
25	0.993	0.993	0.993	0.993	1.007	1.007	1.007	4.6		1.9378	2	0.0002		0.65	0.38	0.13
26	0.993	0.993	0.993	0.993	1.007	1.007	1.007	4.5		1.6117	2	0.0002		0.62	0.37	0.14
27	0.993	0.993	0.993	0.993	1.007	1.007	1.007	4.5		1.4405	2	0.0002		0.59	0.36	0.15
28	0.994	0.994	0.994	0.994	1.006	1.006	1.006	4.5		1.2952	1	0.0001		0.56	0.35	0.15
29	0.994	0.994	0.994	0.994	1.006	1.006	1.006	4.5		1.1022	1	0.0001		0.54	0.34	0.16
30	0.994	0.994	0.994	0.994	1.006	1.006	1.006	4.5		0.9994	1	0.0001		0.51	0.33	0.16
31	0.994	0.994	0.994	0.994	1.006	1.006	1.006	4.5		0.9091	1	0.0001		0.49	0.32	0.16
32	0.994	0.994	0.994	0.994	1.006	1.006	1.006	4.4		0.7872	1	0.0001		0.47	0.31	0.16
33	0.994	0.994	0.994	0.994	1.006	1.006	1.006	4.4		0.7223	1	0.0001		0.45	0.30	0.17
34	0.994	0.994	0.994	0.994	1.006	1.006	1.006	4.4		0.6630	1	0.0001		0.43	0.29	0.16
35	0.994	0.994	0.994	0.994	1.006	1.006	1.006	4.4		0.5826	1	0.0001		0.41	0.28	0.16
36	0.994	0.994	0.994	0.994	1.006	1.006	1.006	4.4		0.5392	1	0.0001		0.39	0.27	0.16
37	0.994	0.994	0.994	0.994	1.006	1.006	1.006	4.4		0.4988	1	0.0001		0.38	0.26	0.16
38	0.994	0.994	0.994	0.994	1.006	1.006	1.006	4.4		0.4631	1	0.0001		0.36	0.26	0.16
39	0.995	0.995	0.995	0.995	1.005	1.005	1.005	4.3		0.2128	0	0.0000		0.25	0.19	0.13

	DM	DN	DO	DP	DQ	DR	DS	DT	DU	DV	DW
1	F	T	is		ib0	ib1	ib2	ib3	ib4	ib5	i6
2	0.70	0.90	gm/cm/s		gm/cm/s	gm/cm/s	gm/cm/s	gm/cm/s	gm/cm/s	gm/cm/s	gm/cm/s
3	0.00	0.00	-0.0972		4.9805	1.3771	-0.0020	-0.8313	-0.2904	0.0105	0.3121
4	0.00	0.00	-0.0845		3.1115	0.9620	0.0000	-0.3195	-0.1566	0.0004	0.0717
5	0.00	0.00	-0.0541		2.7323	0.9173	0.0002	-0.1876	-0.1217	0.0000	0.0243
6	-0.73	0.00	-0.0498		2.4167	0.8741	0.0019	-0.1081	-0.0958	0.0000	0.0059
7	-0.69	0.00	-0.0254		2.0331	0.7890	0.0056	-0.0561	-0.0724	-0.0009	0.0004
8	-0.66	-0.76	-0.0106		1.8374	0.7567	0.0110	-0.0311	-0.0591	-0.0029	0.0000
9	-0.66	-0.79	0.0034		2.2959	0.9542	0.0153	-0.0359	-0.0778	0.0042	0.0000
10	-0.60	-0.75	0.0183		1.9652	0.8625	0.0228	-0.0176	-0.0571	-0.0084	-0.0004
11	-0.56	-0.72	0.0304		1.7801	0.8182	0.0312	-0.0088	-0.0475	-0.0130	-0.0025
12	-2.06	-2.75	1.4297		45.0746	18.1991	0.2554	-0.5457	-0.4763	0.0066	0.4082
13	-1.65	-2.27	1.1810		28.6440	12.3239	0.3137	-0.1764	-0.2785	0.0000	0.0671
14	-1.32	-1.88	1.0175		21.8328	9.6364	0.2978	-0.1048	-0.2382	-0.0006	0.0117
15	-1.06	-1.57	0.8216		15.0275	6.9940	0.3171	-0.0323	-0.1554	-0.0071	-0.0002
16	-0.85	-1.31	0.6991		11.9541	5.6825	0.2931	-0.0193	-0.1392	-0.0171	-0.0066
17	-0.69	-1.11	0.5636		8.6447	4.2941	0.2889	-0.0052	-0.0991	-0.0319	-0.0167
18	-0.56	-0.93	0.4841		7.1416	3.6094	0.2657	-0.0029	-0.0911	-0.0455	-0.0353
19	-0.45	-0.79	0.3962		5.4031	2.8313	0.2525	-0.0005	-0.0692	-0.0594	-0.0509
20	-0.36	-0.67	0.3427		4.5722	2.4298	0.2312	-0.0002	-0.0649	-0.0714	-0.0707
21	-0.29	-0.57	0.2853		3.5892	1.9646	0.2155	0.0000	-0.0517	-0.0807	-0.0855
22	-0.22	-0.48	0.2488		3.0959	1.7141	0.1971	0.0000	-0.0492	-0.0894	-0.1006
23	-0.18	-0.41	0.2099		2.4979	1.4169	0.1815	0.0000	-0.0406	-0.0940	-0.1117
24	-0.13	-0.35	0.1852		2.1951	1.2565	0.1668	0.0000	-0.0389	-0.0995	-0.1213
25	-0.10	-0.30	0.1639		1.9378	1.1189	0.1534	0.0000	-0.0373	-0.1040	-0.1283
26	-0.07	-0.25	0.1407		1.6117	0.9480	0.1403	0.0000	-0.0318	-0.1036	-0.1343
27	-0.04	-0.21	0.1254		1.4405	0.8532	0.1294	0.0000	-0.0306	-0.1058	-0.1389
28	-0.02	-0.18	0.1124		1.2952	0.7722	0.1198	0.0000	-0.0294	-0.1071	-0.1395
29	-0.01	-0.15	0.0979		1.1022	0.6665	0.1094	0.0000	-0.0257	-0.1038	-0.1417
30	0.01	-0.13	0.0882		0.9994	0.6075	0.1014	0.0000	-0.0248	-0.1039	-0.1430
31	0.02	-0.11	0.0797		0.9091	0.5554	0.0942	0.0000	-0.0240	-0.1035	-0.1397
32	0.03	-0.09	0.0702		0.7872	0.4862	0.0859	0.0000	-0.0213	-0.0988	-0.1397
33	0.04	-0.07	0.0639		0.7223	0.4479	0.0801	0.0000	-0.0206	-0.0976	-0.1393
34	0.05	-0.06	0.0581		0.6630	0.4127	0.0747	0.0000	-0.0199	-0.0963	-0.1387
35	0.05	-0.05	0.0517		0.5826	0.3657	0.0682	0.0000	-0.0179	-0.0911	-0.1332
36	0.06	-0.04	0.0474		0.5392	0.3394	0.0639	0.0000	-0.0173	-0.0894	-0.1315
37	0.06	-0.03	0.0434		0.4988	0.3149	0.0597	0.0000	-0.0167	-0.0876	-0.1296
38	0.07	-0.02	0.0399		0.4631	0.2932	0.0560	0.0000	-0.0161	-0.0857	-0.1274
39	0.07	0.02	0.0172		0.2128	0.1382	0.0284	0.0000	-0.0097	-0.0581	-0.0898

	DX	DY	DZ	EA	EB	EC	ED	EE	EF	EG	EH
1	ib7	ib8	ib9	ib10	ib11		Is	Ib	dmass		dz/dt
2	gm/cm/s	gm/cm/s	gm/cm/s	gm/cm/s	gm/cm/s		gm/cm/wave	gm/cm/wave	gm/cm/wave		mm/step
3	0.0105	-0.2904	-0.8313	-0.0020	1.3771		-0.4861	2.4252	-1.1797		-71.8
4	0.0004	-0.1566	-0.3195	0.0000	0.9620		-0.4225	1.7316	-0.0127		-0.6
5	0.0000	-0.1217	-0.1876	0.0002	0.9173		-0.2703	1.6554	0.0662		2.4
6	0.0000	-0.0958	-0.1081	0.0019	0.8741		-0.2489	1.5694	-0.1463		-4.3
7	-0.0009	-0.0724	-0.0561	0.0056	0.7890		-0.1269	1.4016	0.0481		1.2
8	-0.0029	-0.0591	-0.0311	0.0110	0.7567		-0.0531	1.3277	0.4336		9.2
9	-0.0043	-0.0728	-0.0359	0.0153	0.9542		0.0172	1.6704	-0.0555		-1.0
10	-0.0084	-0.0571	-0.0176	0.0228	0.8625		0.0916	1.4873	-0.0363		-0.6
11	-0.0130	-0.0475	-0.0088	0.0312	0.8182		0.1519	1.3906	39.0899		561.9
12	0.0066	-0.4763	-0.5457	0.2554	18.1991		7.1487	33.4838	-12.6125		-161.2
13	0.0000	-0.2785	-0.1764	0.3137	12.3239		5.9048	22.1151	-5.8388		-67.9
14	-0.0006	-0.2382	-0.1048	0.2978	9.6364		5.0873	17.0939	-5.8812		-62.1
15	-0.0071	-0.1554	-0.0323	0.3171	6.9940		4.1082	12.1917	-2.9903		-28.7
16	-0.0171	-0.1392	-0.0193	0.2931	6.6825		3.4956	9.8141	-3.1866		-28.0
17	-0.0319	-0.0991	-0.0052	0.2889	4.2941		2.8182	7.3049	-1.6210		-13.2
18	-0.0455	-0.0911	-0.0029	0.2657	3.6094		2.4204	6.0817	-1.8224		-13.7
19	-0.0594	-0.0692	-0.0005	0.2525	2.8313		1.9809	4.6988	-0.9787		-6.8
20	-0.0714	-0.0649	-0.0002	0.2312	2.4298		1.7133	3.9876	-1.1020		-7.2
21	-0.0807	-0.0517	0.0000	0.2155	1.9646		1.4263	3.1725	-0.6233		-3.8
22	-0.0894	-0.0492	0.0000	0.1971	1.7141		1.2441	2.7315	-0.7072		-4.0
23	-0.0940	-0.0406	0.0000	0.1815	1.4169		1.0497	2.2187	-0.4034		-2.2
24	-0.0995	-0.0389	0.0000	0.1668	1.2565		0.9262	1.9388	-0.3464		-1.8
25	-0.1040	-0.0373	0.0000	0.1534	1.1189		0.8193	1.6994	-0.4031		-1.9
26	-0.1036	-0.0318	0.0000	0.1403	0.9480		0.7033	1.4122	-0.2389		-1.1
27	-0.1058	-0.0306	0.0000	0.1294	0.8532		0.6272	1.2495	-0.2033		-0.9
28	-0.1071	-0.0294	0.0000	0.1198	0.7722		0.5621	1.1113	-0.2442		-1.0
29	-0.1038	-0.0257	0.0000	0.1094	0.6665		0.4894	0.9398	-0.1473		-0.6
30	-0.1039	-0.0248	0.0000	0.1014	0.6075		0.4411	0.8409	-0.1290		-0.5
31	-0.1035	-0.0240	0.0000	0.0942	0.5554		0.3986	0.7543	-0.1555		-0.6
32	-0.0988	-0.0213	0.0000	0.0859	0.4862		0.3510	0.6465	-0.0937		-0.3
33	-0.0976	-0.0206	0.0000	0.0801	0.4479		0.3194	0.5844	-0.0854		-0.3
34	-0.0963	-0.0199	0.0000	0.0747	0.4127		0.2907	0.5277	-0.1018		-0.3
35	-0.0911	-0.0179	0.0000	0.0682	0.3657		0.2586	0.4581	-0.0626		-0.2
36	-0.0894	-0.0173	0.0000	0.0639	0.3394		0.2370	0.4171	-0.0579		-0.2
37	-0.0876	-0.0167	0.0000	0.0597	0.3149		0.2171	0.3791	-0.0507		-0.1
38	-0.0857	-0.0161	0.0000	0.0560	0.2932		0.1996	0.3460	-0.3258		-0.9
39	-0.0581	-0.0097	0.0000	0.0284	0.1382		0.0862	0.1336	-0.2198		-0.4

	EI	EJ	EK	EL	EM	EN	EO	EP	EQ	ER	ES	ET	EU	EV	EW
1	new		x	h m	h m	h m	h m	h m	dz/dt 1	dz/dt 2	dz/dt 3	dz/dt 4	Hm	Hm	Uo m/s
2	profile		m	t=0	t=1	t=2	t=3	t=4	mm/stp	mm/stp	mm/stp	mm/stp	t=0	t=4	t=0
3	0.572		17	-0.50				-0.57	-24	-14	-16	-72	0.24	0.30	1.19
4	0.601		23	-0.60				-0.60	1	5	6	-1	0.27	0.33	1.03
5	0.698		29	-0.70				-0.70	3	-3	-4	2	0.31	0.39	0.99
6	0.804		35	-0.80				-0.80	0	0	0	-4	0.35	0.44	0.96
7	0.899		42	-0.90				-0.90	2	2	2	1	0.39	0.49	0.91
8	0.991		49	-1.00				-0.99	2	2	2	9	0.44	0.55	0.89
9	1.101		56	-1.10				-1.10	1	0	0	-1	0.48	0.60	0.96
10	1.201		64	-1.20				-1.20	1	0	1	-1	0.52	0.65	0.92
11	0.738		72	-1.30				-0.74	167	136	128	562	0.56	0.70	0.89
12	1.561		81	-1.40				-1.56	-37	-12	-9	-161	1.48	1.48	3.58
13	1.568		89	-1.50				-1.57	-15	-19	-16	-68	1.49	1.49	3.08
14	1.662		98	-1.60				-1.66	-14	-8	-8	-62	1.45	1.45	2.81
15	1.729		108	-1.70				-1.73	-7	-8	-8	-29	1.46	1.46	2.48
16	1.828		118	-1.80				-1.83	-6	-4	-4	-28	1.44	1.44	2.30
17	1.913		127	-1.90				-1.91	-3	-3	-3	-13	1.45	1.45	2.07
18	2.014		138	-2.00				-2.01	-3	-3	-3	-14	1.42	1.42	1.94
19	2.107		148	-2.10				-2.11	-2	-2	-2	-7	1.43	1.43	1.76
20	2.207		159	-2.20				-2.21	-2	-2	-2	-7	1.41	1.41	1.67
21	2.304		170	-2.30				-2.30	-1	-1	-1	-4	1.43	1.43	1.54
22	2.404		181	-2.40				-2.40	-2	-2	-2	-4	1.41	1.41	1.47
23	2.502		192	-2.50				-2.50	-2	-2	-2	-2	1.42	1.42	1.36
24	2.602		204	-2.60				-2.60	-2	-2	-2	-2	1.40	1.40	1.31
25	2.702		216	-2.70				-2.70	-2	-2	-2	-2	1.39	1.39	1.25
26	2.801		228	-2.80				-2.80	-1	-1	-1	-1	1.40	1.40	1.18
27	2.901		240	-2.90				-2.90	-1	-1	-1	-1	1.39	1.39	1.14
28	3.001		253	-3.00				-3.00	-1	-1	-1	-1	1.38	1.38	1.10
29	3.101		266	-3.10				-3.10	-1	-1	-1	-1	1.39	1.39	1.04
30	3.200		279	-3.20				-3.20	0	0	0	0	1.38	1.38	1.00
31	3.301		292	-3.30				-3.30	0	0	0	-1	1.37	1.37	0.97
32	3.400		305	-3.40				-3.40	0	0	0	0	1.38	1.38	0.93
33	3.500		319	-3.50				-3.50	0	0	0	0	1.37	1.37	0.90
34	3.600		332	-3.60				-3.60	0	0	0	0	1.37	1.37	0.88
35	3.700		346	-3.70				-3.70	0	0	0	0	1.38	1.38	0.84
36	3.800		361	-3.80				-3.80	0	0	0	0	1.37	1.37	0.82
37	3.901		375	-3.90				-3.90	0	0	0	0	1.36	1.36	0.80
38	4.001		389	-4.00				-4.00	-1	-1	-1	-1	1.36	1.36	0.78
39	5.000		544	-5.00				-5.00	0	0	0	0	1.36	1.36	0.60

COLUMN P Non-dimensional depth = kh

COLUMN Q Non-dimensional depth = h/L_∞

COLUMN R Blank spacer.

L
=ARGUMENT("h")
=ARGUMENT("T")
=1.56*T*T
=SET.NAME("LA",DEREF(A4))
=2*3.1416*h/LA
=(EXP(A6)-EXP(-A6))/(EXP(A6)+EXP(-A6))
=9.81*T*T*A7/(2*3.1416)
=SET.NAME("LB",DEREF(A8))
=IF(LB<LA,GOTO(A12))
=RETURN(LB)
=SET.NAME("LA",LA-1)
=GOTO(A6)

Figure 17.3 WAVELENGTH macro to calculate local L iteratively.

COLUMNS S-X Calculation of the hyperbolic functions from the equations in Appendix I.

COLUMN Y Blank spacer.

COLUMN Z Local wave celerity from Eq.5.1 (AII.8).

COLUMN AA Calculates n from Eq.5.10 (AII.9).

COLUMN AB Blank spacer.

COLUMN AC The formula is IF($H$3="on",$D$9,0) where $H$3 is the control panel for default wave height and $D$9=H_∞. If the default is set on the control panel then H is set equal to H_∞ otherwise it is set equal to zero.

COLUMN AD The formula is IF($H$4="on", $D$9*(($D$7/ (2*AA3*Z3))^0.5),0) where $H$4 is the control panel for shoaling wave height and $D$9=H_∞, $D$7=C_∞, AA3=n and Z3=C. This means that if the local shoaling wave height is chosen on the control panel then H is given by Eq.5.14 (Eq.AII.10) otherwise it is set equal to zero.

COLUMN AE Calculates the frictional wave energy loss in the previous, seaward cell and the formula for AE3 is (4*((AC4 + AD4 - AF4)^2) * ((2*PI()/O4)^2) * M4 * (J4-J3))/(3*PI()*T4*(W4+2*P4)) where (AC4 + AD4 - AF4) is the wave height which may or may not include shoaling, depending upon the control panel setting but which does include the integrated frictional losses. O4 is the wavelength, M4 is ythe coefficient of friction, (J4-J3) is the horizontal width of the cell, T4=sinh(kh),

W4=sinh(2kh) and P4=kh. The formula is equivalent to Eq.5.19 (AII.11).

COLUMN AF Summates the frictional losses for all seaward cells.

COLUMN AG Displays the local wave height as either H_∞ or as H_s minus the frictional losses.

COLUMN AH Calculates the local angle of incidence and is written (ASIN(Z3 + SIN($D$6 + PI()/180)/$D$7))*180/PI() where Z3=C, $D$6=α_∞ and $D$7=C_∞ from Eq.5.22 (AII.12). This assumes shoreparallel, though not necessarily planar, bathymetries.

COLUMN AI Calculates the wave energy loss due to refraction which, for cell AI3 is written IF($H$6="on", $D$9(($D$7/(2*AA3*Z3))^0.5) * (1-((COS($D$6*PI()/180))/(COS(AH3*PI()/180)))^0.5),0) where $H$6 is the control panel setting for refraction losses, D=H_∞, $D$7=C_∞, AA3=n, Z3=C, $D$6=α_∞ and AH3=α. The formula computes the refraction height change from Eq.5.25 (AIV.13).

COLUMN AJ Summates the wave height changes due to shoaling, seabed friction and refraction according to the control panel settings.

COLUMN AK Determines whether the wave has broken using the criterion for γ_b from Eq.5.26 (AII.14).

COLUMN AL Determines the wave height using AJ directly unless both breaking occurs and the depth is shoaling when AIV.14 is used. If breaking has occured but the water is deepening shorewards then the wave height is set equal to the value in the next seaward cell.

COLUMN AM Determines a local surf scaling parameter from Table 5.2 (AII.15).

COLUMN AN determines the Ursell Number from Eq.6.25 (AII.16) and we recall that the shallow water limit for second order wave theory is Ur<26.3.

COLUMN AO Blank spacer.

COLUMNS AP-BA Determine the nearbed flow velocity at 12x30° increments through the wave cycle from first order theory. The formula for cell AP3 is
=IF($H$11 = "off", 0, 100*$D$11*PI()*AL3*COS(2*PI()*0/12)) /($D$4*T3))
where $H$11 is the control panel setting for first order theory. Thus a value of zero is returned if first order theory is not to be used, otherwise the formula is Eq.6.2 (AII.17). $D$11 is the attenuation coefficient, AL3=H, $D$4=T and T3=sinh(kh).

COLUMN BB Blank spacer

COLUMNS BC-BN. Determine the nearbed flow velocity at twelve 30° increments through the wave cycle from second order theory.

COLUMN BO Blank spacer

COLUMN BP Calculates the Stokes drift which for cell BP3 is:

=IF($H$13="off",0,100*$D$11*(5/4)*((PI()*AL3/O3)^2)*Z3*(1/((T3)^2)))

where $H$13 is the control panel setting, and then according to Eq.AII.18.

COLUMN BQ Blank spacer

COLUMNS BR to CC summate the first order, second order and drift flow components each of which return a zero value according to the control panel settings

COLUMN CD Blank spacer

COLUMN CE-CP Calculates the effect of bedslope on the bedload transport rate from Eq.10.34 (AII.20). The formula is
    IF($II$19="off",0,$D$13/(COS(K3)*($D$13+(BR3*K3/(ABS(BR3))))))
where $H$19 is the control panel setting for slope inclusive sediment transport, $D$13=tanϕ, K3=tanβ and BR3 is the resultant flow at the corresponding phase in the wave cycle.

COLUMN CQ Is the bedload calibration coefficient from $D$24 (AIV.21)

COLUMN CR-DC Calculate the effect of bedslope on the bedload threshold from Eq.10.36 (AII.22). The formula is:
    IF($H$19="off",1,((($D$13+(K3*BR3/ABS(BR3))*COS(K3) /$D$13))^0.5)))
where the variables are as in Columns CE-CP above.

COLUMN DD Calculates the threshold from Eq.10.37 (D<0.05cm) or from Eq.10.38 (D>0.005 cm) as shown by Eq.AII.23 and AII.24.

COLUMN DE Blank spacer

COLUMNS DF Calculates the maximum instantaneous bedload transport rate from Eq.10.33 (Eq.AIV.25)

COLUMN DG Calculates the reference concentration in mg/l (Eq.AII.26).

COLUMN DH Calculates the reference concentration in g cm^{-3}.

COLUMN DI Blank spacer.

COLUMN DJ-DN Calculates the drift current at heights of 10, 30, 50, 70 and 90 cms above the bed using the second order solution, Eq.6.29 (Eq.AII.27)

COLUMN DO Calculates the total suspended sediment transport rate by integrating through the bottom 100cm of the drift current. (Eq.AII.28 and Eq.AII.29) with A_c set equal to 0.22.

COLUMN DP Blank spacer

Column DQ-EB Calculates the instantaneous bedload transport rate (AII.30) for twelths of the wave period.

COLUMN EC Blank spacer

COLUMN ED Calculates the nett suspended load transport per wave by simply multiplying the mean rate by the wave period (Eq.AII.31).

COLUMN EE Calculates the net bedload transport per wave by integrating columns DQ to EB (Eq.AII.32).

COLUMN EF Summates the shorewards inputs from offshore and the seawards inputs from onshore with the local accretion and erosion in each cell to calculate the mass change per wave.

COLUMN EG Blank spacer

COLUMN EH Calculates the local change in elevation from the mass continuity equation

COLUMN EI Calculates the new profile by adding the Column EH results to the original profile.

COLUMN EJ Blank spacer

COLUMN EK Initial profile distances offshore

COLUMNS EL-EP Evolving profile form at t=0 to 4 where steps are equal fractions of the total iteration time.

COLUMN EQ-ET Evolving response function values corresponding to profiles in EM-EP

COLUMNS EU and EV Wave height values at the beginning and at the end of the iteration time.

COLUMNS EW and EX Peak onshore nearbed flow velocities at the beginning and at the end of the iteration time.

17.3 Bathymetric input

The model requires, as inputs, both the wave parameters and a pre-existing seabed bathymetry. The former requirements are detailed in the following sections and are achieved by the careful choice of a prototype example. The latter requirement presents a more difficult choice and is not well precedented in the beach literature. Davidson-Arnott (1980) and later Watanabe (1988d) both used planar bathymetries as initial boundary conditions. It is argued here that a better choice involves the use a "typical" orthogonal profile, and that operating the process functions over such an initial bathymetry will more closely represent reality. Therefore we take the results of Dean (1977) who examined a very large number of profiles along the eastern seaboard of the United States (cf Chapter Fourteen) and found that:

$$h = a \, x^{2/3}$$

Eq.17.1

where h is again the water depth at a distance x from the shoreline, and a varied between about 0.04 and 0.11. We choose mid-range values so that the profile is represented by the equation:

$$h = 0.075 \, x^{2/3} \qquad\qquad \text{Eq.17.2}$$

and:

$$x = (13.33h)^{3/2} \qquad\qquad \text{Eq.17.3}$$

The differential of any geomorphological equations of this kind with respect to distance is the gradient of the surface so that Eq.17.3 gives:

$$\tan\beta = 0.05 \, x^{-1/3} \qquad\qquad \text{Eq.17.4}$$

The initial profile for the model is then calculated from Eq.14.10 and 14.11 and the results are shown in Table 17.1 and Figure 17.3.

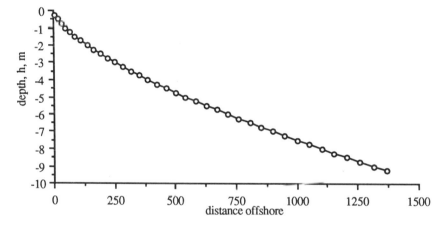

Figure 17. Initial bathymetric profile for the simulations described in this book.

17.4 Wave Parameter Input

Chapter Eighteen approaches the problem of analysing the model through process sensitivity and are therefore not concerned with the absolute magnitude of the input parameters. Nevertheless a reasonable choice of wave height, period and incident direction must be made because the result of the process sensitivity analysis will itself depend upon these parameters.

Spurn Head on the English coast of the North Sea (Plate II) is the location for much of the author's experimental work. The nearshore orthogonal profiles are close to those described above, except for the presence of some tidal channels, and it is therefore not unreasonable to utilise wave data from that site for these simulations. The Institute of Oceanographic Sciences analysed a twelve month wave record from the Dowsing Light vessel some 20 kms offshore from the site. Typical deep water wave heights were found to be 1-2 m with a mean annual zero crossing period of about 5 seconds. The illustrations and examples in this chapter have therefore been generated with T=5 s and H_∞=1.5 m.

253

These will be taken as input parameters for the process sensitivity analysis in the following sections.

17.5 Pseudo constants

In addition to the input parameters, the SLOPES model contains a series of coefficients in the hydrodynamic and sediment dynamic equations. Each is discussed in detail in the appropriate chapter of this book and Table 17.1 summarises the values which will be used for the process sensitivity analysis in the following sections. Although each is given a constant value here, it is in fact a variable, hence the term pseudo-constant. The additional analysis of the effect of each pseudo-constant on the orthogonal system is beyond the scope of this book.

Table 17.2			
Pseudo-constant values in the SLOPES model			
Term	value	units	Chapter
Density of seawater	1025	kg m^{-3}	4
Density of sediment	2650	kg m^{-3}	3
Packing of sediment	0.63	n.d.	3
Bulk density of sediment	1669.5	kg m^{-3}	3
Gravitational acceleration	9.81	m s^{-2}	
Viscosity of seawater	0.01	N s m^{-2}	4
Wave friction factor	0.01	n.d.	5
Attenuation of second-order			
wave coefficient	0.75	n.d.	6

17.5 System characterisation

It is most useful to represent the first-order orthogonal profile, and the response of that profile to time and to changing inputs through time in terms of a few characteristics. This is in part a hankering for the elegance of the analytical solution described earlier, but is also a more practical requirement for inter-comparison. Two sets of choices have been made here. Since the objective of this book is to analyse the geomorphological response of the orthogonal system, the results will be characterised by the response function, $\partial z/\partial t$, which describes the system response to change, and since the input profile is concave, and concavity is a common geomorphic form descriptor, the concavity of the profile as indicated by $\partial^2 z/\partial x^2$ will be considered. The relationships between these system characteristics and the process and parameter sensitivity analyses will be investigated in the following chapters.

chapter eighteen

Process Sensitivity Analysis

18.1 Introduction

It is difficult to choose a routine method by which to report the results from this numerical model of a geomorphological system, and it is undoubtedly true that this problem will become more and more common as other models are developed in the future. In order to make progress, the view is taken here that, in effect, models can be examined in one of two modes. On the one hand a limited system is described to which are then added ever more complex process modules. In this way the effect of each process can be understood individually even if the totality is beyond comprehension. Alternatively the full model can be examined for a variation in input parameters. These two approaches will be called *process sensitivity* and *parameter sensitivity* respectively, and are described separately in the present and in the following chapters.

18.2 Minimal Process Solutions

The hydrodynamic parameters, profiles and response functions for a fixed (deep water) wave height, first order flow and bedload transport with a zero threshold speed and without any bedslope effects for a ninety six hour simulation are shown in Figure 18.1. The simulation is, of course, perculiar, but nevertheless a number of interesting results are predicted by the model, and are discussed in this section.

Firstly, there is, by definition, no change in the wave height but the wavelength decreases from 30.47m at the seaward limit of the model (h=5m) to less than 11.87m at the shoreward limit (h=0.5m). The effect of this is to generate a five-fold change in kh and hence sinh(kh), the former varying from 1.0309 to 0.2646 and the latter from 1.2245 to 0.2678 during shoaling. The result of this is, in turn, to produce nearbed flows which vary considerably from deep water to the shore. The maxima, for example, range from 57.7 cm s^{-1} in 5.0m of water to 263.9 cm s^{-1} in 0.5m of water.

Secondly, since the flows are symmetrical and there are no bedslope effects in the transport function in this simulation, then there is no net bedload transport and hence no geomorphic response. There is however bedload transport and since, with $u_{cr}=0$, the rate varies with the cube of the flow speed. The resulting rates also, therefore, increase

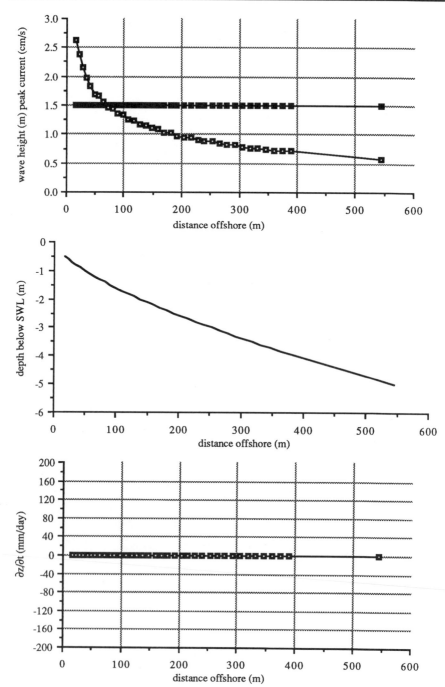

Figure 18.1 Minimal process model solution (a) wave height at t=0 (open squares) and at t=96 hrs (closed diamonds) and peak onshore flow speed at t=0 (closed squares) and t=96 hrs (open diamonds).; (b) surface elevation and (c) the response function.

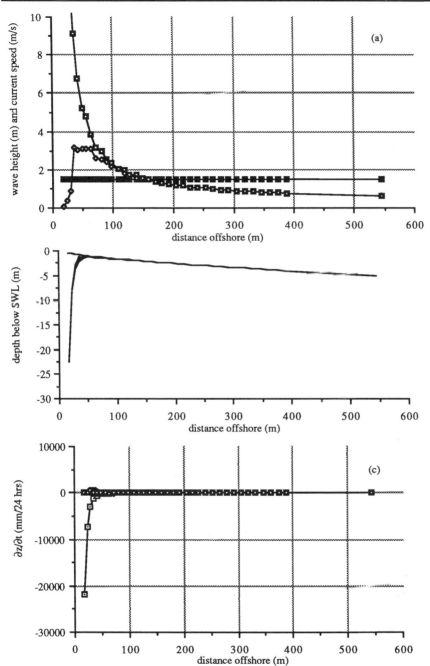

Figure 18.2 Second order model solutions (a) hydrodynamic parameters (all symbols as in Figure 18.1); (b) surface elevation and (c) response function for t=24 (open squares), t=48 hrs (closed diamonds), t=72 hrs (closed squares) and t=96 hrs (open diamonds).

considerably across the profile. The peak rate increases only from 0.1923 gm cm^{-1} s^{-1} in deep water to 18.3813 gm cm^{-1} s^{-1} at the shoreline.

The results are therefore somewhat repetitive, but do provide a yardstick by which the effects of the inclusion of other processes can be examined.

18.3 Flow Process Solutions

The flow processes are first-order oscillatory currents, second-order oscillatory currents and second-order drift currents, although the use of the first two are, of course, mutually exclusive. The flow processes are introduced to the minimal process model before the shoaling transformations because, as will be seen, it is necessary to enable an asymmetric oscillatory flow before the system will evidence any geomorphic response.

18.3(a) Second order solution

The hydrodynamic parameters, profiles and response functions which are produced by the replacement of linear, Airy wave theory with second order, Stokes wave theory for the nearbed orthogonal currents are shown in Figure 18.2. The inclusion of an asymmetric flow function has a marked effect on the net bedload transport which now becomes non-zero, and this is reflected in the geomorphic response of the system.

The flow is now asymmetric and the peak onshore flow velocities range from effectively sinusoidal at 60.2 cms^{-1} in 5.0m water to 182.8 cm s^{-1} in a depth of 1.8m. Since the offshore return flows range over the same distance from 55.3 to 53.6 cm s^{-1}, it is apparent that the flow asymmetry increases from a velocity ratio (Chapter Six) of about unity in deep water to more than three at the shore. Shorewards of the ninth cell (i.e. in depths of less than 1.4m) second order theory has broken down, predicting a "shorewards" return flow and an Ursell parameter of more than 100.

The asymmetry produces net sediment transport and therefore predicts geomorphic change, although under these conditions this is only in the shallowest water where, as described above, the wave theory is inapplicable. This problem disappears when the shoaling transformations are included in the simulation, but for the present it is sufficient to note that the simulation confirms the important intuitive result that second-order theory generates shorewards sediment transport at all depths when there is no bedslope control on the bedload transport formula.

18.3(b) Drift Current Solution

The hydrodynamic parameters, profiles and response functions which are produced by the inclusion of the Stokes nearbed drift current are shown in Figure 18.3. The current is always shoreward at the bed, and produces a considerable effect on the system in the present simulations. Given that the wave height is presently constant, the magnitude of the drift current depends inversely on the wave length and inversely on the square of the sinh(kh) term. Since the wavelength reduces from 30.48 m in 5m of water to 19.12m in about 1.5m of water, whilst the hyperbolic increases only from 1.2247 to 0.5265, then the result is a drift current which increases shorewards from 6.8cm s^{-1} in deep water to 51.4 cm s^{-1} in a depth of 1.5m and to much higher values in the shallower cells.. The combination of second-order oscillatory flow combines with this drift to generate the unrealistically high currents in the shallowest cells as shown in Figure 8.3(a).

The drift current increases the shorewards transport of sediment, and decreases the seawards transport of sediment by an amount which increases towards the shoreline. This positive $\partial i/\approx x$ in a shorewards direction leads necessarily to erosion (Figure 18.3(b)), which ranges from 0.4 mm/24hrs in a depth of 6m to 27.9 mm per day in a depth of 1.5

258

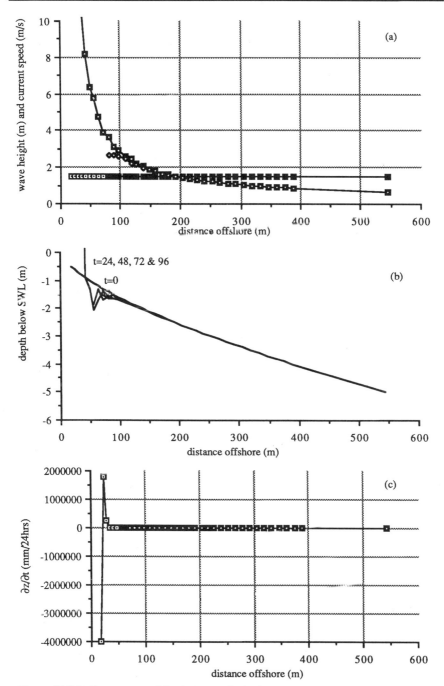

Figure 18.3 Drift current model solution (a) hydrodynamic parameters, (b) profiles and (c) response functions (all symbols in this and subsequent diagrams are as in Figures 18.1 and 18.2).

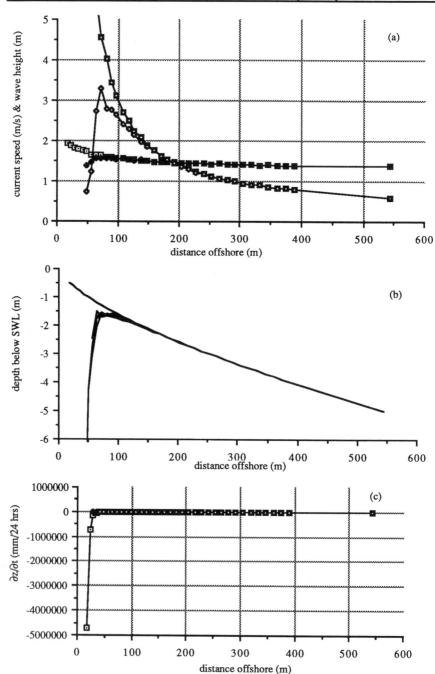

Figure 18.4 Shoaling height model solution (a) H at t=0hrs and t=72 hrs and u_{in} at same times (closed squares and open diamonds), (b) profiles at t=0 hrs, 24 hrs, 48 hrs, 72 hrs and 96 hrs, the last four of which show slow accretion and a large shoreward hollow, (c) response functions showing massive shoreward erosion followed by stability.

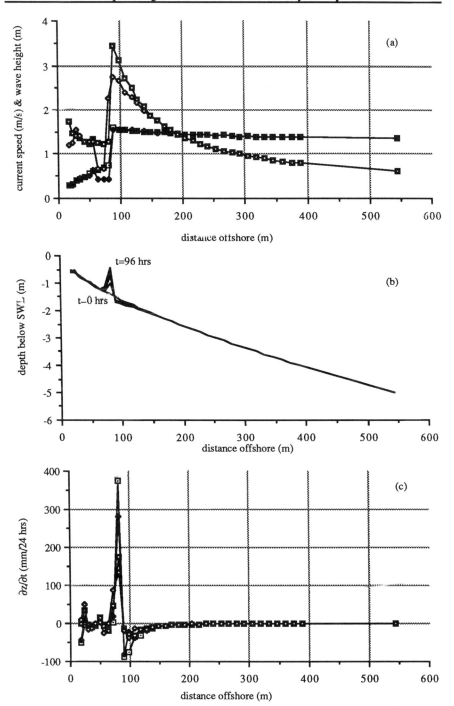

Figure 18.5 Breaking model solution (a) hydrodynamic parameters, (b) profile changes and (c) response function changes through the 96 hour simulation.

m and to values measured in km/24hrs in the shallowest cells at the end of the ninety six hour simulation. The excess sediment is apparently lost shorewards from the first cell but, for the reasons given earlier, the hydrodynamics in this region are not well represented and the fate of this material will remain to be resolved with the inclusion of further processes.

18.4 Tranformation Process Solutions

The processes of wave height change during shoaling which are to be included are the shoaling height transformation, breaking, seabed frictional energy loss, and refraction. The effect of introducing these processes to the system described in section 18.3 is as follows.

18.4(a) Shoaling Height Solution

The hydrodynamic parameters, profiles and response functions which are produced by the replacement of the constant wave height constraint by the wave height increase due to the shoaling transformation are shown in Figure 18.4.

The hashed cells which appeared on the computer screen during this simulation indicated that the numbers have become too ridiculous for even Excel to contemplate, and this is an instructive illustration of the dangers of computer modelling. This has happened because the shoaling transformation equations lead, with mathematical inexorability, to wave heights in very shallow water which are unrealistically large. Figure 18.4(a) shows that, even during the first iteration, the simulated wave height in 50cm of water was 1.94 m. This problem is rectified when the breaking process is introduced in the next section, and we shall confine the discussion here to an identification of the effects of the shoaling transformations in depths of more than two metres.

Most importantly, the increasing wave heights increase the seabed currents: the maximum shorewards flow has increased to more than 300 cm s^{-1} to at a depth of 3 m because of the additional process. As before the result is a positive $\partial I/\partial x$ in the shoreward direction, but the values are now an order of magnitude larger. Again the model becomes unrealistic in the shoreline cell where it suggests that the depth increased to more than 5 km during the first twenty four hours of the simulation, and then stabilised!

18.4(b) Breaking Solution

The hydrodynamic parameters, profiles and response functions produced by the inclusion of the wave breaking process are shown in Figure 18.5. The simulation appears now to be approaching the prototype system.

Hydrodynamically, the wave breaks in cell 10 about 90m from the shoreline, and from that point onwards the wave height is controlled by the water depth and the breaking process. The maximum wave height varies between about 1.54 and 1.56 m during the simulation as the depth of the seaward boundary of the cell varies. The nearbed flow field reflects the surface waves: the peak onshore current increases to about 221 cm s^{-1} at the breakpoint with a corresponding peak offshore flow of about 68 cms^{-1}. The asymmetry at the breakpoint therefore corresponds to a velocity ratio of slightly more than three.

The resulting net sediment transport in each cell is still shorewards, with peak rates that range from 0.2212 g cm^{-1} s^{-1} in 5.0 m up to 20.8964 g cm^{-1} s^{-1} in 1.68m and then down to 1.6815 g cm^{-1} s^{-1} in the shoreline cell at a depth of 50cm. It is this pattern of change in the generally shorewards directed sediment transport ($\partial I/\partial x$) which controls the geomorphological response of the system as shown by Figures 18.6(a) and 18.6(b). Seaward of the breakpoint $\partial I/\partial x$ is positive and increases in a shoreward direction. However, shoreward of the breakpoint $\partial i/\partial x$ is negative but still increases in a shoreward direction. The result, it appears, is extremely instructive and should now be interpreted in the context of the morphodynamic principles which were introduced in the preceeding chapters.

Consider Figure 18.6(b): $\partial I/\partial x$ is positive from offshore to the breakpoint some 100 m from the shoreline, so that the continuity equation (Eq.13.5) necessitates a negative $\partial z/\partial t$, and therefore the seabed is eroding. Furthermore $\partial I/\partial x$ is increasing in a non-linear fashion and therefore, at least in qualitative terms, the seabed is eroding more rapidly as the breakers are approached from offshore. This can be quantified and is confirmed by inspection of Figure 18.5(c) and of the detailed numerical results of the model. These show that, after 24 hrs, $\partial z/\partial t$ was increasing from -1 mm day^{-1} in 4.0 m of water to -89 mm day^{-1} at the breakpoint. The seabed profile which was initially concave therefore becomes less concave. Mathematically this is written $\partial^2 h/\partial x^2$ remains positive but less so. Sediment is moving shoreward at an increasing rate up to the breakpoint.

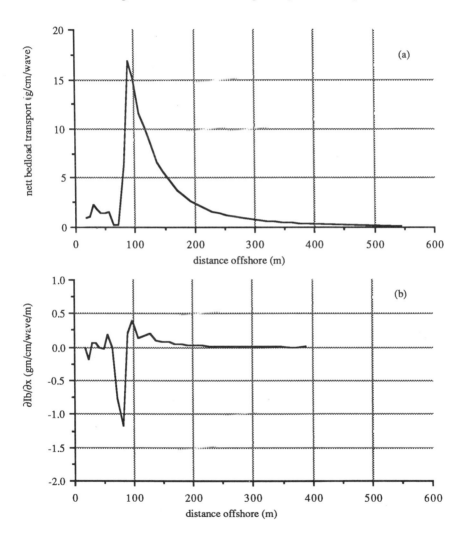

Figure 18.6 (a) cross shore changes in Ib and (b) in di/dx for the SLOPES model as configured in section 18.4(b).

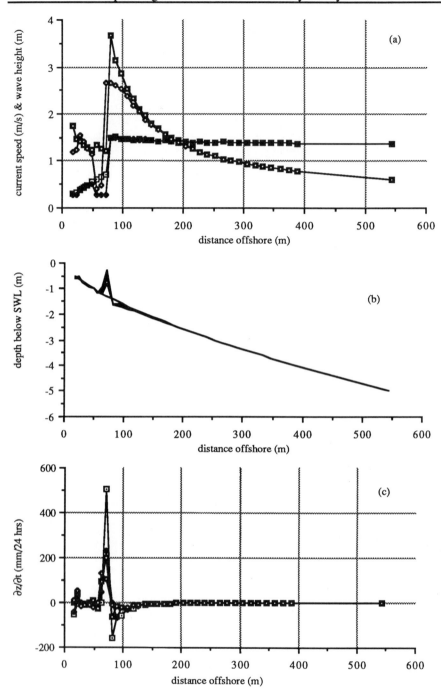

Figure 18.7 Frictional loss model solution (a) hydrodynamic parameters, (b) profile changes and (c) response surfaces for the 96 hr simulation.

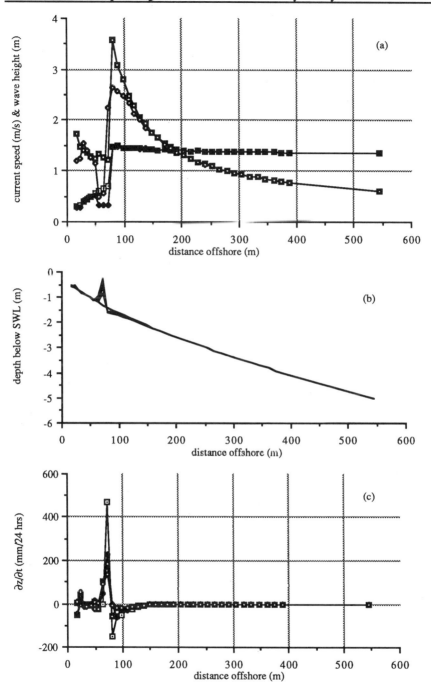

Figure 18.8 Refraction model solution (a) hydrodynamic parameters, (b) profile changes and (c) response functions throughout the 96 hr simulation. Symbols as defined in Figures 18.1 and 18.2.

265

The dynamics change shoreward of the breakpoint, however, because of the scale differences, $\partial I/\partial x$ is now increasing shorewards but is negative. The response function, $\partial z/\partial t$, is therefore negative but decreasingly so in a shorewards sense. The result is an increased concavity. Sediment is moving shoreward from the breakpoint, but at a decreasing rate. The overall result is the construction of a breakpoint bar, but the rate of accretion is decreasing throughout the simulation. This process will be discussed further in later sections of the chapter.

Mathematically we say $\partial^2 h/\partial x^2$ remains positive, but more so. The discontinuity, wherein $\partial I/\partial x$ is large and negative occurs in cell 10, and results in rapid accretion and bar formation.

18.4(c) Frictional Loss Solution

The hydrodynamic parameters, profiles and response functions which are produced by the inclusion of the seabed friction process are shown in Figure 18.7.

The inclusion of this process has but a small effect on the system in this configuration. Maximum wave heights at the seaward boundary of cell nine decrease by 9 cm (compare Figure 18.5(a)) and Figure 18.7(a)) and the associated peak onshore flows reduce to 2.45 m s^{-1}. The transport rates are reduced accordingly, but the net transport and response of the system remains broadly as it was, though reduced off the breakers. Shorewards of the breaker zone, of course, the seabed friction has no effect because height is controlled solely by the water depth and the wave breaking process.

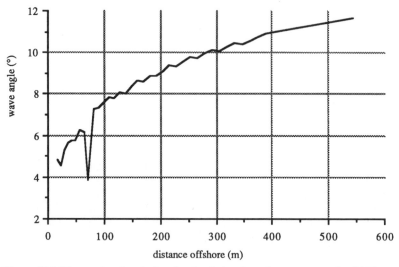

Figure 18.9 Wave refraction during the simulation for a deep water angle of 15°.

18.4(d)   Refraction solution

The hydrodynamic parameters, profiles and response functions which are produced by the inclusion of wave energy loss due to refraction over a shore-parallel (but not planar) bathymetry are shown in Figure 18.8.

The inclusion of this process also has but a slight effect on the system. The wave height during shoaling decreases. The waves immediately seaward of the breakpoint have, for example, lost some 9 cm in height due to the 15° angle of incidence and the subsequent refraction. The result is that the breakpoint moves shoreward by approximately 10m, although the precise location is undefined because of the vertical and thus horizontal

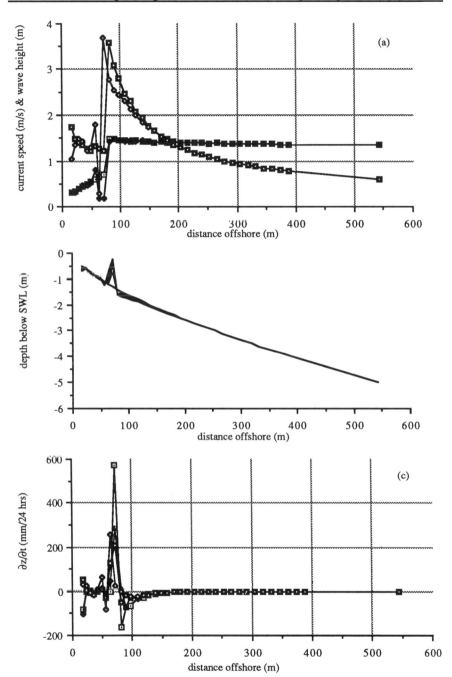

Figure 18.10 Total load model solution. (a) hydrodynamic parameters, (b) profile changes and (c) response functions.

resolution of the model. The whole profile response is effectively shifted shorewards slightly up to the breakpoint, but thereafter, as before, is identical to the breaking and frictional loss solutions. The wave angle for this particular example of shoreparallel, but not planar bathymetric contours is shown in Figure 18.9, where it can be seen that the waves are effectively normal to the shoreline at the breakpoint (x=72 m) and that, therefore, there will be only a weakly developed longshore current. The dip in the wave angle diagram at the breakpoint corresponds to refraction toward and then away from the shoreline over the breakpoint bar which has developed on the profile as shown in Figure 18.8(b).

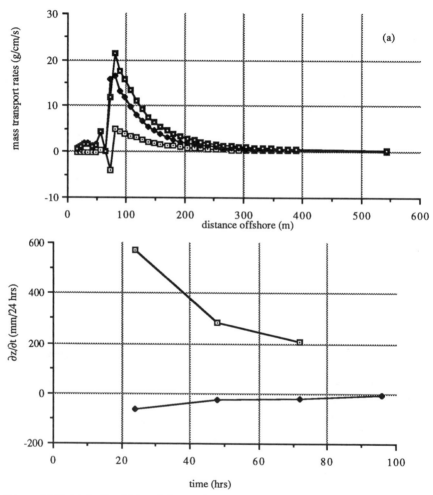

Figure 18.11 (a) Bedload (closed diamonds), suspended load (open squares) and total load transport rates (open squares) at the seawards boundary of each cell in the total load solution and (b) The response function plotted through time for the bar crest (x=72m and open squares) and offshore from the bar (x=98m and closed diamonds). The bar crest moved shoreward during the last 24 hrs of the simulation and therefore the fourth $\partial z/\partial t$ was not plotted.

18.5 Sediment dynamic process solutions

The additional process which control the movement of loose, non-cohesive material in the beach system, and which are included in the sediment dynamic process solutions are the transport of suspended load, thresholds and the effect of bedslope of the thresholds and transport rates. The effect of introducing these processes to the system described in section 18.4 is as follows.

18.5(a) Total load solution

The results screen, response function and profiles which result from the inclusion of the suspended sediment processes are shown in Figure 18.10. Before discussing the results in detail, the absolute values of the suspension concentrations and transport rates will be described because the original equations were found to be inappropriate for the model. In particular, it will be recalled that the suspended load transport rate is calculated by integrating the product of the drift velocity profile and the concentration profile. The concentration profile is calculated from a logarithmic decay equation referenced to the concentration at a height of 1 cm above the seabed. The reference concentration is calculated from the empirical relationship which had been determined between the reference concentration and the bedload transport rate. This approach was chosen because it appears to be physically realistic, but the resulting transport rate depends critically on the relationship between the bedload rate and the reference concentration. The model utilised Vincent *et al.*'s (1982) formula given by Eq.10.30 as:

$$C_a = 330 \frac{i_b}{\rho_s u} \qquad \qquad \text{Eq.18.1}$$

where C_a is measured in mg litre^{-1}, i_b has units of g cm^{-1} s^{-1}, ρ_s has units of g cm^{-3} and u has units of cm s^{-1}. It was found that this formula gave reference concentrations of the order of 1 to 10 mg litre^{-1} which were in agreement with the original Vincent *et al.* (1982) results. However these values are approximately two orders of magnitude smaller than other data sets. For example, Dyer (1986, p. 171) quotes sand concentrations measured at a height of 18 cm in a 96 cm s-1 tidal current of 1,000-2,000 mg litre^{-1}. Again, Sleath (1984, p.273) quotes measurements at a height of 1.5 cm above a rippled sand bed in an oscillatory flow which range from 5 to 10 kg m^{-3}. Since 1 kg m^{-3}=1000 mg m^{-3}=1000 mg litre^{-1}, Sleath's data range from 5 to 10,000 mg litre^{-1}. Sternberg *et al.* (1989) reports OBS results from the NSTS experiment with nearbed concentrations of 2 to 40 kg m^{-3}, i.e. 2,000 to 40,000 mg litre^{-1}. Finally Horikawa (1988, p.188) report further flume channel results with sand concentrations in the 1 to 5 x10^3 ppm which is equivalent to 1,000 to 5,000 mg litre^{-1}. Given these results, and initial simulations with the SLOPES model which showed that Eq.18.1 generated suspended load rates which were about two orders of magnitude less than the bedload rates reported earlier in this chapter, a decision was taken to increase the multiplier in the equation by the somewhat arbitrary value of 100. The results reported below include this change, and further confirmation must await the results of additional field work.

The addition of the suspended load process does not, of course, have any effect on the hydrodynamic regime at the beginning of the simulation, but the considerable change in the geomorphic response of the system which it produces does feedback to alter the shoaling transformations in subsequent iterations of the model. Notice, for example, how the wave height at the break point some 90m from the shoreline changes from the bedload solution (Figure 18.8(a)) to the total load solution (Figure 18.10(a)). This is wholly attrinbutable to the reduced bar development as a result of the changed sediment dynamic regime, and this will now be described.

The changing sediment dynamic regime is well illustrated by considering the total load rate (I) as shown in Figures 18.11(a) for comparison with Figures 18.6(a) and (b). The patterns are similar, for the total load increases shorewards towards the breakers with the suspended component, in this simulation, contributing more than half of the total transport. The spatial change in the transport rate again increases at an increasing rate in a shorewards direction, and the inclusion of the suspended load has increased $\partial I/\partial x$ by almost an order of magnitude. Shoreward of the breakers, rates are much lower at the end of the simulation because the bar creams off the wave energy. However there is now, for the first time, evidence of net seawards sediment transport of suspended load which tends to balance the onshore bedload transport and to result in negligible geomorphic response in shallow water.

The behaviour of the response function, $\partial z/\partial t$, through time is shown at two locations in Figure 18.11(b), where, with increasing bed elevation at the bar crest, $\partial z/\partial t$ is positive but $\partial^2 z/\partial t^2$ is negative indicative of stability. The rates enable an estimation to be made of the relaxation time of the system, R_t (Chapter Thirteen). Given that $\partial^2 z/\partial t^2$ is about 400 mm/24 hr over 72 hours (approximately 0.015 mm/hr/hr), then 72 hrs value of $\partial z/\partial t$ of 208 mm/24hr (9 mm/hr) will reduce to zero in 570 hrs. The relaxation time of the system at the bar crest in this simulation is therefore more than three weeks, although there is some evidence that $\partial^2 z/\partial t^2$ is not constant but is decreasing with time. That is the system is approaching the equilibrium asymtopically in the manner described in Chapter Thirteen. Conversely in the region seawards of the bar where $\partial z/\partial t$ is negative, $\partial^2 z/\partial t^2$ is positive which is again indicative of stability (Chapter Thirteen). These trends are discussed further in a later section and in Chapter Twenty.

18.5(b) Threshold Inclusive Solution

The inclusion of a threshold process was found to make an insignificant difference to the system because, for the sand sized sediment (D=0.05 cm) the threshold flow speed was low. It was found to vary, for example, from 4.5 cm s^{-1} in 9.25 m of water to 3.4 cm s^{-1} in 0.50 m of water. These values are small in comparison with the peak onshore flow speeds which ranged from 57 cm s^{-1} in deep water to more than 2m s^{-1} at the break point.

18.5(c) Slope Inclusive Solution

The hydrodynamic parameters, profiles and response functions which are produced by the inclusion of the slope inclusive processes with an exponent of seven on the A_b term (Chapter Nine) in the simulation are shown in Figure 18.12.

The inclusion of these processes has a marked effect on the system as is best exemplified and explained by the plot of the slope correction factor, A_b, in Figure 18.13 for the system at the end of 72 hrs of the simulation. The diagram shows the correction which the model applies to the peak shorewards transport. A different correction is, of course, applied at each of the $\pi/6$ steps in each wave because it depends upon the sign of the seabed gradient and the direction of the flow. It is apparent that the slope effect decreases the onshore (upslope) transport by a factor which ranges from 0.90 on the flat seabed further offshore to about 0.35 on the steeper seabed immediately seaward of the bar. Conversely the onshore transport rate is enhanced by the same gravitational effects by a factor which ranges up to almost 2.1 on the steep shoreward side of the bar. The movement of the onshore factor above and below the unity line is, quite correctly, caused by the (negative) shorewards sloping face of the inside of the bar.

These results show the manner in which the slope-inclusive solution tands to supress change by reducing the shorewards flux of sediment in the regions offshore from and inshore from the bar, and to flatten the bar by enhancing shorewards transport on its inner face. The effect is well exemplified by again considering the change in $\partial z/\partial t$ over time at,

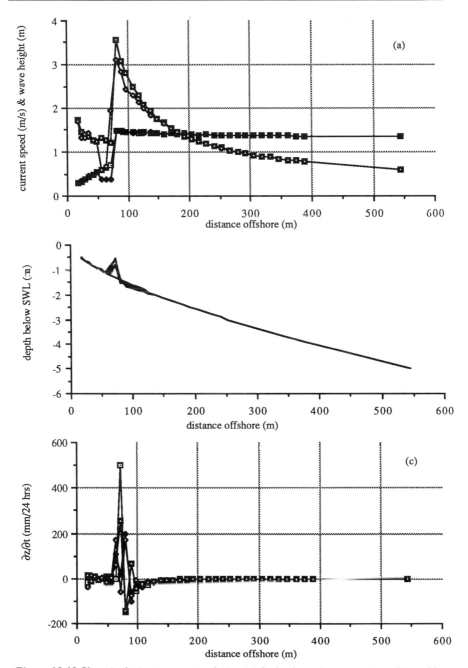

Figure 18.12 Slope inclusive transport solution (a) hydrodynamic parameters, (b) profiles and (c) response functions.

for example, the bar crest. The values are now 501, 110, 62 and 3 mm/24 hr. The value of $\partial^2 z/\partial t^2$ for the system is therefore about 0.08 mm hr^{-1} (50 mm /24hr/24hr), giving a ralaxation time of about four days.

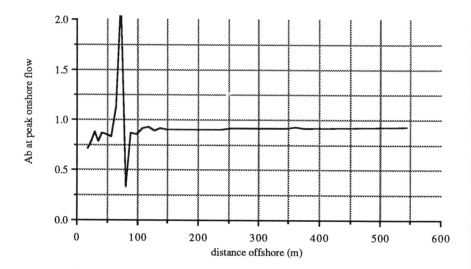

Figure 18.13 The bedload slope correction factor, Ab, which the model applies to the peak shoreward transport atfter seventy two hours in the slope-inclusive simulation.

The system is therefore stable, and responds by more quickly establishing an equilibrium when the slope-inclusive effects are represented in the model, though the absolute values depend upon the initial bathymetry (the geomorphic memory, Chapter One), and a fuller discussion but be reserved for Chapter Twenty. It is none-the-less apparent that the full SLOPES model is generating a system response which is of the same order of magnitude as the prototype values which were discussed in Chapters Fourteen to Sixteen. These results will be summarised in Chapter Twenty, but initially the sensitivity of the model to changes in the input parameters is considered.

chapter nineteen

PARAMETER SENSITIVITY ANALYSIS

19.1 Introduction

The alternative analysis to the preceeding process sensitivity techniques which can be applied to the SLOPES model is that of input parameter sensitivity, whereby the effect of the three independent parameters, H_∞, the deep water wave height, T, the wave period and α_∞ the deep water wave incident angle are examined independently. The results of this analysis are described in the following sections, and are followed by a brief analysis of the effect of varying the initial seabed profile in Section 19.5.

19.2 Wave Height Solutions

The model was run with all of the processes detailed in the preceeding chapter enabled, and with a range of deep water wave heights from 1.0 to 3.0 m in increments of 1.0 m. The wave period was taken as T=5s and the wave incident angle as α_∞=15°. The resulting hydrodynamic parameters, profiles and response functions are shown in Figure 19.1.

The system as a whole operates in a similar manner to that described earlier and the change in wave height produces a predictable and intuitively correct geomorphic response. For low values of H_∞ (1.0 m), all of the flow and sediment transport parameters are low, and the geomorphic response is slow and supressed in comparison with the results from the earlier simulations. A bar begins to form some seventy metres from the shore, constructed largely of material eroded from cells further offshore. The increasing wave height, however, increases the orthogonal currents and the sediment transport parameters. In the final simulation for H_∞=3.0 m, the transport rate reaches a peak value of more than one hundred g cm^{-1} wave^{-1} at a distance of about 130m from the shoreline. Theresult is that larger bars form within the twenty four hour simulation, and they do so in increasingly deep water because the larger waves are breaking further from the shoreline.

The response function appears to correctly reflect these changes, increasing from a maximum of about 300 mm 24 hrs^{-1} for the smallest waves to some 1650 mm 24 hrs^{-1} for the largest. There is however evidence of a problem due to the time steps of the iterations within the model, because the breakpoinbt bar accretes above the still water level. Although practically this response is possible, it is more likely due to the iteration time steps. This problem will be discussed further in the following chapter.

19.3 Wave Period Solutions

The model was run with all of the processes detailed in the preceeding chapter enabled, and with a range of wave periods from 4.0 to 8.0 s in increments of 2.0 s. The wave height was taken as H_∞=1.5 m and the wave incident angle as α_∞=15°. The resulting hydrodynamic parameters, profiles and response functions are illustrated in Figure 19.2.

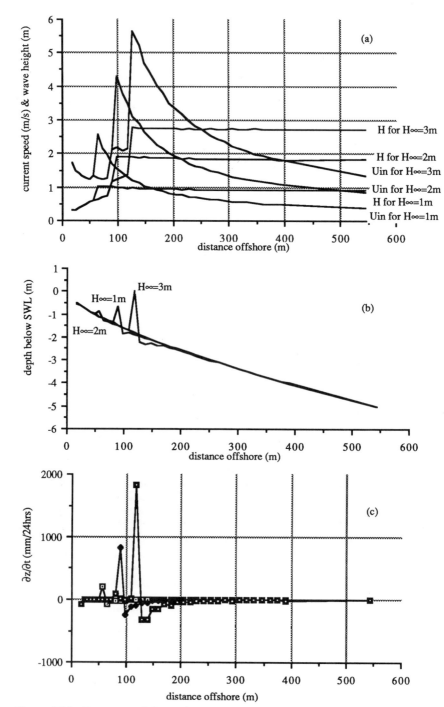

Figure 19.1 Response of the orthogonal profile model to wave height changes. (a) hydrodynamic parameters, (b) profiles and (c) response functions.

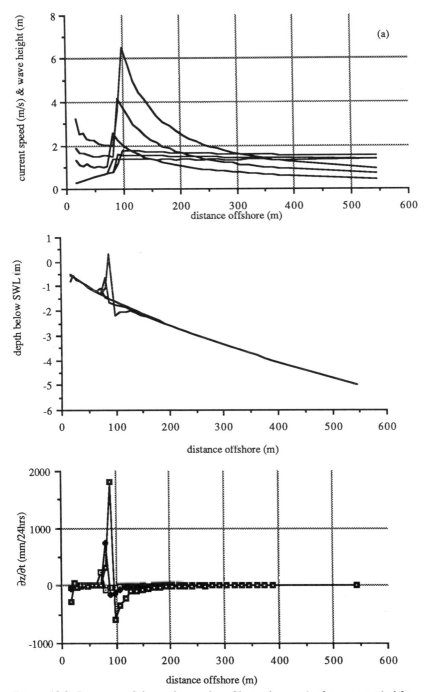

Figure 19.2 Response of the orthogonal profile to changes in the wave period from 4 to 8s. (a) hydrodynamic parameters, (b) profiles and (c) response functions.

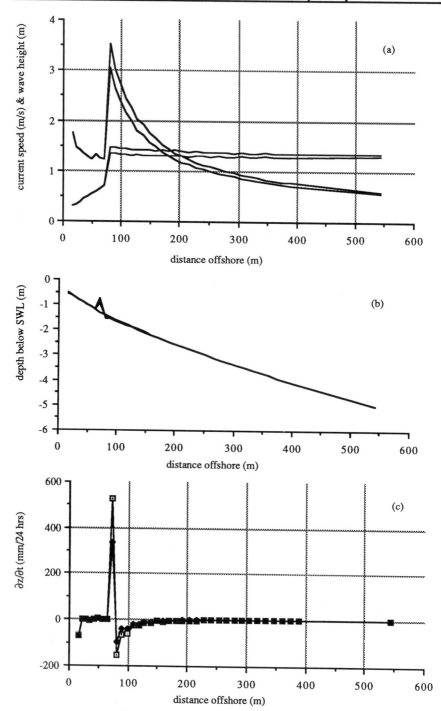

Figure 19.3 Profile response of the system to wave incident angle changes.

The system as a whole operates in a similar manner to that described earlier and the change in wave period produces predictable and intuitively correct geomorphic responses. For low values of T (4.0 s), all the hydrodynamic and sediment transport parameters are low, and the geomorphic response is slow and supressed. In Figure 19.2(a) the upper three curved lines represent the peak onshore currents and the lower three lines the local wave height. In each case the lowest line corresponds to T=4.0 s, the middle line to T=6.0 s and the upper line to T=8.0 s. With increasing wave period, however, the flow and transport parameters are increased and the bar forms further from the shoreline. The peak transport rates are, for example, in excess of 200 g cm^{-1} wave^{-1} during the T=8.0 s simulation. Once again the simulation becomes unrealistc because of the iteration time step problem which was discussed earlier and will be examined in the following chapter.

Finally the response function is shown in Figure 19.2(c) wherein the open circles represent T=4.0 s, the closed diamonds represent T=6.0 s and the closed circles represent T=8.0 s. The response functions confirm that the bar is again constructed largely from sediment eroded further offshore, and the absolute values at the breakpoint increase from about 300 mm 24 hrs^{-1} for T=4.0 s to some 1800 mm 24 hrs^{-1} for T=8.0 s,

19.4 Incident Angle Solutions

The model was run with all of the processes detailed in the preceeding chapter enabled, and with two wave incident angles α_∞=20° and α_∞=40° and for a simulation length of twenty four hours. The wave height was taken as H_∞=1.5 m and the wave period was taken as T=5 s. The resulting hydrodynamic parameters, profiles and response functions are shown in Figure 19.3.

The system as a whole operates in a similar manner to that described earlier and the change in wave incident angle produces a predictable and intuitively correct geomorphic response. This can be understood by recalling that an increased angle of incidence, leads to greater refraction and therefore to a greater divergence of the wave orthogonals. The results is that there is an inverse relationship between the angle of incidence and the local wave height at a particular depth. Thus the upper line in Figure 19.3(a) represents the results from the α_∞=20° whilst the lower line represents the results from the α_∞=40° simulation. In each case the curved lines represent the orthogonal currents, whilst the approximately horizontal lines represent the local wave heights. It is apparent that the change in the angle of incidence has but little effect on the hydrodynamic parameters. The results therefore confirm that, as the angle of incidence increases, the local height, flow and transport parameters decrease and the bar development is similar to but the reverse of the effect of increasing wave height shown in Figure 19.1. The lower values of α_∞ generates the smaller bar in Figure 19.3(b) and the correspondingly smaller response functions in Figure 19.3(c). It is apparent that, were the offshore profiles rather shallower, then an increased angle of wave approach would lead to smaller breakers and therefore to a more shoreward position of bar formation. These results will also be summarised in the final chapter.

chapter twenty

TIME, TRENDS AND CHANGE

20.1 Introduction

With but a few exceptions, all of the preceding work in this book has utilised a monochromatic wave train approaching from a single deep water direction across shore-parallel (but not necessarily regularly spaced) bathymetric contours. Although the analysis has proven fruitful and has led to a number of interesting conclusions about the behaviour of the orthogonal system, the reality is more complex. This chapter is designed to summarise and to conclude the book by investigating that complexity and considering some of the the effects of short term change on the orthogonal profile system.

The preceding chapters have described the design, construction and operation of a version of SLOPES which, in so far as has been possible, utilises precedented theoretical formulae. That is, a computer model has been assembled which initially represents a power law form of the orthogonal profile between depths of 0.50 and 5.0 m. The hydrodynamic processes have been represented by linear shoaling, integrated linear frictional loss, wave refraction following Snell's Law and breaking based upon a γ_b of 0.72. Within the surf zone the wave heights have been taken to be a linear function of the water depth. Oscillatory currents have been driven by Stokes' second order equations, and the second order solutions of Longuet Higgins have been used to compute the profile of the residual currents. The theoretical magnitudes of the oscillatory currents have been reduced in accordance with some laboratory experiments, and second-order theory has been forced into shallow water by attenuating the magnitude of the the first harmonic. The bedload transport rate has been driven by an integrated, instantaneous, Bagnold type of excess stress transport function which has been calibrated with a range of existing and new experimental data. The bedload rate has been used to compute a reference concentration by a method which is analogous to that described by Vincent, Young and Swift, but it was found to be necessary to increase the resulting values in order to more nearly equate bedload and suspended load transport rates and to more nearly replicate alternative laboratory and field measurements. The suspended load concentration profile has been computed from the reference concentration using a logarithmic decay law, and the suspended load transport rates have been computed by depth integration of the vector products of the concentration and drift velocity profiles. The bedload threshold conditions have been modelled using Komar and Miller's formulae, and the effect of bedslope on the threshold has been modelled using Dyer's geometrical relationships, which have been shown to hold in laboratory conditions. The effect of bedslope on the rate of bedload transport has been modelled using Bagnold's theoretical correction factor, but the dependence of the rate on the slope has been increased because of the tentative results of some laboratory and (aeolian) field experiments. Finally the integrated net bedload and suspended loads have been transported between a series of cells on the orthogonal profile, and thus the response of the profile to a variety of wave conditions has been investigated.

279

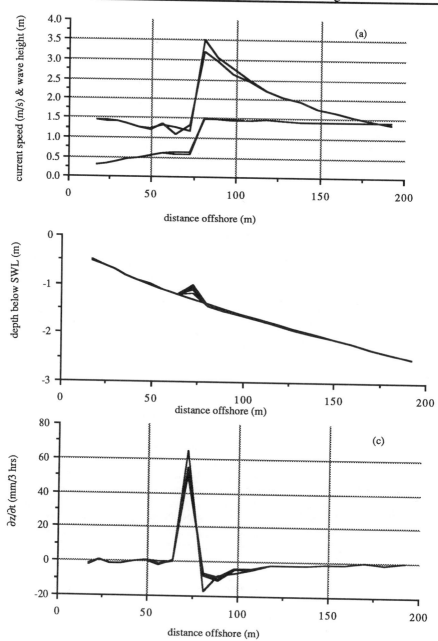

Figure 20.1 Twelve hour simulation in three hourly steps showing (a) hydrodynamic parameters, (b) profiles and (c) response functions. Details in the text.

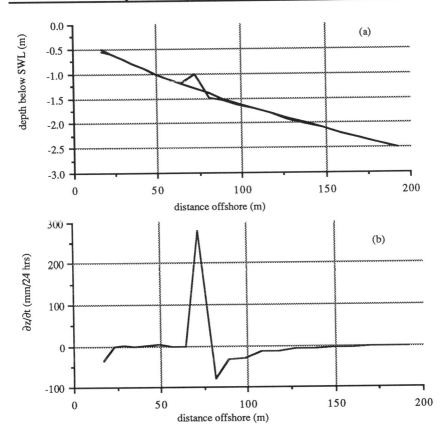

Figure 20.2 Single twelve hour simulation showing (a) profiles and (b) response function. Details in the text.

The results of the simulation work are summarised in Section 20.4, but firstly attention must be given to some of the shortcomings of the model. The hydrodynamic, sediment dynamic and morphodynamic processes have been represented by a series of algorithms and it is reasonable to assume that there may well be errors both in the process elements of the model and in the implementation of the process algorithms within SLOPES. These problems are discussed in section 20.5. In addition, there are two rate related problems which are a result of the operation of SLOPES and of the choice of monochromatic input parameters. The first concerns the iteration time and we have seen that the changes in the response function, that is $\partial^2 z/\partial t^2$, are more pronounced when the model is running with a longer iteration time. This problem is discussed in section 20.2. Secondly, the incident waves on a real beach are not monochromatic and therefore the continually changing wave heights and periods will result in a more complicated profile response than has been described in the preceding chapters. Therefore, a brief analysis of the effect of iteration time and of wave height and period distributions will be presented in section 20.3 before moving to summarise the results.

20.2 Iteration Time

The various simulation results which have been presented in Chapters Eighteen and Nineteen are based upon four sequential sets of twenty four hour iteration times. That is the processes are presumed to operate without feedback for twenty four hours in order to calculate net transports and profile response. The result of this iteration time is that a change in the profile, for example seabed steepening due to bar formation, does not change the transport rate due to the slope inclusive effects until the completion of the iteration period. The result of this is, in turn, that the offshore transport is not enhanced and therefore the response function achieves a larger value than would be the case with a shorter iteration time. The effect is illustrated by Figures 20.1 and 20.2 which show the hydrodynamic parameters, profiles and response functions which result from running SLOPES for twelve hours with H_∞=1.5 m, T=5 s and α_∞=15°. Figure 20.1 illustrates the use of four sequential sets of three hour iteration times and Figure 20.2 illustrates the result of a single, twelve hour iteration time. It is apparent that the response function has, at the breakpoint, a maximum value of 62 mm/3 hrs which reduces to about 45 mm/3 hrs after nine hours in the former case, compared to a value of about 280 mm/12 hrs (70 mm/3 hrs) throughout the latter case. The conclusion is that shorter iteration times reduce the response function for stable equilibria and therefore reduce the relaxation time of the system.

20.3 Orthogonal Response to Change

We have seen, in Chapter Six, that the three wave parameters can exhibit more or less random variation. The wave heights in any wave train are distributed about a mean value, the wave periods are better represented by a continuous rather than a line (monochromatic) spectrum and waves arriving at a particular site on the shore may have originated along different (and intersecting) orthogonal paths. The three dimensional seabed bathymetry will, therefore, be an amalgam of the intersecting, two-dimensional orthogonal profiles because, except for the narrow longshore current region shorewards of the breakers, there is no mechanism for wave induced transport except along each orthogonal. The effects of the distributions of height and period are analysed in the following sections.

Table 20.1		
Rayleigh wave height distribution for H_{rms}=2m		
H (m)	P(H)	%
0.25	0.123	3
0.50	0.235	6
0.75	0.325	8
1.00	0.389	10
1.25	0.422	11
1.50	0.425	11
1.75	0.404	10
2.00	0.365	9
2.25	0.314	8
2.50	0.259	7
2.75	0.205	5
3.00	0.155	4
3.25	0.114	3
3.50	0.080	2
3.75	0.054	2
4.00	0.035	1

20.3 (a) A Distribution of Wave Height

Early conjecture and recent field results have supported the use of the Rayleigh probability distribution (Section 5.8) for wave height in a random sea. The distribution is given by:

$$P(H) = \frac{2H}{H_{rms}^2} \exp - \left(\frac{H}{H_{rms}}\right)^2 \qquad \text{Eq.20.1}$$

where $P(H)$ is the probability of a wave height H, and the distribution has a root mean squared value of H_{rms}. Table 20.1 shows the wave height distribution given by the Rayleigh equation for $H_{rms}=2$ m. The maximum probability (0.425) corresponds to a wave height less than H_{rms} because of the inherent skewness of the Rayleigh distribution. The effect of this height distribution was compared with the same duration of waves having $H_\infty=H_{rms}$ by running two 100 hr simulations on the SLOPES model. The random wave height distribution was investigated by running sixteen sequential simulations with the range of wave heights and the percentage durations given in the table. That is three hours of $H_\infty=0.25$ were followed by six hours of $H_\infty=0.50$ and so on. The constant wave height was investigated by running sixteen sequential simulations of six hours duration and each with $H_\infty=2$ m, the root mean squared value. In all of the simulations the wave period was set to $T=8s$ and the offshore wave angle was set to $\alpha_\infty=15°$.

The results showed that, during the variable wave height simulations, the maximum response function was less than 1cm per sequence for waves of $H_\infty<1$ m, and that these conditions do not therefore contribute significantly to the geomorphic response of the profile over such periods of time. The results for $\partial z/\partial t$ remain small and therefore the profile alters little until the larger waves were introduced into the simulation. The use of the larger wave heights of 3.0 m, 3.5 m, 3.75 m and 4.0 m accounted for only 4%, 2%, 2% and 1% of the total time yet resulted in the majority of the profile response. The results showed that the offshore bar developed, as before, by shoreward transport of material, but that the profile responded at a decreasing rate ($\partial^2 z/\partial t^2$ is negative).

A comparison of the results from the two simulations showed that identical profiles were produced depths of less than 1.30 m and, to within 1cm, in depths of more than 3.70 m. The region between these two limits, within which an offshore bar developed in both simulations, showed a maximum difference between the two profiles of 40cm in about 1.60 m of water and the variable height simulation developed shallower water closer to the shore. This clearly was a result not of the distribution of wave heights, but of the sequence of occurence of particular wave heights. The bar in the variable height simulation developed closer to the shore under the smaller waves, and there was insufficient time for the larger waves to then move it offshore as the regime changed. We are clearly reaching the limits of sensible simulation at this stage, further investigation depending upon the patterns within a random sequence and upon the relaxation time of the beach system on a wave by wave basis. Furthermore it was evident that the relaxation time was not constant along the profile at any particular time.

20.3 (b) A Distribution of Wave Periods

The distribution of wave periods can be represented by the Pierson-Moskowitz spectrum (Section 5.8):

$$E(\omega) = \frac{2\pi A g^2}{\omega^5} \exp \left(-B(\omega/\omega_o)^4 \right) \qquad \text{Eq.20.2}$$

where $E(\omega)$ is called the spectral density function, ω is the wave frequency, A is an empirical constant ≈ 0.0081, B is a second empirical constant ≈ 0.74 and ω_0 is the central frequency which can be related to the wind speed. Taking a wave period of T=8 s giving a central frequency of $\omega_0 = 0.125$ Hz generates the P-M Spectrum shown in Table 20.2.

Table 20.2

Percentage energy in
frequency bands from the Pierson-Moskovitz spectra

$\omega/2\pi$	T	Mean T	Percentage
0.10-0.12	10-8.33	9.0	0.00
0.12-0.14	8.33-7.14	8.2	0.23
0.14-0.16	7.14-6.25	7.6	4.69
0.16-0.18	6.25-5.56	5.9	14.22
0.18-0.20	5.56-5.00	5.3	19.39
0.20-0.22	5.00-4.55	4.8	18.42
0.22-0.24	4.55-4.17	4.3	14.77
0.24-0.26	4.17-3.85	4.0	10.95
0.26-0.28	3.85-3.57	3.7	7.84
0.28-0.30	3.57-3.33	3.4	5.55
0.30-0.32	3.33-3.13	3.2	3.94

The table shows the occurence of wave energy in the different frequency bands, and that the distribution has a mean wave period of about 5 s. The effect of this distribution of wave periods on the orthogonal profile was compared with the effect of the same duration of average waves in the SLOPES model by running one hundred hours of sequential simulations for each case. The deep water wave height was set to $H_\infty = 2$ m and the incident wave angle to $\alpha_\infty = 15°$ for both of the simulations. The variable wave periods given in the table were simulated by 0.23 hours of T=8.2 s followed by 4.69 hours of T=7.6 s and so on. The fixed wave period was simulated by running one hundred hours with a wave period of T=5s.

Once again the results showed that the longer period waves were responsible for the majority of the profile response. Furthermore the longer waves tended to move the profile offshore because they entered shallow water farther from the shoreline. In the variable period simulations, the bar formed about 150 m offshore and tended to migrate shoreward but the shorter waves did not operate for long enough to establish an equilibrium position. Once again, therefore, the resulting profile depended upon the order of occurence of particular waves in a random sequence and upon the pre-existing geomorphology at that time.

20.4 Concluding Remarks

The geomorphological analysis of the response of an orthogonal profile defined initially by the power law:

$$h = ax^b \hspace{4cm} \text{Eq.20.3}$$

to the various process functions and to the monochromatic wave conditions detailed in the present and preceding chapters has suggested a number of conclusions. The process sensitivity analyses reported in Chapter Eighteen suggested that:

1. With the exclusion of shoaling, frictional loss, breaking, refraction, drift

current, suspended load and slope inclusive effects, symmetrical oscillatory current theory generates high bedload transport rates in shallow water, but zero net transport and therefore $\partial z/\partial t=0$ for all water depths.

2. With the same exclusions as in 1, second order, asymmetrical current theory generates net shorewards transport in all water depths, $\partial I/\partial x$ is positive in a shoreward sense and therefore $\partial z/\partial t$ is increasingly negative in a shoreward sense. Profile concavity decreases because $\partial z/\partial t$ increases in a shoreward sense.

3. With the exclusion of frictional loss, breaking, refraction, drift current, suspended load and slope inclusive effects, the shoaling height transformations serve to increase the magnitude of I, $\partial I/\partial x$ in a shoreward sense and profile concavity.

4. Frictional loss, refraction and threshold effects have but a small effect on the simulated system tending, in general, to reduce the magnitude of I, $\partial I/\partial x$ in a shorewards sense and the rate of increase of profile concavity is also reduced because $\partial z/\partial t$ decreases in a shoreward sense.

5. The inclusion of the breaking process introduces a marked discontinuity into the orthogonal system. The magnitude of I and $\partial I/\partial x$ and therefore of $\partial z/\partial t$ increase in a shoreward sense from deep water to the breakpoint as above, but at the breakpoint, $\partial I/\partial x$ decreases very rapidly and becomes negative because of the sudden loss of wave energy and hence reduction in the transport rate. The result is that $\partial z/\partial t$ achieves a locally large, positive value (cf the mass continuity equation in Chapter Thirteen) and the breakpoint bar develops. Shorewards of the breakpoint I and $\partial I/\partial x$ again both increase in a shorewards sense generating a new concavity. The precise location and magnitude of the discontinuity depends upon the pre-existing form of the orthogonal profile and upon the incident wave conditions as discussed below.

6. The inclusion of the drift currents into the purely bedload solutions described above increases the onshore and decreases the offshore transports, thus increasing $\partial I/\partial x$ in a shoreward sense along both the surf zone concavity and the offshore concavity which develop because of the breaking processes.

7. Suspended load is the first process by which, under certain wave and profile conditions, net offshore transport is introduced into the system. The precise values depend upon the form of the concentration profile, which is scalar, and the drift current profile which is generally directed offshore above the bed. Net offshore suspended load transport appears to be usual in the region offshore from the breakers because the large bedload rates generate large reference concentrations resulting in large suspended load concentrations at heights of between 0.10 and 0.70 m above the bed. Drift currents at these elevations are offshore and have rather high values of the order of 0.50 m s^{-1}. The overall result is that suspended load generally suppresses the magnitude of $\partial I/\partial x$ in a shoreward sense, so that $\partial z/\partial t$ remains negative but small and therefore the decrease in profile concavity is less marked. The smaller values of $\partial I/\partial x$ at the discontinuity also therefore suppress bar formation.

8. The inclusion, finally, of the slope dependency in the transport processes

appears to generate a considerable change in the response of the profile. The slope inclusive effects enhance downslope transports and supress upslope transports by up to 30%. The process does not change the direction of the transport rate for any one of the p/6 time increments simulated in each wave, but the net transport, obtained by integrating through the full wave cycle, is less onshore (upslope) dominated and appears frequently to become offshore (downslope) dominated. The effect is, of course enhanced on the steeper slopes in shallower water, and particularly on the steep, seaward face of the bar. The overall result is that the shoreward increase in I on both concavities is decreased, $\partial I/\partial x$ is therefore decreased and therefore the decrease in the profile concavity is less marked. Bar formation is also supressed.

9. The response function, $\partial z/\partial t$, varies markedly in a spatial sense along the profile. However, and in general, $\partial 2z/\partial t2$ is negative when $\partial z/\partial t$ is positive whereas $\partial 2z/\partial t2$ is positive when $\partial z/\partial t$ is negative. That is the rate of both accretion and erosion reduce through time and the orthogonal profile, therefore, appears to exhibit state attractor behaviour. Relaxation times also vary along the profile and are of the order of tens to hundreds of hours, but there is some evidence of exponential decay of the response function. The concept of a half-life relaxation time may therefore be usefully explored.

The parameter sensitivity analyses reported in Chapter Nineteen suggested that:

10. An increased wave height or wave period or a decreased angle of incidence increases the magnitude of the hydrodynamic parameters at each depth, increasing $\partial I/\partial x$ and therefore increasing $\partial z/\partial t$ in a negative sense along the concavities. The profile concavity therefore reduces and the bar forms farther offshore.

11. Overall, the bar formation and subsquent changes in the surf zone processes so that they reflect, but at a smaller scale, the offshore responses suggest that the orthogonal profile might better be described by the combination of two power functions with a point of inflection at the breakpoint. Should conditions prevail which generate, for example, surging breakers at the shoreline, the inner concavity simply assumes zero length. The range of beach states described in Chapter Fourteen might then be reasonably duplicated.

20.5 The Future: A Spectral Approach
Finally, it may be worth considering the type of future developments which are likely to occur in the subject. These may conveniently be divided into the hydrodynamic, sediment dynamic and morphodynamic elements of the system.

It is evident that, for monochromatic waves, second order theory appears appropriate up to a shallow water limit at which the first harmonic becomes dominant. Shoreward of this region either a higher order theory or the cnoidal wave theory would appear to be appropriate and, in the shallowest parts of the system, either bore theory or solitary wave theory may have to be used. However, the predictions of these theories have yet to be tested in detail in the field and, in any case, it may be worthwhile considering the inclusion of a realistically random wave field in the model at an early stage. Furthermore the whole subject of long wave generation in the nearshore has been carefully excluded from the present treatment, because it remains difficult to predict the occurence or magnitude of long wave motions on the beach. It is apparent from, for example, the various Australian papers by Wright and Short and others, that the geomorphological state

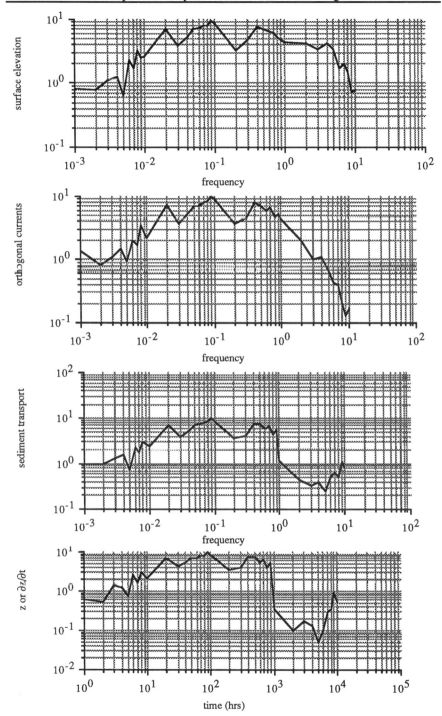

Figure 20.5 Representation of the orthogonal system as detailed in Section 20.7.

of the beach controls the reflection of waves in the nearshore, and therefore the generation of long wave motions. The future developments in the hydrodynamic field should therefore include the formulation and testing of a random wave model which will also have to account for beach state in generating long wave motions in the nearshore. The emphasis should perhaps be placed on the testing of the predictions, because even a relatively unsophisticated theoretical treatment which included these developments would be superior to the use of essentially monochromatic wave theory for the orthogonal system.

The small scale physics of the orthogonal sediment transport processes is not well understood at the present time. It appears that, for sand sized sediment, the Bagnold type of excess stress bedload function offers an adequate predictive capability under sinusoidal or second order flow conditions but that there remains a very considerable scatter in the experimental data. The inclusion of threshold and bedslope effects is important but again requires further experimental analysis and we presently have little appreciation of the effect of the ubiquitous ripples on the transport rates. The prediction of the suspended load rate on the orthogonal profile is still very much a black art and requires considerable refinement. It appears to be physically correct to relate the suspended load concentration profile to a reference concentration which is linked to the bedload rate but there remains considerable uncertainty about the form of the reference concentration or the diffusion coefficients and profiles for the suspensate. Again, recent research is demonstrating that the suspended load concentrations are related to, if not dominated by, long period energy and the comments on shallow water hydrodynamics are therefore once again applicable.

The morphodynamic processes are essentially those of mass continuity and of ensuring that the feedbacks generated by the geomorphological form are properly accounted for in the hydrodynamic and sediment dynamic process elements. Problems of modelling methodology remain unresolved however. It is unlikely that new insights will be developed through new empirical or analytical models because the system is too complex. However the "fond memories of what explanation used to be like in nineteenth-century physics" (page 11) are difficult to eradicate. The future developments in the morphodynamic field should therefore include the formulation of new or improved phase space representations of the system utilising perhaps the response function, $\partial z/\partial t$, to identify and to demonstrate the morphodynamic attractors which result from the SLOPES model or from future numerical solutions. Since the eigen-function analyses of, for example Winant, Clarke, Eliot and others are showing that there appears to be a series of spatial frequencies in the orthogonal profile response, morphodynamic research should erect general, theoretical systems models which predict the appropriate frequencies in order to investigate the nature of the forcing functions.

Finally, it will be apparent that these concluding remarks have been constructed to introduce a new development, and therefore this book will close with a supposition. Given that hydrodynamicists are developing the transfer functions required to predict orthogonal, oscillatory current spectra from surface elevation spectra (Chapter Six), and sediment dynamicists are developing the transfer functions required to predict orthogonal, oscillatory transport rate spectra from the current spectra (Chapter Twelve), and supposing that geomorphologists were to develop the transfer functions required to predict orthogonal profile spectra from the transport spectra, then might it be possible to depict the orthogonal system by the type of diagram shown in Figure 20.5.

APPENDICES

BIBLIOGRAPHY AND AUTHOR INDEX

SUBJECT INDEX

APPENDIX I

Circular and Hyperbolic Functions

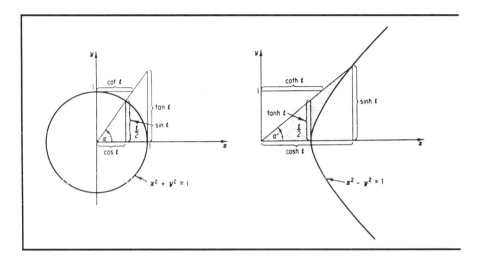

Figure AII.1. The circular finctions referred to a unit circle and the hyperbolic functions referred to a unit hyperbola.

Kinsman (1984) remarks that Longuet-Higgins divides mathematicians into those who understand analytically and those who understand geometrically, and both of these people see themselves as "geometers". Given such a lead, it would be unwise to do other than to define the circular and hyperbolic functions geometrically. Consider Figure AII.1(a) which shows a circle of unit radius given by:

$$x^2 + y^2 = 1$$
Eq.AI.1

The circular functions are all line segments referred to this circle. The argument (α in the present examples) of the circular functions is taken to be the central angle, and the common definition of the cosine is

$$\cos\alpha \equiv \frac{\text{adjacent}}{\text{hypotenuse}}$$
Eq.AI.2

Far more powerfully, it is defined in terms of the average of two exponentials of complex argument:

$$\cos\alpha \equiv \frac{e^{i\alpha} + e^{-i\alpha}}{2}$$
Eq.AI.3

291

Similarly the circular functions sine and tangent are defines as

$$\sin\alpha \equiv \frac{e^{i\alpha} - e^{-i\alpha}}{2}$$

Eq.AI.4

$$\tan\alpha \equiv \frac{\sin\alpha}{\cos\alpha} \equiv x\frac{e^{i\alpha} - e^{-i\alpha}}{e^{i\alpha} + e^{-i\alpha}}$$

Eq.AI.5

It is not suprising, therefore, to find that the hyperbolic functions are defined in a similar way by reference to a unit rectangular hyperbola as shown in Figure AII.1(b):

$$x^2 - y^2 = 1$$

Eq.AI.6

for which the corresponding hyperbolic functions are:

$$\cosh\alpha \equiv \frac{e^{\alpha} + e^{-\alpha}}{2}$$

Eq.AI.7

$$\sinh\alpha \equiv \frac{e^{\alpha} - e^{-\alpha}}{2}$$

Eq.AI.8

$$\tanh\alpha \equiv \frac{e^{\alpha} + e^{-\alpha}}{e^{\alpha} - e^{-\alpha}}$$

Eq.AI.9

These functions are plotted in Figure 5.7. Kinsman (1984) notes his preference for the term circular functions rather than trigonometric functions for those given above because it correctly suggests that they occupy a position in a structure that includes hyperbolic functions, parabolic functions, elliptic functions and cylinder or Bessel functions.

APPENDIX II

Formulae utilised in the SLOPES model

$$L_\infty = 1.56\ T^2 \qquad\qquad \text{AII.1}$$

$$C_\infty = L_\infty / T \qquad\qquad \text{AII.2}$$

$$\rho_b = \rho_s\ C \qquad\qquad \text{AII.3}$$

$$z_t = Z_o + M_2 \sin(28.9841 t^\circ) + S_2 \sin(30.0000 t^\circ + p^\circ) \qquad\qquad \text{AII.4}$$

$$x = (13.33h)^{1.5} \qquad\qquad \text{AII.5}$$

$$\tan\beta = 0.05 x^{-0.33} \qquad\qquad \text{AII.6}$$

$$L = \frac{gT^2}{2\pi} \tanh\left(\frac{2\pi h}{L}\right) \qquad\qquad \text{AII.7}$$

$$C = \frac{L}{T} \qquad\qquad \text{AII.8}$$

$$n = \frac{1}{2}\left(1 + \frac{2kh}{\sinh(kh)}\right) \qquad\qquad \text{AII.9}$$

$$H_s = H_\infty \sqrt{\frac{1}{2n}\frac{C_\infty}{C}} \qquad\qquad \text{AII.10}$$

$$\Delta H_f = \int_{h=L_\infty/4}^{h} \frac{4\ f_e\ k^2\ H^2}{3\pi\ \sinh(kh)(\sinh 2kh + 2kh)}\ dx \qquad\qquad \text{AII.11}$$

$$\alpha = \sin^{-1}\left(\frac{\sin\alpha_\infty\ C}{C_\infty}\right) \qquad\qquad \text{AII.12}$$

$$\Delta H_r = H_{\bullet\bullet} \sqrt{\frac{C_\infty}{2nC}\frac{\cos\alpha_\infty}{\cos\alpha}} \qquad\qquad \text{AII.13}$$

$$H_b = 0.72\ h\ (1 + 6.4\ \tan\beta) \qquad\qquad \text{AII.14}$$

$$\varepsilon = \frac{4\pi^2 H_b}{gT^2 \tan^2\beta} \qquad\qquad \text{AII.15}$$

$$Ur = \frac{HL^2}{h^3} \qquad\qquad \text{AII.16}$$

$$u(t) = A_a \frac{\pi H}{T \sinh(kh)} \cos(\omega t)$$ AII.17

$$u(t) = A_a \left(\frac{\pi H}{T \sinh(kh)} \cos(\omega t) + \frac{3}{4} \left(\frac{\pi H}{L} \right)^2 C \frac{1}{\sinh^4(kh)} \cos(2\omega t) \right)$$ AII.18

$$U = \frac{5}{4} \left(\frac{pH}{L} \right)^2 C \frac{1}{\sinh^2(kh)}$$ AII.19

$$A_b = \frac{\tan\phi}{\cos\beta \left(\tan\phi + \frac{u(t) \tan\beta}{|u(t)|} \right)}$$ AII.20

$$k_b = 1.0 \times 10^{-6}$$ AII.21

$$B_b = \sqrt{\frac{\tan\phi + \frac{u(t) \tan\beta}{|u(t)|}}{\tan\phi}} \cos\beta$$ AII.22

$$u_{cr} = 25.5 \left(\frac{DH}{\sinh(kh)} \right)^{0.25} \qquad \text{for} D < 0.05 \text{ cm}$$ AII.23

$$u_{cr} = D^{0.375} \left(\frac{H}{\sinh(kh)} \right)^{0.125} \qquad \text{for } D > 0.05 \text{ cm}$$ AII.24

$$i_{max} = A_{bmax} k_b (u_{max}^2 - B_{bmax}^2 u_{cr}^2) u_{max}$$ AII.25

$$C_a = 330 \frac{i_b}{\rho_s u}$$ AII.26

$$U_z = \frac{1}{4} \left(\frac{\pi H}{T} \right) \left(\frac{\pi H}{L} \right) \frac{1}{\sinh^2(kh)}$$

$$\{2 \cosh \left(2kh \left(\frac{z}{h} - 1 \right) \right) + 3 + 2kh \left(3 \frac{z^2}{h^2} + 4 \frac{z}{h} + 1 \right) \sinh 2kh$$

$$+ 3 \left(\frac{\sinh 2kh}{kh} + \frac{3}{2} \right) \left(\frac{z^2}{h^2} - 1 \right) \}$$ AII.27

$$C_z = C_a \left(1 - A_c \ln \frac{z}{z_a} \right)$$ AII.28

$$i_s = \frac{1}{T} \int_{t=0}^{t=T} \int_{z=0}^{z=1} U_z C_z \, dz \, dt$$ AII.29

$$i_b(t) = A_b(t) \, k_b \left(u(t)^2 - B_b(t)^2 \, u_{cr}^2 \right) u(t) \qquad\qquad \text{AII.30}$$

$$I_s = T \, \overline{i_s} \qquad\qquad \text{AII.31}$$

$$I_b = \int_{t=0}^{t=T} i_b(t) \ dt \qquad\qquad \text{AII.32}$$

$$\frac{\partial z}{\partial t} = -\frac{1}{\rho_b} \left(\frac{\partial i}{\partial t} + \frac{1}{U} \frac{\partial i}{\partial x} \right) \qquad\qquad \text{AII.33}$$

BIBLIOGRAPHY & AUTHOR INDEX

Square bracketed numbers [ ] refer to text sections which reference the publication.

Abou Seida, M.M., 1965. Bed load function due to wave action. Univ. Calif. Hydraulic. Eng. Lab., HEL-2-11.

Airy, G.B., 1845 [2.2, 5.3, 6.2, 6.3]. Tides and waves. *Encyc. Metrop.*, **192**, 241-396.

Aksoy, S., 1972 [9.4]. Fluid forces acting on a sphere near a boundary. *Proc. 15th Cong. Int. Assoc. Hydraulic. Res..* 217-224.

Allen, J.R.L., 1970a [3.7]. The avalanching of granular solids on dune and similar slopes. *J. Geol.*, **78**. 326-351.

Allen, J.R.L., 1970b [13.4]. *Physical Processes of Sedimentation.* Allen and Unwin, London. 248pp.

Allen, J.R.L., 1982a [9.4, 9.5, 13.5]. Simple models for the shape and symmetry of tidal sand waves. 1, Statically-stable equilibrium forms. *Mar. Geol.*, **48**. 31-49.

Allen, J.R.L., 1982b [9.4, 9.5, 13.5]. Simple models for the shape and symmetry of tidal sand waves. 2, Dynamically-stable symmetrical equilibrium forms. *Mar. Geol.*, **48**. 51-73.

Allen, J.R.L., 1982c [9.4, 9.5]. Simple models for the shape and symmetry of tidal sand waves. 1, Dynamically-stable asymmetrical forms without flow seperation. *Mar. Geol.*, **48**. 321-336.

Allen, J.R.L., 1985 [3.1, 3.3]. *Principles of Physical Sedimentology.* George Allen and Unwin. London. 272pp.

Anderson, M.G., (ed.) 1988 [2.1]. *Modelling Geomorphological Systems.* Wiley, Chichester. 458pp.

Anderson, M.G. and K.M.Sambles, 1988 [1.6]. A review of the bases of geomorphological modelling. In: M.G.Anderson (ed.) *Modelling Geomorphological Systsems.* Wiley, Chichester. 1-32.

Anikouchine, W.A. and R.W.Sternberg, 1973 [4.1]. *The World Ocean: An Introduction to Oceanography.* Prentice-Hall, Englewood Cliffs, New Jersey. 338pp.

Aranuvachapun, S and A.J.Johnson, 1979 [14.6]. Beach profiles at Gorleston and Great Yarmouth. *Coast. Eng.*, **2**. 201-213.

Aubrey, D.G., 1979 [2.10, 14.6]. Seasonal patterns of onshore/offshore sediment movement. *Jnl. Geophys. Res.*, **84**, 6347-6354.

Aubrey, D.G. and R.M.Ross, 1985 [14.6]. The quantitative description of beach cycles. *Mar. Geol.*, **69**. 155-170.

Aubrey, D.G., D.L.Inman and C.D.Winant, 1980 [14.6]. The statistical prediction of beach changes in southern California. *J. Geophys. Res.*, **85**. 3264-3276.

Bagnold, R.A., 1941 [9.3]. *The Physics of Blown Sand and Desrt Dunes.* Chapman and Hall. 265pp.

Bagnold,R.A., 1946 [5.6(a), 17.2(a), 10.2]. Motion of waves in shallow water. Interaction between waves and sand bottoms. *Proc. Roy. Soc. Lon.*, Series A, **187**. 1-18.

Bagnold,R.A., 1947 [5.3]. Sand movement by waves: some small scale experiments with sand at very low density. *J. Inst. Civ. Eng.*, **27**. 447-469.

Bagnold, R.A., 1954 [3.3]. Experiments on the gravity free dispersion of large solid spheres in a Newtonian fluid under shear. *Proc. Roy. Soc. Lon.*, A, **225**. 49-63.

Bagnold, R.A., 1956 [9.5, 9.6, 10.3]. The flow of cohesionless grains in fluids. *Phil. Trans. Roy. Soc. Lon.*, **B249**. 235-297.

Bagnold, R.A., 1963 [9.5, 11.3, 15.3]. Beach and nearshore processes. Part 1, Mechanics of marine sedimentation. In: M.N.Hill (ed.), *The Sea: Vol. 3.* Wiley-Interscience,

New York. 507-528.

Bagnold, R.A., 1966 [9.5, 15.3]. An approach to the sediment transport problem from general physics. U.S. Geol. Surv. Prof. Pap., 422-I. 235-297.

Bagnold, R.A., 1972 [9.4]. Fluid forces on a body in shear-flow; experimental use of 'stationary flow'. *Proc. Roy. Soc. Lon.*, **A340**. 147-171.

Bailard, J.A., 1981 [15.4]. An energetics total load model for a plane beach. J. Geophys. Res., **C86**. 10938-10954.

Bailard, J.A. and D.L.Inman, 1980 [9.5, 15.4]. An energetics model for a plane sloping beach, local transport. *J. Geophys. Res.*, **87**. 2035-2043.

Bisal, F and K.F.Nielsen, 1962 [9.3]. Movement of soil particles in saltation. *Canad. J. Soil Sci.*, **42**. 81-86.

Bascom, 1954 [2.2]. Characteristics of natural beaches. *Am. Soc. Civ. Eng., Proc. 4th Conf. Coast. Eng..* 163-180.

Bassett, A.B., 1888 [12.6]. On the motion of a sphere in a viscous liquid. *Phil. Trans. Roy. Soc.*, **179**. 43-63.

Batchelor, G.K., 1967 [12.6]. *An Introduction to Fluid Dynamics*, Cambridge Univ. Press, London.

Battjes, J.A. 1972 [5.9(a)]. Set-up due to irregular waves. *Proc. 13th Conf. Coast. Eng..* 1993-2004.

Battjes, J.A., 1974a [5.6(c), 5.7]. Computation od set-up, longshore currents, run-up and overtopping due to wind generated waves. Comm. on Hydraulic., Delft University of Technology. Report 74-2.

Battjes, J.A., 1974b [5.6(c)]. Surf similarity. *Proc. 14th Coast. Eng. Conf..* 446-480.

Becker, H.A., H.C.Hottel and G.C.Williams, 1967 [2.8]. The light-scatter technique for the study of turbulence and mixing. *J.Fluid Mechanics*, **30**. 259.

Bessel, F.W., 1826 [12.6]. *On the Incorrectness of the Reduction to a Vacuum Formerly Used in Pendulum Experiments*, Berlin Academy.

Bird (1984) [1.4] *Coasts: An Introduction to Coastal Geomorphology*. 3rd Ed., Blackwell, Oxford. 320pp.

Birkemeier, W.A. and R.A.Dalrymple, 1976 [16.1]. Numerical Models for the Prediction of Wave Set-Up and Nearshore Circulation. Tech. Rpt. 1 ONR Contract N00014-76-C-0342, University of Delaware. 127pp.

Bisal, F. and K.F.Nielson, 1962 [9.3]. Movement of soil particles in saltation. *Canad. J. Soil Sci.*, **42**. 81-86.

Bishop, C.T. and M.A.Donelan, 1987 [2.7]. Measuring waves with pressure transducers. *Coast. Eng.*, **11**, 309-328.

Bodge, K.R., 1989 [11.2, 11.4]. A literature review of the distribution of longshore sediment transport across the surf zone. *J. Coastal Res.*, **5**. 307-328.

Bonnefille, R. and L.Pernecker, 1966 [10.2]. Le debut d'entrainment des sediments sous action de la houile. *Bull. C.R.E.C.*, **15**. 27-32.

Bourke, P.J., D.J.Pulling, L.E.Gill and W.H.Denton, 1967 [2.8]. The measurement of turbulent velocity fluctuations and turbulent temperature fluctuations in the supercritical region by a hot wire anemometer and cold wire resistance thermometer. *Proc. Inst. Mech. Engnrs.*, **182**. Part 31.

Boussinesq, J., 1872 [2.2, 5.6(c)]. Theories des ondes et des remous qui se propagent le long d'un canal rectangulaire horizontal, en communicant au liquide contenu dans ce canal des vitesses sensibles paralleles a la surface au fond. *J. Math. Pures Appliques*. Ser. 2, **17**, 55-108.

Bowden, K.F. and R.A.White, 1966 [6.5]. Measurements of the orbital velocities of sea waves and their use in determining the directional spectrum. *Geophys. J. Roy. Astron. Soc.*, **12**. 33-54.

Bowen, A.J., 1969a [7.3]. The generation of longshore currents on a plane beach. *J.Mar. Res.*, **27**. 206-215.

Bowen, A.J., 1969b [7.3, 7.4]. Rip Currents, 1: theoretical investigations. *J. Geophys. Res.*, **74**, 5467-5478.

Bowen, A.J., 1980 [6.3, 15.3, 15.5]. Simple models of nearshore sedimentation; beach profiles and longshore bars. In: *The Coastline of Canada*, S.B.McCann (ed.). Geological Survey of Canada, Paper 80-10. 1-11.

Bowen, A.J. and D.L.Inman, 1969 [7.4]. Rip Currents, 2: laboratory and field observations. *J. Geophys. Res.*, **74**. 5479-5490.

Bowen, A.J., D.L.Inman and V.P.Simmons, 1968 [7.2] Wave set-up and set-down. *J.Geophys. Res.*, **73**. 2569-2577.

Bowen, A.J. and D.A.Huntley, 1984 [5.6(c), 7.4]. Waves, long waves and nearshore morphology. *Mar. Geol.*, **60**. 1-13.

Bowman, D., 1981 [14.6]. Efficiency of eigenfunctions for discriminant analysis of subaerial non-tidal beach profiles. *Mar. Geol.*, **39**. 243-258.

Bradbury, L.J.S. and T.P.Castro, 1971 [2.8]. A pulsed wire technique for velocity measurements in highly turbulent flows. *J.Fluid Mech.*, **49**. 657.

Bretschneider, C.L., 1954 [5.6(a)]. Field investigation of wave energy loss of shallow water ocean waves. U.S.Army, Beach Erosion Board, Tech. Memo., 46.

British Admiralty Tide Tables, 1988 [8.1]. Volume 1: European Waters. N.P.201-88. Hydrographer of the Navy. 438pp.

Brown, C.B., 1950 [9.4]. Sediment transportation. In: H.Rouse (ed.) *Engineering Hydraulics*. Wiley, New York.

Bruun, P., 1954 [2.2]. Coast erosion and the development of beach profiles. U.S.Army, *Beach Erosion Board, Tech. Memo.*, 44. 79pp.

Caldwell, J.M., 1956 [11.2]. Wave action and sand movement near Anaheim Bay, California. *U.S.Army Corps Eng., Beach Erosion Board, Tech. Memo.*, **68**. 21pp.

Carson, M.A. and M.J.Kirkby, 1972 [13.4]. *Hillslope Form and Process*. University Press, Cambridge. 475pp.

Carr, A.P., M.W.L.Blackley and H.L.King, 1982 [2.3]. Spatial and seasonal aspects of beach stability. *Earth Surf. Proc. and Landf.*, **7**. 267-282.

Carstens, M.R., F.M.Nielson and H.D.Altinbilek, 1969 [5.6(a), 10.2]. Bed forms generated in the laboratory under an oscillatory flow: analytical and experimental study. *C.E.R.C. Tech. Memo.* 28.

Carter, R.W.G., 1989 [1.7, 17.2]. *Coastal Environments*. Academic Press, London. 617pp.

Carter,T.G., Lui,P.L. and Mei,C.C., 1972 [17.2(b)]. Mass transport by waves and offshore bedforms. *Jnl Waterways, Harbors and Coastal Eng Div*, A.S.C.E., 99, 165-184.

Cavaleri, L., J.A.Ewing and N.D.Smith, 1978 [6.5]. Measurement of the pressure and velocity field below surface waves. In: *Turbulent Fluxes through the Sea Surface Wave Dynamics and Predictions*. NATO Conf. Serv. V., Plenum Press, New York. 257-272.

Chakrabarti, A., 1977 [14.2]. Mass-spring-damper system as the mathematical model for the pattern of sand movement for an eroding beach around Digha, West Bengahl, India. *J. Sedim. Pet.*, **47**. 311-330.

Chan, K.W., M.H.I.Baird and G.F.Round, 1972 [10.2]. Behaviour of dense particles in a horizontally oscillating liquid. *Proc. Roy. Soc. Lon.*, **A330**. 537-559.

Chappell, J and I.G.Eliot, 1979 [14.2, 16.1, 16.7]. Surf-beach dynamics in time and space. An Australian beach study and elements of a predictive model. *Mar. Geol.*, **32**. 231-250.

Chepil, W.S., 1958 [9.4]. The use of evenly spaced hemispheres to evaluate aerodynamic forces on a soil surface. *Trans. Am. Geophys. Union*, **39**. 397-404.

Chorley, R.J., 1967 [2.1]. Models in geomorphology. In: R.J.Chorley and P.Haggett (eds.) *Models in Geography*. Methuen, London. 59-96.

Clarke, D.J. and I.G.Eliot, 1982 [14.6]. Description of littoral, longshore sediment movement from empirical eigenfunction analysis. *Geol. Soc. Aust. J.*, **29**. 327-341.

Clarke, D.J. and I.G.Eliot, 1983 [14.6]. Onshore-offshore patterns of sediment exchange in the littoral zone of a sandy beach. *Geol. Soc. Aust. J.*, **30**. 341-351.

Clarke, D.J., I.G.Eliot and J.R.Frew, 1984 [14.6]. Variation in subaerial beach sediment volume on a small sandy beach over a monthly lunar tidal cycle. *Mar. Geol.*, **58**. 319-344.

Clarke, D.J. and I.G.Eliot, 1988a [14.6]. Low-frequency variation in seasonal intensity of coastal weather systems and sediment movement on the beachface of a sandy beach. *Mar. Geol.*, **79**. 23-39.

Clarke, D.J. and I.G.Eliot, 1988b [14.6]. Low-frequency changes of sediment volume on the beachface at Warilla Beach, New South Wales, 1975-1985. *Mar. Geol.*, **79**. 189-211.

Clift, R., Grace, J.R., and Weber, M.E., 1978 [12.1]. *Bubbles, Drops and Particles*. Academic Press, New York.

Clifton, H.E., 1976 [6.3]. Wave-generated structures - a conceptual model. In: R.A.Davis and R.L.Ethington (Eds.) *Beach and Nearshore Processes*. Soc. Econ. Pal. Min., Special Publication **24**. 126-148.

Clifton, H.E. and J.R.Dingler, 1984 [6.2, 6.3]. Wave-formed structures and palaeo-environmental reconstruction. *Mar. Geol.*, **60**. 165-198.

Coastal Engineering Research Centre, 19?? [1.7]. *Shore Protection Manual*. U.S.Army Corps of Engineers, Washington. 2 Volumes.

Coleman, N.L., 1967 [9.4]. A theoretical and experimental study of drag and lift forces acting on a sphere resting on a hypothetical stream bed. *Proc. 12th Cong. Int. Assoc. Hydraulic. Res.*, **3**. 185-192.

Collins, J.I., 1970 [5.9(a)]. Probabilities of breaking wave characteristics. *Proc. 13th Conf. Coast. Eng.*. 399-412.

Cook, P.A., 1986 [13.10]. Nonlinear Dynamical Systems. Prentice Hall International, Englewood Cliffs, New Jersey. 216pp.

Cote, L.J., 1960 [2.7]. The directional spectrum of wind generated sea as determined from data obtained from S.W.O.P.. *Meteor. Pap.*, **2**, New York. 88pp.

Culling, W.E.H., 1960 [13.9] Analytical theory of erosion. *J.Geol.*, **69**. 336-344.

Dally, W.R., and R.G.Dean, 1984 [16.1]. Suspended sediment transport and beach profile evolution. *J. Waterway, Port, Coast. and Ocean Eng.*, ASCE, **110**. 15-33.

Dalrymple, R.A., (ed.) 1985 [2.1]. *Physical Modelling in Coastal Engineering*. Balkema, Rotterdam.

Dalrymple, R.A. and W.W.Thompson, 1976 [14.3, 14.6, 15.4]. Study of equilibrium beach profiles. *Proc. 15th Coast. Eng. Conf.*, Am. Soc. Civ. Eng.. 1277-1296.

Dalrymple,R.A. and J.T.Kirkby, 1986 [17.2(d)]. Water waves over ripples. *Jnl Waterways, Port, Coastal and Ocean Eng*, 112, 309-319.

Das, M.M., 1971 [11.1, 11.2]. Longshore sediment transport rates: a compilation of data. *U.S. Army Corps. Eng., C.E.R.C.* Misc. Paper 1-71.Vicksburg MS 39180-0631

Davidson-Arnott, R.G.D., 1981 [16.1, 16.3, 17.3]. Computer simulation of nearshore bar formation. *Earth Surf. Proc. Landf.*, **6**, 23-34.

Davies,A.G., 1982 [17.2(d)]. The reflection of wave energy from undulations on the seabed. *Dynamics of Atmospheres and Oceans*, 6, 207-232.

Davies, A.G., 1983 [6.2]. Wave interaction with a rippled bed. In: B.Johns (ed.) *Physical Oceanography of Coastal and Shelf Seas*. Elsevier, Amsterdam. 1-66.

Davies,A.G. and A.D.Heathershaw, 1984 [17.2(d)]. Surface wave propogation over sinusoidally varying topography. *Jnl Fluid Mech*, **144**, 419-443.

Davies, T.R.H. and M.F.A.Samad, 1978 [9.4]. Fluid dynamic lift on a bed particle. *J.Hydraulic. Div., ASCE*, **104**. 1171-1182.

Davies, T.V., 1951 [5.6(c)]. Symmetrical, finite amplitude gravity waves. In: *Gravity Waves*, Nat. Bur. of Standards, **521**. 55-60.

De, S.E., 1955 [6.3]. Contributions to the theory of Stokes waves. *Proc. Cam. Phil. Soc.*, **51**. 713-736.

Deacon, G.E.R., 1946 [2.7]. Ocean waves and swell. *Oceanogr. Pap. Challenger Soc.*. **1**. 13pp.

Dean, R.G., 1973 [14.5, 15.3]. Heuristic models of sand transport in the surf zone. *Proc. Conf. Eng. Dynamics in the Surf Zone, Sydney*. 208-214.

Dean, R.G., 1977 [14.4, 14.6, 15.3, 17.3]. Equilibrium beach profiles. U.S.Atlantic and Gulf Coasts, Ocean Engineering Rpt. 12. Dept. Civ. Eng., Univ. Delaware, Newark, Delaware.

Dean, R.G. and M.Perlin, 1986 [6.3]. Intercomparison of nearbottom kinematics by several wave theories nd field and laboratory data. *Coatal Eng.*, **9**. 399-437.

Defant, A., 1961 [2.7, 8.3]. *Physical Oceanography*. Pergamon, Oxford. 598pp.

Dittmar, C., 1884 [4.3]. Challenger Report. Physics and Chemistry Volume I, No. 11.

Dolan, R., B.P.Hayden and W.Felder, 1977 [14.6]. Systematic variations in inshore bathymetry. *J. Geol.*, **85**. 129-141.

Dolan, R., B.P.Hayden and J.Heywood, 1978a [14.6]. A new photogrammetric method for determining shoreline erosion. *Coast. Eng.*, **2**. 21-39..

Dolan, R., B.P.Hayden and J.Heywood, 1978b [14.6]. Analysis of coastal erosion and storm surge hazards. *Coast. Eng.*, **2**. 41-53.

Dolan, R., B.P.Hayden and W.Felder, 1979a [14.6]. Shoreline periodicities and edge waves. *J. Geol.*, **87**. 175-185.

Dolan, R., B.P.Hayden and W.Felder, 1979b [14.6]. Shoreline periodicities and linear offshore shoals. *J. Geol.*, **87**. 393-402.

Donat, J., 1929 [9.4]. Uber sohlangriff und geschiebetrieb. *Wasserwirtschaft* **26, 27**.

Doodson, A.T. and H.D.Warburg, 1941 [8.1]. *Admiralty Manual of Tides*. H.M.S.O., London.270pp (Reprinted 1973)

Doornkamp, J.C. and C.A.M.King, 1971 [14.3, 15.4]. *Numerical Analysis in Geomorphology: An Introduction*. Arnold, London.

Draper, L., 1957 [6.3]. Attenuation of sea waves with depth. *La Houille Blanche*, **6**. 1-6.

Draper, L., 1970 [2.7]. Routine sea-wave measurement - a survey. *Underwater Sci. and Tech.*, **2**. 81-86.

Driver, J.S., 1980 [2.7, 6.3]. A guide to sea wave recording. *Institute of Oceanographic Sciences Report* **103**. 51pp.

Du Boys, M.P., 1879 [9.4]. Le Rhone et les rivieres a lit affouillable. *Mem. Doc. Ann. Ponts Chaussees*. Ser 5, **18**. 141-195.

Dubuat, C., 1786 [12.6]. *Principes d'hydraulique*, Paris.

Dyer, K.D., 1980 [9.7]. Velocity profiles over a rippled sand bed and the threshold of movement of sand. *Estuar. Coast. Mar. Sci.*, **10**. 181-199.

Dyer,K.D., 1986 [1.7, 5.1, 6.1, 6.6, 9.1, 9.4, 9.7, 10.4]. *Coastal and Estuarine Sediment Dynamics*. Wiley, Chichester. 342pp.

Eagleson, P.S., 1965 [7.3]. *Theoretical Study of Longshore Ciurrents on a Plane Beach*. M.I.T. Hydrodynamic Lab., Tech. Rpt., **82**. Cambridge, Mass.. 31pp.

Eagleson, P.S., R.G.Dean, 1959 [10.2]. Wave-induced motion of bottom sediment particles. *Proc. A.S.C.E. J. Hydraulic. Div.*, **85**. 53-79.

Eagleson, P.S., B.Glenne and J.A.Dracup, 1963 [14.3, 14.6]. Equilibrium characteristics of sand beaches. Proc. A.S.C.E., J. Hydraulic. Div., **89**. 35-57.

Ebersole, B.A. and R.A.Dalrymple, 1979 [16.6]. A numerical model for nearshore circulation including convective accelerations and lateral mixing. Dept. Civ. Eng., Univ. Delaware, Tech. Rpt. 4, Off. of Naval Res. Geography Programs, Ocean Eng. Rpt., 21.

Eckart, C., 1951 [6.4, 6.8, 7.4] Surface waves on water of variable depth. Wave Rpt., 100. Scripps Inst. Oceanog., University of California. 99pp.

Einstein, H.A., 1950 [9.4, 9.7]. The bedload function for sediment transport in open channel flows. U.S. Dept. Agric. Soil Conservations Service. *Tech. Bull.* 1026.

Einstein, H.A. and E.-S.A. El-Samni, 1949 [9.4]. Hydrodynamic forces on a rough wall. *Rev. Modern Physics*, **21**. 520-524.

Eliot, I.G. and D.J.Clarke, 1982 [14.6]. Temporal and spatial variability of the sediment budget of the subaerial beach at Warilla, New South Wales. *Aust. J. Mar. Freshw. Res.*, **33**. 945-969.

Embleton, C. and J.B.Thornes (eds.) 1979 [1.1]. *Process in Geomorphology*. Arnold, London. 436pp.

Ewing, H., 1986 [2.7]. Presentation and interpretation of directional wave data. *Underwater Technology*, **2**, 17-23.

Exner, F.M., 1920 [13.4]. Sitzungsber. Akad. Wiss. Wein., Math-Naturw. Kl., Abt. 2a, **129**, 929-952.

Flemming, R.H. and R.Revelle, 1939 [4.6]. Physical processes in the ocean. In: P.D.Trask (ed.) *Recent Marine Sediments*: Annal. Ass. Pet. Geol., Tulsa, Okla..

Flemming, C.A. and J.N.Hunt, 1976a [16.1, 16.2]. A mathematical sediment transport model for unidirectional flow. *Proc. Instn. Civ. Eng.*, 61. 297-310.

Fleming, C.A. and J.N.Hunt., 1976b [16.1, 16.2]. Application of a sediment transport model. *Proc. 15th Conf. Coast. Eng.*, ASCE. 1184-1202.

Flick, R.E., R.T.Guza and D.L.Inman, 1981 [5.9(b)]. Elevation and velocity measurements of laboratory shoaling waves. *J. Geophys. Res.*, **86**. 4149-4160.

Folk, R.L., 1966 [3.2]. A review of grain size parameters. Sedimentology. **6**, 73-93.

Fox, W.T., 1985 [2.1, 14.6]. Modelling coastal systems. In: R.A.Davids (ed.) *Coastal Sedimentary Environments*, 2nd Ed.. Springer-Verlag, New York. 665-705.

Fox, W.T. and R.A.Davis, 1973 [16.1]. Simulation model for storm cycles and beach erosion on Lake Michigan. *Geol. Soc. Amer. Bull.*, **84**. 1769-1790.

Fox, W.T. and R.A.Davis, 1978 [2.2]. Seasonal variation in beach erosion and sedimentation on the Oregon coast. *Geol. Soc. Am. Bull.*, **89**. 1541-1549.

Francis, J.R.D., 1973 [9.6]. Experiments on the motions of solitary grains along the bed of a water stream. *Phil. Trans. Roy. Soc. Lon.*, **A332**, 443-471.

Freilich, M. and R.T.Guza, 1984 [5.9] Nonlinear effects on shoaling surface gravity waves. *Phil. Trans. Roy. Soc. Lon.*, **A-311**. 1-41.

Froude, W., 1862 [6.2]. On the rolling of ships. *Trans. Inst. Nav. Archs.*, **3**. 45-62.

Gadd, P.E., J.W.Lavelle and D.J.P.Swift, 1978 [9.5]. Estimates of sand transport on the New York Shelf using near-bottom current-meter observations. *J.Sedim. Petrol.*, **48**. 239-252.

Galvin, C.J. and P.S.Eagleson, 1965 [7.3]. Experimental study of longshore currents on a plane beach. *U.S.Army, C.E.R.C. Tech. Memo.*, **10**. 80pp.

Galvin, C.J., 1968 [5.6(c)]. Breaker type classification on three laboratory beaches. *J. Geophys. Res.*, **73**. 3651-3659.

Galvin, C.J., 1972 [5.6(c)]. Wave breaking in shallow water. In: R.E.Meyer (ed) *Waves on Beaches*, Academic Press, New York. 413-456.

Galvin, C.J. and P.S.Eagleson, 1965 [7.3]. Experimental study of longshore currents on a plane beach. *U.S.Army, C.E.R.C. Tech. Memo.*, **10**. 80pp.

Gerstner, F., 1802 [2.2, 5.3, 6.2, 6.7] *Theorie der Wellen.* Abhandlungen der koniglichen bomischen Gessellschaft der Wissenschaften, Prague.

Gilbert, G.K., 1885 [2.2]. The topographic features of lake shores. *U.S. Geol. Surv. 5th Ann. Rpt..* 69-123.

Glossary of Geology and Related Sciences, 1972 [1.4]. Amer. Geol. Inst. under Nat. Acad. Sci., Nat. Res. Counc. Washington, D.C.. 325pp plus 72pp supplement.

Goda, Y., 1975 [5.9(a)]. Irregular wave deformation in the surf zone. *Coast. Eng. Japan*, **18**. 13-26.

Goddet, J., 1960 [10.2]. Etude du debut d'entrainment des materiaux mobiles sous l'action de la houle. *La Houille Blanche*, **15**. 122-135.

Gordon, G., 1978 [2.1]. *System Simulation.* Prentice-Hall, Englewood Cliffs, New Jersey.

Graf, W.H., 1971 [9.1, 9.5]. Hydraulics of Sediment Transport. McGraw-Hill, New York. 514pp.

Grace, R.A. and R.Y.Rocheleau, 1973 [6.3]. Near-bottom velocities under Waikiki swell. University of Hawaii, Ocean Eng., Tech. Rpt., **31**.

Grass, A.J., 1970 [9.4]. Initial instability of a fine sand bed. *J. Hydraulic. Div. A.S.C.E.*, **96**. 619-632.

Greer, M.N. and O.S.Madsen, 1978 [11.1]. Longshore sediment transport data: a review. *Proc. 16th Int. Conf. Coast. Eng.*, A.S.C.E..1563-1576.

Gresswell, R.K., 1937 [14.2]. The geomorphology of the south-west Lancashire coastline. *Geogr. J.*, **90**. 335-348.

Guy, H.P., D.B.Simmons and E.V.Richardson, 1966 [9.5]. Summary of alluvial channel data from flume experiments 1955-1961. *U.S.Geol. Surv. Prof. Pap.*, **462-I**. 92pp.

Guza, R.T. and A.J.Bowen, 1975 [5.6(c)]. The resonant instabilities of long waves obliquely incident on a beach. *J. Geophys. Res.*, **80**. 4529-4534.

Guza, R.T. and A.J.Bowen, 1977 [5.6(c)]. Resonant interactions from waves breaking on a beach. *Proc. 15th Coast. Eng. Conf..* 560-579.

Guza, R.T. and D.L.Inman, 1975 [5.6(c)]. Edge waves and beach cusps. *J. Geophys. Res.*, **80**. 2997-3012.

Guza, R.T. and E.B.Thornton, 1980 [2.8, 5.9(b), 6.5]. Local and shoaled comparisons of sea surface elevations, pressures and velocities. *J. Geophys. Res.*, **85**. 1524-1530.

Guza, R.T. and E.B.Thornton, 1989 [2.7, 2.8]. General Measurements. In: R.J.Seymour (ed.) *Nearshore Sediment Dynamics.* Plenum, New York. 51-60.

Gwyther, R.F., 1900 [5.6(c)]. The classes of long progressive waves. *Phil. Mag.*, **50**(5). 213.

Haines-Young, R and J.Petch, 1986 [1.3]. *Physical Geography : Its Nature and Methods.* Harper & Row, London, 230pp

Hallermeir, R.J., 1980 [10.2]. Sand motion initiation by water waves: two asymptotes. *Proc. A.S.C.E., J. Waterw. Port. Coastal. Div.*, **106**. 299-318.

Hallermeier, R.J., 1981 [14.5]. A profile zonation for seasonal sand beaches from wave climate. *Coast. Eng.*, **4**. 253-277.

Hallermeier, R.J., 1982a [10.3]. Oscillatory bedload transport: data review and simple formulation. *Cont. Shelf Res.*, **1**. 159-190.

Hallermeier, R.J., 1982b [11.1]. Bedload and wave thrust computations of alongshore sand transport. *J. Geophys. Res.*, **87**. 5741-5751.

Halliwell, G.R. and C.N.K.Moores, 1979 [14.6]. The space-time structure and variability of the shelf water-slope water and Gulf Stream surface temperature fronts and associated warm core eddies. *J. Geophys. Res.*, **84**. 7707-7725.

Hammond, F.D.C., A.D.Heathershaw and D.N.Langhorne, 1984 [9.3]. A comparison between Shield's threshold criterion and the movement of loosely packed gravel in a tidal channel. *Sedimentology*, **31**, 51-62.

Hanes, D.M. and D.A.Huntley, 1986 [2.9]. Continuous measurements of suspended sand concentration in a wave dominated nearshore environment. *Continental Shelf Research*, **6**. 585-596.

Hanson, H., 1989 [16.1]. Genesis - A generalised shoreline change numerical model. *Jnl. Coastal Res.*, **5**. 1-27.

Hanson, H. and N.C.Kraus, 1986 [16.1]. Seawall boundary conditions in numerical models of shoreline change. Tech. Rpt., CERC-86-3. U.S.Army Engrs. Waterways Experiment Station. 59pp.

Harbaugh, J. and G.Bonham-Carter, 1970 [16.1]. *Computer Simulation in Geology.* Wiley, New York. 575pp.

Hardisty, J., 1983 [9.5]. An assessment and calibration of formulations for Bagnold's bedload equation. *J. Sedim. Petrol.*, **53**. 1007-1010.

Hardisty, J., 1984 [6.3, 9.5]. A dynamic approach to the intertidal profile. In: M.W.Clark (ed.) *Coastal Research: U.K.Perspectives.* Geo Books, Norwich. 17-33.

Hardisty, J., 1986a [2.9]. A new optoelectronic technique for the measurement of seabed sand transport. *Oceanology*, **6**. 169-174.

Hardisty, J., 1986b [6.3, 9.5, 15.3]. A morphodynamic model for beach gradients. *Earth Surface Processes and Landforms*, **11**. 327-333.

Hardisty,J., 1987 [5.2, 5.3, 15.4] The transport response function and relaxation time in geomorphic modelling. In: F.Ahnert (ed.) *Geomorphological Models; Theoretical and Empirical Aspects.* Catena Suppl., **10**. 171-179.

Hardisty, J., 1989a [5.8]. Laboratory and site measurements of nearbed currents and sand movement beneath waves: a spectral approach. In: M.H.Palmer (ed) *Advances in Water Modelling and Measurement.* B.H.R.A., Cranfield. 271-290.

Hardisty, J., 1989b [6.3]. Morphodynamic and experimental assessments of wave theories in intermediate water depths. *Earth Surface Processes and Landforms*, **14**. 107-118.

Hardisty, J., 1990a [5.3, 5.4]. Monochromatic wave surface profiles. *Geosystems Library.* WM-1.

Hardisty,J., 1990b [5.4, 5.5]. First order shoaling transformations. *Geosystems Library.* WM-2.

Hardisty, J., 1990c [5.6(a)]. Wave height changes due to seabed friction. *Geosystems Library* WM-3.

Hardisty,J., 1990d [5.6(b)]. Simple effects of wave refraction. *Geosystems Library.* WM-4.

Hardisty,J., 1990e [5.7]. The Shoreline and Orthogonal Process Emulation System (SLOPES). *Geosystems Library.* WM-5.

Hardisty,J., 1990f [5.8(a)]. Pierson-Moskowitch spectra. *Geosystems Library.* WM-6.

Hardisty,J., 1990g [5.8(a)]. JONSWAP spectra. *Geosystems Library.* WM-7.

Hardisty,J., 1990h [5.8(b)]. The Rayleigh probability distribution for wave height. *Geosystems Library.* WM-8.

Hardisty,J., 1990i [6.2(a)]. Orthogonal wave induced currents. *Geosystems Library.* WM-9.

Hardisty,J., 1990j [6.4]. Stokes wave orthogonal and drift currents. *Geosystems Library.* WM-10.

Hardisty,J., 1990k [6.4]. Solitary wave orthogonal and drift currents. *Geosystems Library.* WM-11.

Hardisty,J., 1990l [7.2]. Wave set-up. *Geosystems Library.* WM-12.

Hardisty,J., 1990m [7.3]. Distribution of longshore current across the surf zone. *Geosystems Library*. WM-13.

Hardisty,J., 1990n [8.6]. Tidal elevations. *Geosystems Library*. WM-14.

Hardisty,J., 1990o [9.4]. Subaqueous sediment thresholds under uni-directional currents. *GeoSystems Library*. SD-1.

Hardisty,J., 1990p [9.5]. The Bagnold bedload function for unidirectional, subaqueous transport. *GeoSystems Library*. SD-2.

Hardisty,J., 1990q [9.6]. Subaqueous suspended load thresholds under uni-directional currents. *GeoSystems Library*. SD-3.

Hardisty,J., 1990r [9.7]. Subaqueous suspended load transport rates using the Rouse equation. *GeoSystems Library*. SD-4.

Hardisty,J., 1990s [9.8]. An empiricised total load function for unidirectional subaqueous sediment transport. *GeoSystems Library*. SD-5.

Hardisty, J., 1990t [10.2]. Sediment thresholds under waves. *GeoSystsems Library*. SD-6.

Hardisty, J., 1990u [10.3]. The Bagnold bedload function for sand transport under waves. *GeoSystsems Library*. SD-7.

Hardisty, J., 1990v [10.5]. Sediment suspension profiles under waves. *GeoSystsems Library*. SD-8.

Hardisty, J., 1990w [10.6]. Total load sediment transport under waves. *GeoSystsems Library*. SD-9.

Hardisty, J., 1990x [11.2]. Empirical longshore transport equations. *GeoSystsems Library*, SD-10.

Hardisty,J., 1990y [11.4]. Longshore transport across the breaker zone. *GeoSystems Library*, SD-11.

Hardisty, J., 1990z [12.4]. Terminal velocity of a falling sphere. *GeoSystems Library*, SD-12.

Hardisty,J., 1990aa [8.1]. *British Seas: An Introduction to the Oceanography and Resources of the North-West European Continental Shelf*. Routledge, London. 272pp.

Hardisty, J., 1990ab [12.5]. Acceleration of a sphere in air. *GeoSystems Libary*, SD-13.

Hardisty, J., 1990ac [12.6]. The particle momentum equation. *GeoSystems Library*, SD-14.

Hardisty, J., 1990ad [2.1]. *An Introduction to Wave Recording : With Special Reference to Pressure Transducers and Geomorphological Applications*. Brit. Geom. Res. Group. Tech. Pub. 76pp.

Hardisty, J., 1990ae [14.3]. Empirical beach gradient models. *GeoSystems Library*, SD-15.

Hardisty, J., 1990af [14.4]. Empirical orthogonal profile models. *GeoSystems Library*, SD-16.

Hardisty, J., 1990ag [15.4]. A note on suspension transport in the beach gradient model. *Earth Surface Processes and Landforms*, **15**. 91-96.

Hardisty, J. and D.Hamilton, 1984 [2.9]. Measurements of sediment transport on the seabed southwest of England. *Geo-Marine Letters*, **4**. 19-23.

Hardisty, J., D.Hamilton and J.C.Collier, 1984 [15.4]. A calibration of the Bagnold beach equation. *Mar. Geol.*, 61. 95-101.

Hardisty, J. and R.J.S.Whitehouse, 1988 [9.5]. Evidence for a new sand transport process from experiments on Saharan dunes. *Nature*, **332**. 532-534.

Harrison, W. and W.C.Krumbein, 1964 [14.4, 14.5]. Interactions of the beach-ocean-atmosphere system at Virginia Beach, Virginia. *U.S. Army Coprs of Eng., Coastal Eng. Res. Center, Tech. Memo.*, **7**. 102pp.

Harrison, W., N.A.Pore and D.R.Tuck, 1965 [14.4, 14.5]. Predictor equations for beach processes and responses. *J. Geophys. Res.*, **70**. 6101-6109.

Harrison, W., 1969 [14.5]. Empirical equations for foreshore changes over a tidal cycle. *Mar. Geol.*, **7**. 529-551.

Harvey, H.W., 1955 [4.1]. *The Chemistry and Fertility of Sea Water*. Cambridge University Press, Cambridge. 224pp.

Hattori, M. and R.Kawamata, 1980 [14.5]. Onshore-offshore transport and beach profile change. *Proc. 17th Conf. Coast. Eng., Sydney*. 1175-1194.

Hayden, B., R.Dolan and W.Felder, 1979 [14.6]. Spatial and temporal analysis of shoreline variations. *Coast. Eng.*, **2**. 351-361.

Highland, H.J., 1973 [2.1]. A taxonomy of models. *Simuletter*, IV, **12**. 10-17.

Hill, P.S., A.R.M.Nowell and P.A.Jumars, 1988 [9.7, 9.8]. Flume evaluation of the relationship between suspended sediment concentration and excess boundary shear stress. *J.Geophys. Res.*, **93**. 12499-12509.

Hijum, van E., 1974 [14.3]. Equilibrium profiles of coarse material under wave attack. *Proc. 14th Conf. on Coastal Eng., Am.Soc. Civ. Eng.* Copehagen. 939-957.

Hinze, J.O., 1975 [12.6]. *Turbulence*, 2nd ed, McGraw-Hill, New York.

Homma, M. and K.Horikawa, 1963 [10.4]. A laboratory study on suspended sediment due to wave action. *Proc. 10th Cong. I.A.H.R.*, Lon.. 213-220.

Homma, M., K.Horikawa and R.Kajima, 1965 [10.4]. A study on suspended sediment due to wave action. *Coastal Eng. Japan.*, **8**. 85-103.

Horikawa, K., 1978 [1.7, 5.4, 5.6, 5.7, 11.2, 12.6]. *Coastal Engineering*. University of Tokyo Press. 402pp.

Horikawa, K. (ed.), 1988 [1.7, 2.10, 9.5, 10.3, 16.5]. *Nearshore Dynamics and Coastal Processes*. University of Tokyo Press. 522pp.

Horikawa, K. and A.Watanabe, 1967 [10.2, 10.4]. A study on sand movement due to wave action. *Coastal Eng. Japan*, **10**. 39-57.

Horikawa, K., Harikai, S and Kraus, N.C., 1979 [16.1]. A physical and numerical modelling of waves, currents and sediment transport near a breakwater. *Ann. Rpt. Eng. Res. Inst.*, University of Tokyo, **14**. 49-57.

Hotta, S., M.Mizuguchi and M.Isobe, 1981 [2.8]. Observations of long period waves in the nearshore zone. *Coast. Eng. in Japan*, **24**. 41-76.

Howd, P.A. and R.A.Holman, 1987 [16.1, 16.7]. A simple model for beach foreshore response to long-period waves. *Mar. Geol.*, **78**. 11-22.

Huggett, R., 1980 [2.1]. *Systems Analysis in Geography*. Clarendon Press, Oxford, England.

Hughes, S.A. and T.T.Chiu, 1978 [14.4] The variations in beach profiles when approximated by a theoretical curve. Report UFL/COEL/TR-039, Department of Coastal and Oceanographic Engineering, University of Florida. 136pp

Hughes, M.G. and P.J.Cowell, 1987 [14.3]. Adjustment of reflective beaches to waves. *J. Coast. Res.*, **3**, 153-167.

Huntley, D.A. and A.J.Bowen, 1975 [5.6(c), 6.2]. Comparison of the hydrodynamics of steep and shallow beaches. In: J.R.Hails and A.Carr (eds.) *Nearshore Sediment Dynamics and Sedimentation*. Wiley, London. 69-109.

Huntley, D.A., R.T.Guza and E.B.Thornton, 1981 [7.4]. Field observations of surf beat, 1, progressive edge waves. *J. Geophys. Res.*, **83**, 1913-1920.

Huthnance, J.M., 1982 [9.5]. On the formation of sand banks of finite extent. *Estuarine Coastal and Shelf Sci.*, **15**. 277-299.

Ichikawa, T., O.Ochiai, K.Tomita and K.Murobuse, 1961 [11.2]. Waves and coastal sediment characteristics at Tagono-ura coast, Suruga Bay. *Proc. 8th Conf. on Coast. Eng. Japan*. 161-167.

Ijima, T., S.Sato, H.Aono and K.Ishii, 1960 [11.2]. Wave and coastal sediment characteristics at Fukue coast, Atsumi Bay. *Proc. 7th Conf. Coast. Eng. Japan.* 69-79.

Ijima, T., S.Sato and N.Tanaka, 1964 [11.2]. On the coastal sediment at Kashima Harbour coast. *Proc. 11th Conf. Coat. Eng. Japan.* 175-180.

Ingle, J.C., 1966 [2.7]. *The Movement of Beach Sand.* Elsevier, New York. 221pp.

Inman,D.L., 1952 [3.2]. Measures for describing the size distribution of sediments. *J. Sedim. Petrol.*, **22**, 125-145.

Inman,D.L., 1953 [2.2]. Aereal and seasonal variations in beach and nearshore sediments at La Jolla, California. *U.S.Army Corps Engnrs., Beach Erosion Board, Tech Memo.* **39**. 134pp.

Inman, D.L., 1957 [10.4]. Wave generated ripples in nearshore sands. U.S. Army Corps of Eng., Beach Erosion Board, Tech. Memo., 100. 42pp.

Inman, D.L. and R.A.Bagnold, 1963 [1.5, 7.3, 9.5, 10.3, 11.3, 15.2]. Littoral Processes. Chapter 21 Part II. In: M.N.Hill (ed) *The Sea.* Vol 3. Interscience. 529-554.

Inman, D.L. and W.H.Quinn, 1952 [7.3]. Currents in the surf zone. *Proc. 2nd. Conf. Coast. Eng..* 24-36.

Inman, D.L. and G.A.Rusnak, 1956 [2.2, 2.10]. Changes in sand level on the beach and shelf at La Jolla, California. *U.S.Army Corps Engnrs., Beach Erosion Board, Tech Memo.* 82. 30pp.

Ippen, A.T. and G.Kulin, 1955 [5.6(c)]. The shoaling and breaking of the solitary wave. *Proc. 5th Conf. Coast. Eng..* 27-49.

Iribarren, R and C.Nogales, 1949 [5.6(c)]. Protection des ports. *17th Int. Naval Cong., Lisbon, Section* **II-4.** 31-82.

Ishihara, T. and T.Sawaragi, 1962 [10.2]. Laboratory studies on sand drift the critical velocity and the critical water depth for sand movement and the rate of transport under wave action. *Coastal Eng. Japan*, **5**. 59-65.

Iverson, H.W., and Balent, R., 1951 [12.6]. A correlating modulus for fluid resistance in accelerated motion. *J. Appl. Phys.*, **22**. 324-328

Iwagaki, Y. and T.Sakai, 1970 [6.3]. Horizontal water particle velocities of finite amplitude waves. *Proc. 12th Conf. Coast. Eng.*, **1**. 309-326.

Iwagaki, Y and T.Kakinuma, 1967 [5.6(a)]. On the bottom friction factors off five Japanese coasts. *Coast. Eng. Japan*, **10**. 13-22.

Jago, C.F. and J.Hardisty, 1984 [3.2, 13.5, 15.4]. Sedimentology and morphodynamics of a macrotidal beach, Pendine Sands, S.W.Wales. *Mar. Geol.*, **60**, 123-154.

Jensen, J.K. and T.Sorensen, 1972 [2.9]. Measurement of sediment suspension in combinations of waves and currents. *Proc. 13th Coastal Eng. Conf.*, A.S.C.E.. 1097-1104.

Johns,B., 1970 [17.2(b)]. On the mass transport induced by oscillatory flow in a turbulent boiundary layer. *Jnl Fluid Mech*, **43**, 177-185.

Johnson, D.W., 1919 [2.2]. *Shore Processes and Shoreline Change.*

Jonsson, I.G., 1966 [6.6, 10.3]. Wave boundary layers and friction factors. *Proc. 10th Conf. Coast. Eng..* 127-148.

Kachel, N. and J.D.Smith, 1986 [9.7]. Geological impact of sediment transporting events on the Washington continental shelf. In: R.J.Knight and J.R.McClean (eds.) *Shelf Sands and Sandstones.* Can. Soc. Petrol. Geologists.,, Calgary. 145-162.

Kajima, R., K.Maruyama, T.Shimizu, T.Sakakiyama and S.Saito, 1982 [10.3]. Experimental study of waves on beaches with prototype wave flume. *Proc. 29th Japanese Conf. on Coastal Eng.*, J.S.C.E.. 213-217.

Kajiura, K, 1968 [5.6(a)]. A model of the bottom boundary layer in water waves. *Bull. Earthquake Res. Inst.*, **46**. 75-123.

Kalinske, A.A., 1942 [9.4]. Criteria for determining sand transportation by surface creep and saltation. *Trans. Americ. Geophys. Union.* 639-643.

Kalkanis, G., 1964 [10.3]. Transportation of bed material due to wave action. *U.S.Army, C.E.R.C., Tech. Memo.*, 2.

Kana, T.W., 1976 [2.9]. A new apparatus for collecting simultaneous water samples in the surf-zone. *J. Sedim. Petrol.*, **46**. 1031-1034.

Kemp, P.H., 1960 [5.6(c)]. The relation between wave action and beach profile characteristics. *Proc. 7th Coast. Eng. Conf.*.262-276.

Kemp, P.H., 1975 [5.6(c), 6.3]. Wave asymmetry in the nearshore zone and breaker area. In: J.R.Hails and A.Carr (eds): *Nearshore Sediment Dynamics and Sedimentation.* Wiley, London. 47-67.

Kemp, P.H. and D.T.Plinston, 1968 [5.6(c), 14.3]. Beaches produced by waves of low phase difference. *J. Hyd. Div. Am. Soc. Civ. Eng.*, **94**. 1183-1195.

Kemp, P.H. and D.T.Plinston, 1974 [5.6(c)]. Internal velocities in the uprush and backwash zone. *Proc. 14th Coast. Eng. Conf.*. 575-585.

King, C.A.M., 1972 [1.5, 1.7, 2.5, 11.1, 14.3, 15.4]. *Beaches and Coasts.* 2nd Ed.. Arnold, London. 570pp

King, C.A.M. and F.A.Barnes, 1964 [14.2]. Changes in the configuration of the intertidal beach zones of part of the Lincolnshire coast since 1951. *Zeitschr. Geomorphol.*, NF **8**. 105-126.

King, C.A.M. and M.H.McCullagh, 1971 [16.1]. A simulation model of a complex recurved spit. *J. Geol.*, **79**. 22-36.

Kinsman, B., 1984 [5.1]. *Wind Waves.* Dover Publications, New York. 676pp

Kirkby,J.T., 1986 [5.3]. A general wave equation for waves over rippled beds. *Jnl Fluid Mech*, **162**, 171-186.

Kishi, T. and H.Saeki, 1967 [5.6(c)]. The shoaling breaking of the solitary wave on impermeable rough slopes. *Proc. 10th Conf. Coast. Eng.*. 27-49.

Klebba, A.A. ,1945 [2.7, 5.3]. A summary of shore recording wave meters. *Prel. Rpt. Woods Hole Ocean. Inst.*.

Knudsen, M., 1901 [4.4]. *Hydrographical Tables.* Bianco Luno. Second Edition, 1931.

Koonitz, W.A. and D.L.Inman, 1967 [2.7]. A multipurpose data aquisition system for field and laboratory instrumentation of the nearshore environment. *U.S.Army Coastal Engineering Research Centre, Tech. Memo.*, **21**.

Komar, P.D., 1971 [11.4] The mechanics of sand transport on beaches. *J. Geophys. Res.*, **30**. 5914-5927.

Komar, P.D., 1973 [16.1]. Computer models of delta growth due to sediment input from trivers and longshore transport. *Geol. Soc. Am. Bull.*, **84**. 2217-2226

Komar,P.D., 1975 [7.3, 7.4, 7.5]. Nearshore currents: generation by obliquely incident waves and longshore variations in breaker heights. In: J.Hails and A.Carr (Eds.) *Nearshore Sediment Dynamics and Sedimentation.* Wiley, London. 17-45.

Komar, P.D., 1976 [1.7, 2.1, 2.2, 5.1, 5.4, 5.5, 6.1, 6.2, 6.4, 7.2, 7.3, 7.5, 7.6, 10.2, 11.1, 11.3, 15.1]. *Beach Processes and Sedimentation.* Prentice-Hall, New Jersey, 429pp.

Komar, P.D., 1977 [16.1]. Modelling of sand transport on beaches and the resulting shoreline evolution. In: E.D.Goldberg, I.N.McCave, J.J.O'Brien and H.H.Steele (eds.), *The Sea, Vol.6, Marine Modelling.* Wiley Interscience, New York. 499-513.

Komar, P.D., 1983 [7.1, 7.3, 11.1]. Nearshore currents and sand transport on beaches. In: B.Johns (ed.): *Physical Oceanography of Coastal and Shelf Seas.* Elsevier, Amsterdam. 67-109.

Komar, P.D. and D.L.Inman, 1970 [11.2, 11.3]. Longshore sand transport on beaches. *J.Geophys. Res.*, **76**. 713-721.

Komar, P.D. and M.C.Miller, 1973 [10.2]. The threshold of sediment movement under oscillatory water waves. *J. Sedim. Petrol.*, **43**. 1101-1110.

Komar, P.D. and M.C.Miller, 1974 [10.2]. Sediment threshold under oscillatory waves. *Proc. 14th Conf. Coastal Eng., Copenhagen.* 756-775.

Korteweg, D.J. and G.de Vries, 1895 [5.3, 6.2, 6.5]. On the change of form of form of long waves advancing in a rectangular canal, and on a new type of long stationary wave. *Phil. Mag.*, series 5, **39**. 422-443.

Kraus, N.C., 1983 [16.1]. Applications of a shoreline prediction model. *Proc. Coastal Structures '83.* A.S.C.E.. 632-645.

Krumbein, W.C., 1968 [2.1]. Statistical models in sedimentology. *Sedimentology*, **10**, 7-23.

Krumbein, W.C. and G.D. Monk, 1942 [3.8]. Permeability as a function of the size parameters of unconsolidated sand. *Am. Inst. Mining. Eng. Tech. Memo.* **1492**. 11pp.

Krumbein, W.C. and F.A.Graybill, 1965 [14.3]. *An Introduction to Statistical Models in Geology.* McGraw-Hill, New York. 475pp.

Kuo, C.T. and S.T.Kuo, 1974 [5.9(a)]. Effect of wave breaking on statistical distribution of wave heights. *Proc. Civ. Eng. Oceans.*, **3**. 1211-1231.

Lai, R.Y., and Mockros, L.F., 1972 [12.6]. The Stokes-Flow drag on prolate and oblate spheroids during axial translatory accelerations. *J. Fluid Mech.*, **52**. 1-15.

Laitone, E.V., 1959 [5.6(c)]. Water Waves IV: Shallow Water Waves. Univ. of Calif., Inst. of Eng. Res. Tech. Rpt., 82-11.

Laitone, E.V., 1962 [6.3] Limiting conditions for cnoidal and Stokes waves. *J. Geophys. Res.*, **67**. 1555-1564.

Lakhan, C.V., 1989 [16.1]. Computer simulation of the characteristics of shoreward propogating deep and shallow water waves. In: Lakhan, C.V. and A.S.Trenhaile (eds.) *Applications in Coastal Modelling.* Elsevier, Amsterdam. 107-158.

Lakhan, C.V. and A.S.Trenhaile, 1989a [2.1]. Models and the coastal system. In: Lakhan, C.V. and A.S.Trenhaile (eds.) *Applications in Coastal Modelling.* Elsevier, Amsterdam. 1-16.

Lakhan, C.V. and A.S.Trenhaile (eds.), 1989b [2.1]. *Applications in Coastal Modelling.* Elsevier, Amsterdam. 387pp.

Lamb,H., 1975 [7.4]. *Hydrodynamics.* 6th Ed. Cambridge University Press. 738pp.

Landau, L.D., and Lifshitz, E.M., 1959 [12.6]. *Fluid Mechanics*, Pergamon Press, London.

Langhorne, D.N., 1982 [2.9]. A study of the dynamics of a marine sandwave. *Sedimentology*, **29**. 571-594.

Larras, J., 1956 [10.2]. Effets de la houle et du clapotis sur les fonds de sable. IV Journ. Hydraulic Rep., 9. Paris.

Leeder, M.R., 1979 [9.3]. Bedload dynamics: grain-grain interactions in water flows. *Earth Surface Processes and Landforms.* **4**. 229-240.

Leeder, M.R., 1982 [3.1]. *Sedimentology: Process and Product.* George Allen and Unwin, London. 344pp.

Leeder, M.R., 1983 [9.6]. On the dynamics of sediment suspension by residual Reynolds stresses - confirmation of Bagnold's theory. *Sedimentology*, **30**. 485-491.

Lenau, C.W., 1966 [5.6(c)]. The solitary wave of maximum amplitude. *J. Fluid. Mech.*, **26**. 309-320.

Lenhoff, L., 1982 [10.2]. Incipient motion of sediment particles. Proc. 18th Conf. Coast. Eng., Cape Town.

Leont'ev, I.O., 1985 [15.5]. Sediment transport and beach equilibrium profile. *Coastal Engineering*, 9. 277-291.

Linklater,E., 1972 [2.2]. *The Voyage of the Challenger.* John Murray, London. 288pp.

Lins, H.F., 1985 [14.6]. Storm-generated variations in nearshore beach topography. *Mar. Geol.*, **62**. 13-29.

Lofquist, K.E.B., 1980 [5.6(a)]. Measurements of oscillatory drag on sand ripples. Proc. 17th Conf. Coast. Eng., Sydney.

Longuet-Higgins, M.S., 1952 [5.8(b)]. On the statistical distribution of the heights of sea waves. *J. Mar. Res.*, **11(3)**. 245-266.

Longuet-Higgins,M.S., 1953 [6.4, 17.2(d)]. Mass transport in water waves. *Phil Trans Roy Soc Lon*, **A245**, 535-581.

Longuet-Higgins, M.S., 1956 [6.3]. The refraction of sea waves in shallow water. *J. Fluid Mech.*, **1**. 163-176.

Longuet-Higgins, M.S., 1970a [7.3, 7.4, 11.4, 16.2]. Longshore currents generated by obliquely incident sea waves, 1. *J. Geophys. Res.*, **75**. 6778-6789.

Longuet-Higgins, M.S., 1970b [7.3, 7.4, 11.4, 16.2]. Longshore currents generated by obliquely incident sea waves, 2. *J. Geophys. Res.*, **75**. 6790-6801.

Longuet-Higgins, M.S., 1972 [5.6(c)]. Recent progress in the study of longshore currents. In: R.E.Meyer (ed.) *Waves on Beaches*. Academic Press, New York. 203-248.

Longuet-Higgins, M.S., 1981a [5.3, 6.2]. Trajectories of particles at the surface of steep solitary waves. *J. Fluid Mech.*, **110**. 239-247.

Longuet-Higgins, M.S., 1981b [5.6(a)]. Oscillating flow over steep sand ripples. *J. Fluid Mech.*, **107**. 1-35.

Longuet-Higgins, M.S. and R.W.Stewart, 1964 [7.2, 16.6]. Radiation stress in water waves, a physical discussion with applications. *Deep-Sea Res.*, **75**. 6790-6801.

Losada, M.A. and L.A.Gimenez-Curto, 1981 [5.6(a)]. Flow characteristics on rough permeable slopes under wave action. *Coast. Eng.*, **4**, 187-206.

Lyles, L. and R.K.Krauss, 1971 [9.3]. Threshold velocities and initial particle motion as influenced by air turbulence. *Trans. Am. Soc. Agric. Eng.*, **14**. 563-566.

Madsen, O.S., 1971. On the generation of long waves. J. Geophys. Res., 76. 8672-8683.

Madsen, O.S. and W.D.Grant, 1976 [10.3]. Sediment transport in the coastal environment. M.I.T. Ralph M. Parsons Lab. Report, 209.

Manohar, M., 1955 [10.2]. Mechanics of bottom sediment movement due to wave action. U.S.Army Coprs. Eng., Beach Erosion Board, Tech. memo., 75.

Martinez, P.A. and J.W.Harbaugh, 1989 [16.1, 16.6]. Computer simulation of wave and fluvial-dominated nearshore environments. In: Lakhan, C.V. and A.S.Trenhaile (eds.) *Applications in Coastal Modelling*. Elsevier, Amsterdam. 297-340.

McCammon, R., 1962 [3.2]. Efficiency of percentile measures describing the mean size and sorting of sediment particles. *J. Geol.*, **70**, 453.

McCave, I.N., 1970 [9.6]. Deposition of fine grained suspended sediment from tidal currents. *J. Geophys. Res.*, **75**. 4151-4159.

McCowan, J., 1894 [5.6(c)]. On the highest wave of permanent type. *Phil. Mag.*, **38**. 351-357.

MacDonald, T.C., 1973 [10.5]. Sediment transport due to oscillatory waves. Univ. Calif. Hyd. Lab. Tech. Rep., HEL-2-39.

MacDougall, C.H., 1933 [9.4]. Bed-sediment transportation in open channels. *Trans. Americ. Geophys. Union*, **14**. 491-495.

Mehaute, B. Le, D.Divoky and A.Lin, 1968 [6.3]. Shallow-water waves: a comparison of theories and experiments. *Proc. 11th Conf. Coast. Eng.*, **1**. 86-107.

Mehaute, B. Le, and M.Soldate, 1977 [16.1]. Mathematical modelling of shoreline evolution. *Misc. Rpt. 77-10, U.S. Army Coprs. Eng., Coat. Eng. Res. Center.*. 56pp

Mehaute, B. Le, and M.Soldate, 1980 [16.1]. A numerical model for predicting shoreline changes. *Misc. Rpt., 80-6. U.S. Army Corps Eng., Coast. Res. Center.* 72pp.

Meyers, R.D., 1933 [14.3]. A model of wave action on a beach. Unpub. M.Sc. thesis referenced by King (1972).

Meyer-Peter, E and R.Muller, 1933 [9.4]. Formulas for bed-load transport. *Proc. 2nd Cong. I.A.H.R., Stockholm.* 39-64.

Miles, J.W., 1980. [6.2]. Solitary waves. *Ann. Rev. Fluid Mech.*, **12**. 11-43.

Miller, M.C., I.N.McCave and P.D.Komar, 1977 [9.4]. Threshold of sediment motion under unidirectional currents. *Sedimentology*, **6**. 303-314.

Mimura, N., T.Shimizu and K.Horikawa, 1983 [16.1]. Laboratory study on the influence of a detatched brteakwater on coastal change. Proc. Coastal Structures '83. A.S.C.E.. 740-752.

Morison, J.R. and R.C.Crooke, 1953 [6.3]. The mechanicxs of deep water, shallow water and breaking waves. *U.S.Army, Beach Erosion Board, Tech. Memo.*, **40**. 17pp.

Morrison, D.F., 1967 [14.6]. Multivariate Statistical Methods. McGraw-Hill, New York.

Moss, A.J., 1972 [3.2]. Bed load sediments. *Sedimentology*, **18**. 159-219.

Muir-Wood, A.M., 1969 [8.1, 8.3]. Coastal Hydraulics. Macmillan, London. 187pp.

Munch-Peterson, 1938 [15.1]. Littoral drift formula. *U.S.Army Corps of Engrs., Beach Erosion Board Bull.*, **4**. 1-38.

Nayak, I.V., 1970 [14.3]. Equilibrium profiles of model beaches. *Proc. 12th Conf. Coast. Eng.*, Washington. 1321-1339.

Neumann, G. and W.J.Pierson, 1966 [4.1, 4.4, 8.1]. *Principles of Physical Oceanography*. Prentice-Hall, Englewood Cliffs, New Jersey. 545pp.

Nielson, P., 1979 [10.5]. Some basic concepts of wave sediment transport. Inst. Hydrodynamics and Hydraulic Eng. Tech. Univ. Denmark. Paper 20.

Noda, H. and Y.Matsubara, 1980 [10.3]. Experiments on onshore-offshore sediment transport by waves. Proc. 27th Japanese Conf. on Coastal Eng., J.S.C.E.. 197-201.

Noda, E.K., C.J.Sonu, V.C.Rupert and J.I.Collins, 1974 [16.6]. Nearshore circulations under seabreeze conditions and wave-current interactions in the surf zone. Tetra Tech., Inc., Pasadena, Calif.. Tech. Rpt., TC-149-4.

Nordstrom, C.E. and D.L.Inman, 1975 [2.10]. Sand level changes on Torrey Pines Beach, California. *U.S. Army Corps of Eng., Beach Erosion Board, Eng. Misc. Pap.*, 11-75. 166pp.

O'Brien, M.P. and B.D.Rindlaub, 1934 [9.4]. The transportation of bed-load by streams. *Trans. Americ. Geophys. Union.* **15**. 593-603.

Odar, F., and Hamilton, W.S., 1964 [12.6]. Forces on a Sphere Accelerating in a Viscous Fluid, *J.Fluid Mech.*, **18**. 302-314.

Packham, B.A., 1952 [5.6(c)]. The theory of symmetrical waves of finite amplitude. *Proc. Roy. Soc. Lon.*, A, **213**. 238-249.

Pawka, S.S., D.L.Inman, R.L.Lowe and L.C.Holmes, 1976 [14.2]. Wave climate at Torrey Pines Beach, California. *U.S. Army Corps of Eng., Beach Erosion Board, Tech. Pap.*, 76-5. 372pp.

Perlin, M. and R.G.Dean, 1979 [16.1]. Prediction of beach planform in the lee of a breakwater. *Proc. Coastal Structures'79.* Am. Soc. Civ. Eng..792-808.

Pethick, J.S., 1984 [1.7]. *An Introduction to Coastal Geomorphology.* Arnold, London. 260pp

Pierson, W.J. and L.Moskowitz, 1964 [5.8] A proposed spectral form for fully developed wind seas based upon the similarity theory of S.A.Kitaigorodskii. *J. Geophys. Res.*, **69**. 5181-5190.

Powers, M.C., 1953 [3.6]. A new roundness scale for sedimentary particles. *J. Sedim. Pet.*, **23**. 117-119.

Price, W.A., D.W.Tomlinson and D.H.Willis, 1973 [16.1]. Prediction changes in the plan shape of beaches. *Proc. 13th Coast. Eng. Conf.*, Am. Ass. Civ. Eng.. 1321-1329.

Pugh, D., 1987 [8.1]. *Tides, Surges and Mean Sea-Level.* Wiley, Bath. 427pp.

Putnam, J.A., 1949 [5.7]. Loss of wave energy due to percolation in a permeable sea bottom. *Trans. Am. Geophys. Union.*, **30**, 349-356.

Putnam, J.A., W.H.Munk and M.A.Traylor, 1949) [7.3]. The prediction of longshore currents. *Trans. Am. Geophys. Union*, **30**. 337-345.

Putnam, J.A. and J.W.Johnson, 1949 [5.6(a)]. The dissipation of wave energy by bottom friction. *Trans. Amer. Geophys. Union*, 30, 67-74.

Quick, M.C. and B.C.Har, 1985 [14.5, 14.6]. Criteria for onshore-offshore sediment transport movement on beaches. *Proc. Canad. Coast. Conf.*, St.Jobus, Newfoundland. 257-269.

Raman, H and J.J.Earattupuzha, 1972 [14.3]. Equilibrium conditions in beach wave interaction. Proc. 13th Conf. Coast. Eng., Vancouver. 1237-1256.

Rance, P.J. and N.F.Warren, 1969 [10.2]. The threshold movement of coarse material in oscillatory flow. *Proc. 11th Conf. Coast. Eng..* 487-491.

Rankine, W.J.M., 1863 [6.2]. On the exact form of waves near the surface of deep water. *Phil. Trans. Roy. Soc. Lon.* 127-138.

Raudkivi, A.J., 1976 [9.1, 9.5]. *Loose Boundary Hydraulics.* Pergamon Press, Oxford. 397pp.

Rayleigh, L., 1876 [2.2, 5.6(c)]. On waves. *Phil. Mag.*, **5**, 1. 257-279.

Rayleigh, L., 1877 [2.2, 5.5]. On progressive waves. *Proc. Lond. Math. Soc.*, **9**, 21-26.

Rea, C.C. and P.D.Komar, 1975 [16.1]. Computer simulation models of a hooked beach shoreline configuration. *J. Sedim. Pet.*, **45**, 866-872.

Rector, R.L., 1954 [14.3, 14.4]. Laboratory study of equilibrium profiles of beaches. *U.S.Army Corps Eng., Beach Erosion Board, Tech. Memo.*, **41**. 38pp.

Rescher, N., 1962 [1.6]. The stochastic revolution and the nature of scientific explanation. *Synthese*, **14**. 200-215.

Reynolds, O., 1877 [2.2, 5.5]. On the rate of progression of groups of waves and the rate at which energy is transmitted by waves. *Nature*, **36**, 343-344.

Riley, J.P. and G.Skirrow, 1965 [4.1]. *Chemical Oceanography*, Volumes I and II. Academic Press, New York. 712pp.

Rivett, P., 1972 [2.1]. *Principles of Model Building.* Wiley, London.

Rohrbough, J.D., E.Koehr and W.C.Thompson, 1964 [2.2]. Quasi-weekly and daily profile changes on a distinctive sand beach. *Proc 9th Conf. Coast. Eng., Am. Soc. Civ. Eng.*, Lisbon. 249-258.

Rouse, H., 1938 [9.7, 16.4]. Experiments on the mechanics of sediment suspension. *Proc. Int. Congr. Appl. Mech. 5th.* 550-554.

Rubey, W.W., 1933 [9.4]. Settling velocities of gravel, sand and silt particles. *Am. J. Sci.*, **25**. 325-338.

Russell, J.S., 1838 [6.2]. Report of the Committee on Waves. *Proc. 7th Meeting of the Brit. Ass. Adv. Sci.*, Liverpool. 417-496.

Russell, J.S., 1844 [2.2, 6.2, 6.6]. Report on Waves. *14th Meet. Brit. Ass. Adv. Sci..*311-390.

Russell, R.C.H. and J.D.C.Osario, 1958 [6.4]. An experimental investigation of drift profiles in a closed channel. *Proc. 6th Conf. Coast. Eng..* 171-183.

Sallenger, A.H., P.C.Howard, C.H.Fletcher and P.A.Howd, 1983 [2.8]. A system for measuring bottom profile, waves and currents in the high-energy nearshore environment. *Mar. Geol.*, **51**. 63-76.

Sasaki, T.O. and H.Sakamoto, 1978 [16.1]. Field verification of a shoreline simulation model. *Int. Conf. on Water Resources Eng*, I.A.H.R., 501-518.

Sato, S. and T.Kishi, 1954 [10.2]. Shearing force on sea bed and movement of bed material due to wave motion. *J. Res. Public Works Res. Inst.*, **1**. 1-11.

Sato, S., T.Ijima and N.Tanaka, 1962 [10.2]. A study of the critical depth and mode of sand movement using radioactive glass sand. *Proc. 8th Conf. Coastal Eng., Mexico*. 304-323.

Savage, R.P., 1959 [11.2]. Laboratory study of the effect of groins in the rate of littoral transport. *U.S.Army Corps Eng., Beach Erosion Board, Tech. Memo.*, **75**. 121pp.

Savage, R.P., 1962 [11.1]. Laboratory determination of littoral transport rates. *J. Waterways and Harbors Div., A.S.C.E.*, **88**. 69-92.

Scheidegger, A.E., 1961 [13.9]. Mathematical models of slope development. *Bull. Geol. Soc. Am.*, **72**. 37-49.

Scheidegger, A.E., 1970 [1.1, 1.2, 13.9]. *Theoretical Geomorphology*. 2nd Ed.. Springer-Verlag, Berlin. 435pp.

Schuum, S.A., M.P.Mosley and W.E.Weaver, 1987 [1.2]. *Experimental Fluvial Geomorphology*. Wiley, New York. 413pp.

Schoklitsch, A., 1934 [9.4]. Geschiebetrieb und die geschiebefracht. *Wasserkr. Wasserwirtsch*, **39**.

Seelig, W.N. and J.P.Ahrens, 1981 [5.7]. Estimation of wave reflection and energy dissipation coefficients for beaches, revetments and breakwaters. U.S.Army, C.E.R.C., Rpt. TP 81-1.

Seymour, R.J., 1986 [14.5]. Results of cross-shore transport experiments. *J Waterway, Port, Coastal and Ocean Eng.*, Am.Soc. Civ. Eng., **112**.168-173.

Seymour, R.J. (ed.), 1989 [1.7, 2.10, 11.1]. *Nearshore Sediment Transport*. Plenum Press, New York. 418pp.

Seymour, R.J. and D.B.King, 1982 [14.5]. Field comparison of cross-shore transport models. *J. Waterways, Port, Coastal and Ocean Div.*, Am. Ass. Civ. Eng., 108. 163-173.

Seymour, R.J. and D.Castel, 1989 [14.1, 14.4, 14.5, 14.6]. Modelling cross-shore transport. In: Seymour, R.J. *Nearshore Sediment Transport*. Plenum Press, New York. 387-401.

Shepard, F.P., 1950 [2.2]. Beach cycles in southern California. U.S. Army Corps of Engineers, *Beach Erosion Board Tech. Memo.*, 20, 26pp.

Shepard, F.P. 1973 [1.4]. *Submarine Geology*. 3rd ed.. Harper and Row, New York. 517pp

Shepard, F.P. and E.C.LaFond, 1940 [2.3]. Sand movement near the beach in relation to tides and waves. *American Journal of Science*, **238**. 272-285.

Shepard, F.P. and D.L.Inman, 1950 [2.2, 7.4]. Nershore water circulation related to bottom topography and wave refraction. *Eos Trans.*, AGU, **31**(2), 196-212.

Shepard, F.P. and D.L.Inman, 1951a [2.2]. Sand movement on the shallow inter-canyon shelf at La Jolla, California. *U.S.Army Corps Engnrs., Beach Erosion Board, Tech Memo*. 32. 28pp.

Shepard, F.P. and D.L.Inman, 1951b [7.4]. Nearshore circulation. *Proc. 1st Conf. Coast. Eng,*. 50-59.

Shields, A., 1936 [9.4]. Anwendung der aehnlichkeitsmechanik und der turbulenz forschung auf die geschiebebewegung. *Mitt. Preuss. Versuchsandstalt Wasserbau Schiffbaue. Berlin*, **26**.

Shore Protection Manual, 1984 [1.4, 5.1, 6.1, 11.2]. Coastal Engineering Research Center. Dept of the Army. Vicksburg, Mississippi. Two Volumes.

Short, A.D., 1979a [14.2, 14.5]. Wave power and beach stages - a global model. *Proc. 16th Int. Conf. Coastal Eng.* 1145-1162.

Short, A.D., 1979b [14.2]. Three dimensional beach stage model. *J.Geol.*, **87**. 553-571.

Short, A.D., 1981 [2.3]. Beach response to variations in breaker height. *Proc 17th Int. Conf. Coast. Eng.*, Sydney. 1016-1035.

Short, A.D. and L.D.Wright, 1981 [2.3, 14.6]. Beach systems of the Sydney Region. *Aust. Geogr.*, **15**. 8-16.

Silvester, R. 1974 [1.7, 14.4]. *Coastal Engineering*. Elsevier, Amsterdam. 2 Volumes.

Silvester, R. and Mogridge, G.R., 1970 [10.2]. Reach of waves to the bed of the continental shelf. *Proc. 12th Coastal Eng. Conf., Washington.* 487-491.

Simpson, J.H., 1969 [6.5]. Observations of the directional characteristics of waves. *Geophys. J. Roy. Astron. Soc.*, **17**, 93-120.

Sitarz, J.A., 1963 [14.4]. Contribution a l'etude de l'evolution des plages a partir de la consistence des profile d'equilibre. Trav. Centre. Etude. Rech. Oceanogr., 12-20.

Skjelbreia, L., 1959 [6.2]. *Gravity Waves, Stokes Third Order Approximation: Tables of Functions.* Council on Wave Research, The Engineering Foundation, University of California.

Skjelbreia, L. and J.A.Hendrickson, 1962 [6.2]. *Fifth Order Gravity Wave Theory.* National Engineering Science Co..

Skovgaard, O., I.A.Svendsen, I.G.Jonnson and O. Brink-Kjaer, 1974 [5.3]. *Sinusoidal and cnoidal gravity wave formulae and tables.* Inst. Hydrodyn. and Hydraul. Eng., Technical University of Denmark.

Sleath, J.F.A., 1976 [5.6(a)]. Forces on a rough bed in oscillatory flow. *J.Hydraulic. Res.*, **14**(2). 155-164.

Sleath, J.F.A., 1982 [10.3, 10.5]. The suspension of sand by waves. *J. Hydraulic Res.*, **20**. 439-452.

Sleath, J.F.A., 1984 [1.7, 2.2, 5.1, 5.5, 5.6, 6.1, 6.5, 9.1, 9.4, 9.5, 10.1, 10.2, 10.4, 15.4]. *Sea Bed Mechanics.* Wiley, New York. 335pp.

Smith, J.D. and S.R.McClean, 1977a [9.7]. Spatially averaged flow over a wavy surface. *J. Geophys. Res.*, **82**. 1735-1746.

Smith, J.D. and S.R.McClean, 1977b [9.7]. Boundary layer adjustments to bottom topography and suspended sediment. *Mem. Soc. R. Sci. Liege*, **11**. 123-151.

Snyder et al 1958 [5.5].

Sonu, C.J., 1973 [14.2]. Three-dimensional beach changes. *J. Geol.*, **81**. 42-64.

Sonu, C.J. and J.L.van Beek, 1971 [14.2]. Systematic beach changes on the Outer Banks, North Carolina. *J.Geol.*, **79**. 416-425.

Southgate, H.N., 1988 [16.1]. The nearshore profile model. Rpt SR 157. Hydraulics Research, Wallingford.

Stokes, G.G., 1847 [2.2, 5.3, 6.2, 6.3, 6.4, 7.4]. On the theory of oscillatory waves. *Trans. Camb. Phil. Soc.*, 8.

Stokes, G.G., 1880 [2.2]. On the theory of oscillatory waves. *Math. Phys. Papers I.* C.U.P., London. 197-229.

Strahler, A.N., 1966 [2.3]. Tidal cycle of changes on an equilibrium beach. *J. Geol.*, **74**. 247-268.

Suhayda, J.N. and N.R.Pettigrew, 1977 [2.7, 5.6]. Observations of wave height and celerity in the surf zone. *J. Geophys. Res.*, **82**. 1419-1429.

Sunamura, T., 1980 [10.3]. A laboratory study of offshore transport of sediment and a model for eroding beaches. *Proc. 17th Coastal Eng. Conf.*, A.S.C.E.. 1051-1070.

Sunamura, T., 1984 [10.3]. Onshore-offshore sediment transport rate in the swash zone of laboratory beaches. *Coastal Eng. in Japan*, **27**. 205-212.

Sunamura, T. and K.Horikawa, 1974 [14.5]. Two-dimensional beach transformation due to waves. *Proc. 14th Conf. Coat. Eng, Copenhagen.* 920-938.

Sunamura, T. and I.Takeda, 1984 [10.3]. Landward migration of inner bars. *Mar. Geol.*, **60**. 63-78.

Sunamura, T., K.Bando and K.Horikawa, 1978 [10.3, 10.4]. An experimental study of sand transport mechanisms and rate over asymmetrical ripples. *Proc. 25th Japanese Conf. Coastal Eng., J.S.C.E..* 250-254.

Svendsen, I.A., 1985 [2.1]. Physical modelling of water waves. In: Dalrymple, R.A. (ed.). Balkema, Rotterdam. 13-47.

Sverdrup, H.U. and W.H.Munk, 1946 [5.6(c)]. Theoretical and empirical relations in forcasting breakers and surf. *Trans. Am. Geophys. Union,* **27**. 828-836.

Swain, A., 1984 [14.6]. Additional results of a numerical model for beach profile development. *Proc. Ann. Conf. CSCE,* Halifax, Nova Scotia.

Swain, A., 1989 [14.6]. Beach profile development. In: V.C.Lakhan and A.S.Trenahaile (eds.): *Applications in Coastal Modelling. Elsevier,* Amsterdam. 215-232.

Swain, A. and J.R.Houston, 1983 [14.6]. A numerical model for beach profile development. *Canadian J. of Coast. Eng.,* **12**. 231-234.

Swart, D.H., 1974 [14.6]. A schematization of onshore-offshore transport. Proc. 14th Conf. Coast. Eng.. 884-900.

Swart, D.H., 1976 [14.6]. Predictive equations regarding coastal transports. *Proc. 15th Coast. Eng. Conf., Honolulu.* 1113-1132.

Swift, D.J.P. and J.C.Ludwick, 1976 [13.4]. Substrate response to hydraulic process: grain-size frequency distributions and bed forms. In: Stanley, D.J. and D.J.P.Swift (eds.) *Marine Sediment Transport and Environmental Management.* John Wiley and Sons, New York. 159-196.

Swift, D.J.P., A.W.Niederoda, C.E.Vincent and T.S.Hopkins, 1985 [1.4]. Barrier island evolution, Middle Atlantic Shelf, U.S.A.. Part I: shoreface dynamics. *Mar. Geol.,* **63**, 331-361.

Tanner, W.F., 1974 [1.5]. Advances in near-shore sedimentology: a selective review. *Shore and Beach.* 42.

Tetzlaf, D.M. and J.W.Harbaugh, 1989 [16.6]. *Simulating Clastic Sedimentation.* Van Nostrand-Reinhold.

Thorn, C.E., 1988 [13.1]. *Introduction to Theoretical Geomorphology.* Unwin Hyman, Boston. 247pp.

Thornes, J.B., 1983 [13.10]. Evolutionary Geomorphology. *Geography,* **68**. 225-235.

Thornes, J.B. and D.Brunsden, 1977 [13.1, 13.9]. *Geomorphology and Time.* Methuen, London. 208pp.

Thornton, E.B., 1971 [7.3]. Variation of longshore current across the surf zone. *Proc. 12th Conf. Coast. Eng..* 291-308.

Thornton, E.B., 1973 [11.4]. Distribution of sediment transport across the surf zone. *Proc. 13th Conf. Coast. Eng..* 1049-1068.

Thronton, E.B. and R.F.Krapohl, 1974 [6.5]. Water particle velocities measured under ocean waves. *J. Geophys. Res.,* **79**, 847-852.

Thornton, E.B. and R.T.Guza, 1982 [5.6]. Energy saturation and phase speeds measured on a natural beach. *J. Geophys. Res.,* **87**. 9499-9508.

Thornton, E.B. and R.T.Guza, 1983 [5.8, 5.9(a)]. Transformation of wave height distribution. *J. Geophys. Res.,* **88**. 5925-5938.

Thornton, E.B. and R.T.Guza, 1989 [5.9]. Wind wave transformation. In: R.J.Seymour (ed.) *Nearshore Sediment Dynamics.* Plenum Press, New York. 137-171.

Trenhaile, A.S., 1978 [8.8]. The shore platform of Gaspe, Quebec. , **68**. 95-114.

Trenhaile, A.S., 1980 [8.8]. Sea level oscillations and the development of rock coasts. In: V.C.Lakhan and A.S.Trenhaile (eds.) Applications in Coastal Modelling. Elsevier, Amsterdam. 271-296.

Trenhaile, A.S. and M.G.J. Layzell, 1981 [8.8]. Shore platform morphology and the tidal duration factor. *Trans. Inst. Brit. Geogr.,* **6**. 82-102.

Tsuchiya, Y, H.Yoshida, M.Tanahashi, and R.Tsuchiko, 1983 [2.8]. Long-term observations of currents in coastal area with ultrasonic current meter. *Proc. 30th Japanese Conf. on Coastal Eng.*. 500-504.

Tucker, M.J., A.P.Carr and E.G.Pitt, 1983 [5.6(c)]. The effect of an offshore bank in attenuating waves. *Coastal Engineering*, **7**. 133-144.

University of California, 1951 [2.5].

Ursell, F., 1952 [7.4]. Edge waves on a sloping beach. *Proc. Roy. Soc. Lon., A,* **214**. 79-97.

Ursell, F., 1953 [6.3] The long-wave paradox in the theory of gravity waves. *Proc. Cam. Philos. Soc.*, **49**. 685-694.

Vincent, G.E., 1957 [10.2]. Contribution to the study of sediment transport on a horizontal bed due to wave action. *Proc. 6th Conf., Coastal Eng.*, Miami. 326-355.

Vincent, L., R.Dolan, B.P.Hayden, and D.Resio, 1976 [14.6]. Systematic variations in barrier island topography. *J. Geol.*, **84**. 583-594.

Vincent, C.E., R.A.Young and D.J.P.Swift, 1981 [10.3, 10.5]. Bed-load transport under waves and currents. *Mar. Geol.*, **39**. 71-80.

Vincent, C.E., R.A.Young and D.J.P.Swift, 1982 [10.5]. On the relationship between bedload and suspended sand transport on the Inner Shelf, Long Island, New York. *J. Geophys. Res.*, **87**. 4163-4170.

Visher, G.S., 1969 [3.2]. Grain size distribution and depositional processes. *J. Sedim. Petrol.*, **39**. 1074-1106.

von Karman, T., 1934 [9.7]. Some aspects of the turbulence problem. *Proc. Int. Congr. Appl. Mech. 4th.* 54-91.

Waddell, H., 1932 [3.6]. Volume, shape and roundness of rock particles. *J. Geol.*, **40**. 443-451.

Wang, H. R.A. Dalrymple and J.Shiau, 1975 [16.1]. Computer simulation of beach erosion and profile modification due to waves. Modelling Symposium, V.2. 1369-1384.

Watanabe, A., 1988a, [16.1, 16.5]. Chapter 1. Introduction. In: Part III: Numerical Model of Beach Topography Change. Horikawa, K (ed.): *Nearshore Dynamics and Coastal Processes*. University of Tokyo Press, Tokyo. 241-244.

Watanabe, A., 1988b, [16.1, 16.5]. Chapter 2. Computation of Nearshore Wave Field. In: Part III: Numerical Model of Beach Topography Change. Horikawa, K (ed.): *Nearshore Dynamics and Coastal Processes*. University of Tokyo Press, Tokyo. 245-270.

Watanabe, A., 1988c, [16.1, 16.5]. Chapter 4. Modelling of Sediment Transport and Beach Evolution. In: Part III: Numerical Model of Beach Topography Change. Horikawa, K (ed.): *Nearshore Dynamics and Coastal Processes*. University of Tokyo Press, Tokyo. 292-302.

Watanabe, A., 1988d, [16.1, 16.5, 17.3]. Chapter 5. Application of the Three-Dimensional Beach Evolution Model. In: Part III: Numerical Model of Beach Topography Change. Horikawa, K (ed.): *Nearshore Dynamics and Coastal Processes*. University of Tokyo Press, Tokyo. 303-319.

Watanabe, A., K.Maruyama, T.Shimizu and T.Sakakiyama, 1986 [16.1, 16.5]. Numerical prediction model of three-dimensional beach deformation around a structure. *Coast. Eng. in Japan*, **29**. 179-194.

Watts, G.M., 1953 [11.2]. A study of sand movement at South Lake Worth Inlet, Florida. U.S.Army Corps of Eng., Beach Erosion Board, Tech. Memo., 42. 24pp.

Watts, G.M. and R.F.Dearduff, 1954 [14.2]. Laboratory study of the effect of tidal action on wave-formed beach profiles. *U.S.Army Corps of Eng., Beach Erosion Board,*

Tech. Memo., **52**. 21pp.

Wentworth, C.K., 1922 [3.2]. A scale of grade and class terms for clastic sediments. *J. Geol.*, **30**, 377-392.

White, S.J., 1970 [9.4]. Plane bed threholds of fine grained sediments. *Nature*, **228**. 152-153.

Whitehouse, R.J.S. and J.Hardisty, 1988 [9.4] Experimental assessment of two theories for the effect of bedslope on the threshold of bedload transport. *Mar. Geol.*, **79**. 135-139.

Wiegel, R.L., 1964 [1.7, 2.3, 5.1, 5.3, 5.4, 6.1, 6.2, 14.3]. *Oceanographical Engineering.* Prentice-Hall, Englewood Cliffs, New Jersey. 532pp.

Wilberg, P.L. and J.D.Smith, 1983 [9.7]. A comparison of field data and theoretical models for wave-current interactions at the bed on the continental shelf. *Cont. Shelf. Res*, **2**. 147-162.

Williams, G.P., 1967 [9.5]. Flume experiments on the transport of coarse sand. *U.S. Geol. Surv. Prof. Pap.*, **562-B**. 31pp.

Willis, D.H., 1977 [16.1]. Evaluation of alongshore transport models. *Proc. Coast. Sediments'77.* Am. Ass. Civ. Eng.. 350-365.

Winant, C.D., D.L.Inman and C.E.Nordstrom, 1975 [14.6]. Description of seasonal beach changes using empirical eigenfunctions. *J. Geophys. Res.*, **80**. 1979-1986.

Woldenberg, M.J. (ed.), 1985 [2.1]. *Models in Geomorphology.* Allen and Unwin, London. 434pp.

Wright, L.D., 1976 [14.2]. Nearshore wave power dissipation and the coastal energy regimeof the Sydney-Jervis Bay region New South Wales: a comparison. *Austral. Jnl. Mar. Freshwater Res.*, **32**. 105-140.

Wright, L.D., 1981a [14.2]. Modes of beach cut in relation to surf-zone morphodynamics. *Proc 17th Int. Conf. Coastal Eng.*, Sydney 1980.

Wright, L.D., 1981b [14.2]. Field observations of long-period surf-zone standing waves in relation to contrasting beach morphodynamics. *Aust. J. Freshwater Res.*, **33**. 181-201.

Wright, L.D., J.Chappell, B.G.Thom, M.P.Bradshaw and P.Cowell, 1979a [14.2]. Morphodynamics of reflective and dissipative beach and inshore systems: Southeastern Australia. *Mar. Geol.*, **32**. 105-140.

Wright, L.D., B.G.Thom and J.Chappell, 1979b [14.2]. Morphodynamic variability of high energy beaches. *Proc 16th Int. Conf. Coast. Eng.*. 1180-1194.

Wright, L.D., R.T.Guza and A.D.Short, 1982a [14.2]. Dynamics of a high energy dissipative surf zone. *Mar. Geol.*, **45**. 41-62.

Wright, L.D., P.Nielson, A.D.Short and M.O.Green, 1982b [14.2]. *Morphodynamics of a macrotidal beach.* Mar. Geol., **50**. 97-128.

Wright, L.D. and A.D.Short, 1983 [14.2]. Beach cut in relation to surf zone morphodynamics. *Proc. 17th Conf. Coast. Eng.*, Am. Soc. Civ. Eng., New York. 978-996.

Wright, L.D. and A.D.Short, 1984 [14.2]. Morphodynamic variability of surf zones and beaches: a synthesis. *Mar. Geol.*, **56**. 93-118.

Wright, L.D., A.D.Short, J.D.Boone, B.Hayden, S.Kimball and J.H.List, 1987 [14.6]. The morphodynamic effects of incident wave groupiness and tide range on an energetic beach. *Mar. Geol.*, **74**. 1-20.

Yalin, M.S., 1963 [9.7]. An expression for bed-load transportation. *J. Hydraulic Div. A.S.C.E.*, **89**. 221-250.

Yalin, M.S., 1977 [9.1, 9.3, 9.5]. *Mechanics of Sediment Transport.* 2nd Ed.. Pergamon Press, Oxford. 298pp.

Yamada, H., 1957 [5.6]. On the highest solitary wave. *Rpt. Res. Inst. App. Math.*,

5(18). 53-155.

Yamashita, T., M.Sawamoto, and H.Yokoyama, 1984 [10.3]. Experimental study on the sand transport rate and the mechanism of sand movement due to waves. *Proc. 31st Japanese Conf. on Coastal Eng.*, J.S.C.E.. 281-285.

Zenkovich, V.P., 1967 [1.7]. *Processes of Coastal Development*. Translated by D.G.Fry and edited by J.A.Steers. Oliver and Boyd, Edinburgh. 738pp.

Zhukovets, A.M., 1963 [5.6(a)]. The influence of bottom roughness on wave motion in a shallow body of water. *Bull. Acad. Sci., USSR Geophys. Ser.* **10**. 933-948.

Ziegler, J.M., C.R.Hayes and S.D.Tuttle, 1959 [2.2]. Beach changes during storms on outer Cape Cod, Massachusetts. *J. Geol.*, **67**. 318-336.

SUBJECT INDEX

Acceleration number, 171
Added mass term, 170, 182
Advective terms, 179
Airy wave theory, 49
Asymmetry of currents, 73, 80, 218, 219
Atmosphere, 3

Backshore, 8
Backwash, 218
Bagnold, 113, 124, 127, 218 (see also
      Author Index)
Bar, 8, 15, 236
Barrier, 8
Basset history integral, 170
Bathymetry, 20
Beach, beaches, beach's
      cycles, 15
      definition of, 7
      dissipative, 203
      face, 8
      gradient models, 201, 207, 217
      profile models, see orthogonal
      system
      reflective, 203
      ridge, 8
      states, 201
      tides, 111
Bedload, 28, 115 (and see sediment
      transport)
Bedload spectrum, 186, 288
Bedslope, 20, 32, 113, 219
Bed shear stress see shear stress
Berm, 8, 15
Bernoulli's equation, 118, 165
Bore classification, 61
Breaking, see waves
Buoyancy, 43

Celerity
      group, 49
      phase, 49, 59
Cell circulation, 97
Challenger expedition, 15, 39

Clays, 32
Cnoidal wave theory, 51 (see also waves)
Coastline
      length of, 7
Computer modelling, 229
Continuity, 18
Convective terms, 179
Co-ordinate axes, 4
Cornaglia, 217
Co-tidal and co-range lines, 111
Coulomb's Law, 38
Covalent bonding, 37
Creeping flow regime, 167
Critical flow etc., 115
Cubic packing, 29
Currents
      speed of, 4
      meters, 16
Cusps, 8

Darcy's Law, 3
Descriptive model, 10
Destructive waves, 16
Denudation geochronology, 5
Developing profile analysis, 214
Dielectric constant, 37
Do San, tides at, 107
Drag coefficient, 114, 116, 117, 127,
      137, 167, 168, 182
Drift currents, 18, 258

Earth dimensions, 103
Eddy coefficient for mixing, 94
Eddy diffusivity, 133, 153
Edge waves, 98
Eigenfunctions, 212
Empirical orthogonal function (EOF),
      212
Empirical model, 10, 14
Endogenetic features, 3
Energy flux, 92
Equilibrium (see also stable etc.)
      of beaches, 10, 20, 196, 220,